Neurotransmitters in Plant Life

VICTORIA V. ROSHCHINA

Russian Academy of Sciences
Institute of Cell Biophysics
Pushchino, Moscow Region
Russia

Science Publishers, Inc.

Enfield (NH), USA   Plymouth, UK

SCIENCE PUBLISHERS, INC.
Post Office Box 699
Enfield, New Hampshire 03748
United States of America

Internet site: *http://www.scipub.net*

sales@scipub.net (marketing department)
editor@scipub.net (editorial department)
info@scipub.net (for all other enquiries)

© 2001, Copyright Reserved

```
        Library of Congress Cataloging-in-Publication Data
Roshchina, V. V. (Viktoriia Vladimirovna)
   [Biomediatory v rasteniiakh. English]
   Neurotrasmitters in plant life/Victoria V. Roschina.
      p. cm.
   Includes bibliographical references (p. ).
   ISBN 1-57808-142-4
   1. Biogenic amines. 2. Acetylcholine. 3. Plant cells and
   tissues. 4. Plant physiology. 5. Neurotrasmitters. I. Title.

   QK898.B56 R6713 2001
   572'.5482--dc21
                                                          2001020173
```

All rights reserved. No part of this publication may be reproduced, stored in a retrieval system, or transmitted in any form or by any means, electronic, mechanical, photocopying or otherwise, without the prior permission from the publisher. The request to produce certain material should include a statement of the purpose and extent of the reproduction.

Published by Science Publishers Inc., Enfield, NH, USA
Printed in India.

Preface

The neutrotransmitters acetylcholine and biogenic amines dopamine, noradrenaline, serotonin, and histamine are present not only in animals, but also in plants. What are the functions of these compounds in plant organisms? How many plant reactions are sensitive to neurotransmitters? What is their significance in the field of medicine? This book is a first attempt to answer these questions.

The synthesis, content, and role of acetylcholine, catecholamines, serotonin, and histamine in plants have been reviewed in this work on the basis of world literature and the author's data. The papers describe participation of the components of animal cholinergic and aminergic regulatory systems in the functioning of many plant processes within and outside the cell: from changes in ion permeability of membranes, energetics, and metabolism to complex processes such as a fertilization, motility and finally germination, growth, and morphogenesis. In plants, enzymes of the neurotransmitters' synthesis and catabolism are present as well as some functional analogs of cholino- and aminoreceptors.

In plants, neurotransmitters serve as chemical signallers, trigger seed and pollen germination, regulate ion permeability, energetics, metabolism, and movements (of roots, leaves, and stomata cells), protect against stresses, and participate in intercellular and intracellular relations. The stimulatory effects of some neurotransmitters on plant growth reactions may be commercially useful both in laboratories and in agriculture. Medicinal plants enriched in some neuromediators could be used in pharmacology and medicine as alternatives to purified chemical drugs.

An evolutionary approach to the role of neurotransmitters reveals the presence of the compounds at all steps of evolutionary development, from unicellular organisms to highly organized animals. This shows that non-synaptic functions of substances have arisen earlier than synaptic functions in animals with developed nervous system, and so may play a universal role as elementary molecular agents of irritation in any living cell.

Contents

Preface	iii
Introduction	1
1. Neuromediators, Their Synthesis and Metabolism	4
1.1. Acetylcholine	4
1.1.1. Occurrence of Acetylcholine in Plants	4
1.1.2. Metabolism of Acetylcholine	13
1.1.3. Effect of External Factors on Metabolism of Acetylcholine	17
1.2. Catecholamines	20
1.2.1 Occurrence in Plants	20
1.2.2. Metabolism in Plants	25
1.3. Serotonin	34
1.3.1. Occurrence in Plants	34
1.3.2. Metabolism in Plants	39
1.4. Histamine	43
1.4.1. Occurrence in Plants	43
1.4.2. Metabolism in Plants	48
2. Response of Plants to Acetylcholine and Biogenic Amines	52
2.1. Growth, development and movements of plants	52
2.1.1. Growth	52
2.1.2. Germination of Seeds, Pollen, and Spores	54
2.1.3. Development of Plants	56
2.1.4. Influence of Environmental Factors on the Activity of Neurotransmitters	57
2.1.5. Mechanism of Action on Growth and Development	58
2.1.6. Movements and Motility Reactions of Plants	60
2.2. Membrane Processes	63
2.2.1. Ion Permeability and Membrane Potential of Whole Cells	63
2.2.2. Ion Permeability and Membrane Potential of Organelles	67

	2.3. Energetic and Metabolic Reactions	69
	2.3.1. Electron Transport and Coupled Phosphorylation	70
	2.3.2. Redox Reactions	71
	2.3.3. Fluorescence of Individual Cellular Components and Intact Cells	76
	2.3.4. Effects on Metabolic Processes	80
3.	**The Regulatory Systems with Participation of Neuromediators**	**82**
	3.1. Cholinergic System of Regulation	84
	3.1.1. Cholinoreception	85
	3.1.1.1. Main Features of Cholinoreception	85
	3.1.1.2. Cholinoreception in Plants	88
	3.1.2. Choline Acetyltransferase	98
	3.1.3. Cholinesterase	99
	3.1.3.1. Main Characteristics of Animal Cholinesterases	99
	3.1.3.2. Occurrence of Cholinesterase Activity in Plants	100
	3.1.3.3. Kinetic Parameters of Cholinesterases Found in Plant Extracts	115
	3.1.3.4. Localization of Cholinesterase in Plant Cells	120
	3.1.3.5. Isolation and Purification of Cholinesterases	126
	3.1.3.6. Regulation of Cholinesterase Activity	139
	3.1.3.7. Effects of Pesticides and Growth Regulators with Anticholinesterase Activity on Plant Reactions	146
	3.1.3.8. Special Roles of Cholinesterases	147
	3.2. Other Regulatory Systems (Adrenergic, Dopaminergic, Serotoninergic, Histaminergic)	150
	3.2.1. Reception of Catecholamines, Serotonin and Histamine	150
	3.2.1.1. Main Features of Aminoreception	150
	3.2.1.2. Aminoreception in Plants	152
	3.2.2. Enzymes Participating in Biosynthesis of Biogenic Amines	159
	3.2.3. Enzymes Participating in Catabolism of Biogenic Amines	164
	3.3. Systems of Secondary Messengers Coupling with Receptors	168
4.	**Physiological Role of Neuromediators in Plants**	**178**
	4.1. Mediatory Function	179
	4.1.1. Intracellular and Intercellular Chemical Signaling in Plants	183

4.1.2. Electric Potentials in Plants and Possible Role of Neuromediators	187
4.1.3. Neuromediators as Possible Triggers of Secondary Messenger Systems	191
4.2. Non-mediatory Functions	192
4.2.1. Mediators as Regulators of Energetic and Metabolic Processes	194
4.2.2. Role of Neuro Mediators in Growth and Morphogenetic Reactions	196
4.2.3. Protective Role of Neuro Mediators	197
4.3. Evolution of Regulatory System with Participation of Neuromediators	199
4.4. Perspectives in Practical Use of Neurotransmitters in Plants	205
4.4.1. Medicine and Pharmacology	205
4.4.2. Agriculture and Protection of the Environment	205
Conclusion	207
Appendix: 1. Pharmacological Effects of Neurotransmitters	211
Lists of some pharmacologically important plants, containing neuromediators and their derivatives	
Appendix: 2. Methods for Plants Assays	215
Bibliography	221
Subject Index	257
Latin Index	276

Introduction

The problem of information transmission within and between living cells is as significant as the problem of genetic code because it is a major aspect of irritability, one of the important features of life. Just as genetic code has a common basis in all living organisms in the form of the sequence and combination of several purine and pyrimidine bases, the mechanism of irritability appears to have a common basis in the form of chemical signals, that are uniform for every cell.

In the 19th century, the animal physiologist Claude Bernard, when considering irritability (excitability) as one of the main characteristics of all living things, predicted the occurrence of common mechanisms of perception and fast reaction of organisms to external factors. In his book (Bernard, 1878), he has written that there exists common basis of excitation in all living organisms. He reached this conclusion after experiments on the rapid closing of mechanically irritated *Mimosa* leaves. He demonstrated that anaesthetics block the conducting of excitation impulses in plants, as also observed in animals. However, the molecular mechanism of excitability, including the perception of external stimulus, transmission of information about this stimulus, and responses, were studied only in the 20th century. Electric events in the form of changes of electric potential have been demonstrated to be the basis of the impulse spreading in both animals and plants. Moreover, Jagadish Chandra Bose (1926) proposed even the existence of the nervous mechanisms in plants.

Further investigations were motivated by the search for anaesthetics and sedative drugs. Investigation of the mechanisms by which excitation is transmitted from cell to cell in animals having a nervous system led to the discovery of the low-molecular chemical transmitters (mediators) cholinic ester acetylcholine, the biogenic amines dopamine, noradrenaline, adrenaline, serotonin, histamine, and other compounds at the beginning of the 20th century. In the nerve cell, these compounds are stored in special secretory vesicles and on irritation are liberated in a narrow space (1 nm) between contacting cells: the synaptic chink or synapse. The free mediator binds with proteins-receptors of the neighbouring cell, that results in opening of ion channels in the plasmatic membrane. The ions move according to their concentration gradient, and the cellular electric potential changes. Thus, chemical information is transformed into electric impulse. The interaction of mediator with receptor can also occur through other mechanisms, by triggering of systems of intracellular secondary messengers that regulate the activity of enzymes in cells.

Besides specialized mediator function in organisms having nervous systems, acetylcholine and biogenic amines also play other roles. The evolutionary animal physiologist Koshtoyantz (1963) hypothesized that physiological and biochemical regulatory processes are similar in all animals, from unicellular and primitive multicellular to higher vertebrates. This hypothesis was experimentally confirmed in the work of G.A. Buznikov and coworkers at the Koltsov Institute of Developmental Biology of the Russian Academy of Sciences. They showed that mediators are synthesized and function as regulators in development of embryos and organisms lacking a nervous system (Buznikov, 1967, 1987, 1989, 1990; Buznikov and Turpaev, 1988). The common biological role of substances that are chemical transmitters of excitation in higher animals became evident after the discovery of acetylcholine and biogenic amines in plants. Since the discovery of acetylcholine in plants (Ewins, 1914; Boruttau and Cappenberg, 1921; Emmelin and Feldberg, 1947), dopamine, noradrenaline (Buelow and Gisvold, 1944; Waalkes *et al.*, 1958; Udenfriend *et al.*, 1959), adrenaline (Askar *et al.*, 1972), serotonin (Bowden *et al.*, 1954; Collier and Chesher, 1956), and histamine (Barger and Dale, 1910; Kutcher, 1910; Kandle, 1916; Werle and Raub, 1948; Werle and Roewer, 1950), the general biological significance of these compounds has become evident. It should be noted that acetylcholine (Ewins, 1914) and histamine (Barger and Dale, 1910; Kutcher, 1910) were first found in ergot fungi and then in animals (Loewi, 1937).

In the 1970s, the role of acetylcholine in plants was studied systematically in laboratories in the United States (Jaffe, Fluck and Riov), Germany (Hartmann), India (Gupta and Kasturi), Japan (Oata and Hoshino), and elsewhere. In the beginning more attention was paid to the compound in connection with its possible regulation of growth and development. Later the focus was on methodically finding and identifying acetylcholine and enzymes of its synthesis and hydrolysis. Besides the abovementioned authors, a fundamental contribution to this line of study has been made by researchers in the laboratories of Kutacek and Vackova (Chechoslovakia), Smallman (Great Britain), Kopsewicz, Tretyn and Bednarska (Poland), Kendrick (the Netherlands), Momonoki and Momonoki (Japan), and Madhavan (USA). The role of biogenic amines has been studied less than that of acetylcholine. A few reviews have been devoted to neurotransmitters in plants that summarize the data on acetylcholine (Fluck and Jaffe, 1974a, 1976; Roshchina and Mukhin, 1986; Tretyn and Kwiatkowska, 1987; Hartmann and Gupta, 1989; Tretyn and Kendrick, 1991). Some data on the identification of histamine and catecholamines were reported in reviews devoted to plant amines (Smith, 1977, 1980, 1981), serotonin (Grosse, 1982), and specially catecholamines and serotonin (Roshchina, 1991b). Hence there is the need for a monograph that reviews the data of mediators in plants. The Russian edition of this book by V.V. Roshchina, *Biomediators in Plants. Acetylcholine and Biogenic Amines*, was published in Pushshino by the Biological Center of USSR in 1991 (Roshchina, 1991a). This revised edition in English takes into account new data and concepts that were developed later.

The author faced some difficulties with respect to terminology. The term "neuromediator" is used only for animals with a nervous system, whereas the mediatory function of acetylcholine and biogenic amines in plants was only supposed. The term "regulator" is wide and is not limited to the group of substances considered in this book. Taking into account all of this, author believes it possible to use also the relative term "mediator" or "biomediator", pointing to the common biological significance of the components studied and their possible role as signaling substances and mediators. However, our experience shows that the term "neurotransmitters" or "neuromediators" is more understood by a wider audience of both animal and plant physiologists.

Another difficulty was the need to use analogies based on concepts that refer to animal tissue, mainly synaptic contacts between nervous cells, whereas plants lack similar structures: at present their existence is only supposed. Perhaps the similarity in the mechanism of neurotransmitter or neuromediator action can be manifested only at membranous and molecular levels. In this monograph, experimental data about mediators in plants have been summarized and their probable interpretation is given.

Neuromediators, Their Synthesis and Metabolism

1.1. ACETYLCHOLINE

1.1.1. Occurrence of Acetylcholine in Plants

Acetylcholine was first discovered, extracted, isolated, and identified by Ewins (1914) in preparations of ergot spur fungus *Claviceps purpurea*. Other investigators were unable to reproduce his results and interpreted them as arising from bacterial confirmation.

A reawakening of interest in acetylcholine contained in the plant cell was noted in 1947 when Emmelin and Feldberg (Emmelin and Feldberg, 1947) found this substance in stinging trichomes and leaves of common nettle by biological method, based on muscle contraction. Acetylcholine was then investigated in many plants by the methods of biotests (Ladeira *et al.*, 1982a), paper and thin-layer chromatography, electrophoresis (Hartmann, 1971), gas-liquid chromatography (Hartmann and Kilbinger, 1974a) and its combination with mass-spectroscopy (Miura and Shih, 1984a; Tretyn *et al.*, 1987a) as well as nuclear magnetic resonance (NMR) spectroscopy (Tretyn *et al.*, 1987a). This compound was identified in the composition of conserved cabbage and silage (Keil and Portner, 1934, 1935; Marquardt and Spitznagel, 1959), as well as in bacteria that were able to accumulate a large amount of acetylcholine (Stephenson and Rowatt, 1947; Rowatt, 1948; Girvin and Stevenson, 1954; Marquardt and Falk, 1957; Marquardt and Spitznagel, 1959). Since the quantity of acetylcholine in bacteria inducing the fermentation can be higher than the level of this compound in plant tissues, its non-bacterial origin had to be proved. The identification and the quantitative determination of the substance was done by various methods, biological, chemical, chromatographic, electrophoretic, and biochemical (Whittaker, 1963). Initially, the biological method was used: nerve-muscle preparations of frog (Emmelin and Feldberg, 1947) or muscle of leech (Ladeira *et al.*, 1982a) served as biotest on acetylcholine.

However, the use of plant extracts in which, besides acetylcholine, there are other biologically active substances, including mediators serotonin and histamine, made the quantitative determination difficult owing to the superposition or imitation of effects of various compounds. The detection of acetylcholine in most biotests was possible only with the addition of physostigmine, known inhibitor of animal cholinesterases. In the opposite case enzymic hydrolysis of cholinic esters took place, and the concentration of acetylcholine decreased sharply. Other factors also influence the synthesis of acetylcholine. Its amount decreased with increased heat or at pH 10. To get around the shortcomings of the biological method, the compound was purified by paper and thin-layer chromatography as well as electrophoresis (Hartmann, 1971) and, following identification, markers and reagents were used on functional groups. Gas and gas-liquid chromatography were also used (Hartmann and Kilbinger, 1974a, b). These methods were less sensitive to acetylcholine, which was determined only in concentrations of 10^{-6}–10^{-10} M. On the contrary, biotests react on 10^{-13}–10^{-14} M of this compound (Fluck and Jaffe, 1976).

During the 1980s, gas and gas-liquid chromatography were combined with identification by mass-spectroscopy (Miura and Shih, 1984a, b; Tretyn et al., 1987a) and NMR-spectroscopy (Tretyn et al., 1987a). Detailed analysis, using a combination of gas chromatography and NMR-spectroscopy, was done on plant parts by Tretyn with coworkers (1987a). The amount of acetylcholine in organs and tissues varies significantly from species to species (Table 1). Acetylcholine was found in 65 species of 33 families of multicellular plants and in blue-green bacterium *Oscillatoria* (Smallman and Maneckjee, 1981). Analysis of the table shows that this cholinic ester is present in 3 species of fungi of the families Agaricaceae and Hypocreaceae, in one species of moss of the family Funariaceae, and one species of gymnosperm plants of the family Pinaceae. Other acetylcholine-containing species are represented by angiosperms from 27 families. The acetylcholine content depends often on the plant species.

Acetylcholine is particularly abundant in secretory cells of common nettle stinging hairs, where its concentration reaches 10^{-1} M or 120–180 nmol g^{-1} of fresh mass. Together with histamine contained in the secretion, acetylcholine may provoke a pain response and formation of blisters when the plant comes in contact with human skin (Emmelin and Feldberg, 1947). Strong burns are induced by stinging hairs of Australian species *Laportea moroides* (Webb, 1948; Robertson and Macfarlane, 1957; Macfarlane, 1963; Chew, 1965; Oelrich and Robertson, 1970), a poisonous plant of Queensland, because of its pungent and pain-inducing substance (Webb, 1948). As early as in the 19th century, burns provoked by this plant were reported

Table 1. Plant families and species in which acetylcholine is found

Family and species	Part of plant and amount of acetylcholine (nmoles g^{-1} of fresh or *dry mass in parentheses, followed by $\mu g\ g^{-1}$ of fresh mass)	Reference
PLANTS		
Alangiaceae *Alangium lamarckii* Thw.	leaves (ND)	Dasgupta, 1966
Altingiaceae *Liquidambar styraciflua* L.	leaves (0.36) 0.058	Miura and Shih, 1984a
Amaranthaceae *Amaranthus caudatus* L.	stems, shoots (3.1) 0.51	Hartmann and Kilbinger, 1974b
Anacardiaceae *Rhus copallina* L.	leaves (0.78) 0.13	Miura and Shih, 1984a
Apocynaceae *Amsonia angustifolia* Michx.	leaves, seeds, roots (ND)	Tanada, 1972
Aquifoliaceae *Ilex opaca* Ait.	leaves (0.14) 0.023	Miura and Shih, 1984a
Asteraceae (Compositae) *Helianthus annuus* L.	stems, shoots (7.9) 1.29 roots (3.5) 0.57	Hartmann and Kilbinger, 1974b
Porophyllum lanceolatum DC.	leaves (100) 16.32	Horton and Felippe, 1973
Porophyllum lanceolatum	leaves (1.31–19.0) 0.21–3.1	Ladeira *et al.*, 1982b
Porophyllum lanceolatum	stems (2.4–1.02) 0.39–0.47	Ladeira *et al.*, 1982b
Porophyllum lanceolatum	roots (0.91–4.5) 0.15–0.73	Ladeira *et al.*, 1982b
Taraxacum officinale L.	pollen (ND)	Marquart and Vogg, 1952
Xanthium strumarium L.	roots (1.3–10.2) 0.21–1.66 leaves (1.32–6.0) 0.22–0.98 stems (4.5–12.0) 0.73–2.0	Ladeira *et al.*, 1982b
Betulaceae *Betula pendula* Roth.	leaves (11.4) 1.9	Miura and Shih, 1984a
Caprifoliaceae *Lonicera japonica* Thunb	leaves (0.93) 0.14	Miura and Shih, 1984a
Viburnum dilatatum Thunb	leaves (0.28) 0.046	Miura and Shih, 1984a
Chenopodiaceae *Spinacia oleracea* L.	leaves (ND) shoots (6.8) 1.13	Fielder *et al.*, 1953 Miura and Shih, 1984a
Convolvulaceae *Ipomaea abutiloides* (Carnea)	leaves (ND), stems (14) 2.28 flowers (ND), seeds (85) 13.87	Villalobos *et al.*, 1974
Pharbitis nil L.	seedlings (11, 7–109) 1,5–15	Tretyn *et al.*, 1997
Cruciferae (Brassicaceae) *Brassica campestris* var. *napobrassica* L.	above-ground parts (ND)	Holtz and Janisch, 1937

(Contd.)

Table 1. (*Contd.*)

Family and species	Part of plant and amount of acetylcholine (nmoles g^{-1} of fresh or *dry mass in parentheses, followed by $\mu g\ g^{-1}$ of fresh mass)	Reference
B. oleracea var. *gongylodes* L.	above-ground parts (ND)	Holtz and Janisch, 1937
Capsella bursa pastoris L.	above-ground parts (ND)	Boruttau and Cappenberg, 1921
Raphanus sativus L.	leaves (*112–329) 18.28–53.68 petioles (*237–545) 38.67–88.92 roots (*327–510) 53.36–83.22	Momonoki and Momonoki, 1991
Sinapis alba L.	stems (1.8) 0.29	Hartmann and Kilbinger, 1974b
Sinapis alba L.	leaves (3.1–5.0) 0.51–0.82	Ladeira *et al.*, 1982b
Cucurbitaceae		
Cucumis anguria L.	above-ground parts (2.6) 0.42	Ladeira *et al.*, 1982a
C. sativus L.	leaves (ND)	Holtz and Janisch, 1937
C. sativus L.	leaves (*31–332) 5.06–54.17 stems (*476–2991) 77.67–488.04 roots (*62–247) 10.12–40.3 nodes (*12–284) 1.96–46.34	Momonoki and Momonoki, 1991
Cucurbita pepo L.	stems (10.5) 1.71 roots (3.3) 0.54	Hartmann and Kilbinger, 1974b
Euphorbiaceae		
Cnidoscolus texanus (Muell. Arg)	stinging trichomes (ND)	Lookado and Pollard, 1991
Codiaeum variegatum (L.) Blume	leaves (20.6) 3.36	Miura and Shih, 1984a
Fabaceae (Leguminosae)		
Phaseolus aureus Roxb.	leaves (0.32) 0.052	Jaffe, 1970
Phaseolus aureus Roxb.	leaves (17.7) 2.89	Miura and Shih, 1984a
Phaseolus aureus Roxb.	leaves (0.015–50) 0.002–8.16	Roshchina, 1991a
Phaseolus aureus Roxb.	roots (0.36–0.72) 0.059–0.12	Jaffe, 1970
P. vulgaris L.	leaves (2.0) 0.326 stems (7.4) 1.21;	Hartmann and Kilbinger, 1974b
Pisum sativum L.	roots (0.4) 0.07 leaves (2.2) 0.36	Roshchina and Mukhin, 1985a

(*Contd.*)

Table 1. (*Contd.*)

Family and species	Part of plant and amount of acetylcholine (nmoles g^{-1} of fresh or *dry mass in parentheses, followed by $\mu g\, g^{-1}$ of fresh mass)	Reference
Pisum sativum L.	stems (8.2) 1.34 roots (1.4) 0.23	Hartmann and Kilbinger, 1974b
Pisum sativum L.	roots (17) 2.77	Kasturi and Vasantharajan, 1976
Robinia pseudoacacia L.	leaves (traces)	Roshchina, 1991a
Trifolium sp. L.	pollen (ND)	Marquart and Vogg, 1952
Vicia faba L.	pollen (ND)	Marquart and Vogg, 1952
Vigna sesquipedalis (L.) Fruw	etiolated seedlings (ND), seeds (ND)	Hoshino, 1983
Vigna unguiculata (L.) Walp	leaves (*43–103) 7.02–16.81 stems (*67–132) 10.93–21.54 roots (*16–44) 2.61–7.18 primary pulvini (*11–390) 1.80–63.64 secondary pulvini (*63–125) 10.28–20.40	Momonoki and Momonoki, 1991
Funariaceae		
Funaria hygrometrica Hedw. × *Physcomitrium piriforme* Brid. (hybrid)	callus (0.11–124.12) 0.018–20.25	Hartmann, 1971; Hartmann and Kilbinger, 1974a
Gramineae		
Avena fatua L.	seeds and seedlings (ND)	Tretyn *et al.*, 1985
A. sativa L.	green seedlings (10) 1.63	Tretyn and Tretyn, 1988a, 1990
Stipa tenacissima Linne.	leaves (34) 5.55	Antweiler and Pallade, 1972
Triticum vulgare L.	seeds, seedlings (ND)	Tretyn *et al.*, 1985
Zea mays L.	leaves (6.0) 0.98	Roshchina, 1991a
Lemnaceae		
Lemna gibba (G_3) L.	coleoptiles (ND)	Hoshino and Oata, 1978
Loranthaceae		
Viscum album L.	shoots (ND)	Winterfeld (cited in Fluck and Jaffe, 1976); Krzaczek, 1977 Volinskii *et al.*, 1983
Mimosaceae		
Albizzia julibrissin Durazz.	leaves (ND) seeds (ND)	Satter *et al.*, 1972

(*Contd.*)

Table 1. (*Contd.*)

Family and species	Part of plant and amount of acetylcholine (nmoles g^{-1} of fresh or *dry mass in parentheses, followed by $\mu g\ g^{-1}$ of fresh mass)	Reference
Macroptilium atropurpureum (DC) Urban	leaves (*21–75) 3.43–12.24 petioles (*23–89) 3.75–14.52 primary pulvini (*41–70) 6.69–11.42 secondary pulvini (*35–164) 5.71–26.76 stems (*30–92) 4.90–15.01 roots (*15–20) 2.45–3.26	Momonoki and Momonoki, 1992
Moraceae		
Artocarpus champeden Merr.	seeds, leaves; stems (20) 3.26	Lin, 1955
A. integra Merr.	seeds, leaves (55–3800), 8.97–620.05	Lin, 1955; 1957
A. integra Merr.	seeds, latex (ND)	Vogel *et al.*, 1964
Pinaceae		
Pinus sylvestris L.	above-ground shoots (ND)	Kopcewicz *et al.*, 1977
Plantaginaceae		
Plantago lanceolata L.	seeds, seedlings (ND)	Tretyn *et al.*, 1985
P. rugelli Decne.	leaves (1.65) 0.27	Miura and Shih, 1984a
Polygonaceae		
Rumex obtusifolius L.	leaves (2.8) 0.46, seeds, seedlings (ND)	Ladeira *et al.*, 1982a; Tretyn *et al.*, 1985
Populaceae		
Populus grandidentata Michx.	leaves (13.7) 2.24	Miura and Shih, 1984a
Rosaceae		
Crataegus oxyacantha L.	leaves, flowers, fruits (100–1000) 16.32–163.17	Fielder *et al.*, 1953
Prunus serotina Ehrh.	leaves (1.72) 0.28	Miura and Shih, 1984a
Salicaceae		
Salix caprea L.	pollen (ND)	Marquardt and Vogg, 1952
Scrophulariaceae		
Digitalis ferruginea L.	leaves (220) 35.90	Tulus *et al.*, 1961
D. lanata L. (Ehr.)	leaves (ND)	Neuwald, 1952
D. lanata L. (Ehr.)	leaves (0.175) 0.028	Chkhve Tkhesop, 1987
D. purpurea L.	leaves (ND)	Chkhve Tkhesop, 1987
Smilacaceae		
Smilax hispida Muhl.	leaves (0.78) 0.13	Miura and Shih, 1984a
Solanaceae		
Solanum ambrosiacum L.	fruits (360) 3.76	Fonteles *et al.*, 1993
S. tuberosum L.	tubers (ND)	Marquardt *et al.*, 1952
S. nigrum L.	fruits (ND)	de Melo *et al.* 1978

(*Contd.*)

Table 1. (Contd.)

Family and species	Part of plant and amount of acetylcholine (nmoles g^{-1} of fresh or *dry mass in parentheses, followed by $\mu g\ g^{-1}$ of fresh mass)	Reference
Umbelliferae		
Daucus carota var. *sativa* L.	leaves (ND)	Holtz and Janisch, 1937
Carum copticum Benth.	above-ground shoots (ND)	Devasankaraiah et al., 1974
Urticaceae		
Girardinia heterophylla Gandich (Decne)	leaves (273) 44.55	Saxena et al., 1966
Laportea moroides Wedd.	leaves 0.011–0.029 stinging hairs (0.07–0.175 per hair)	Robertson and Macfarlane, 1957
Urtica dioica L.	leaves, hairs, stems, roots (ND)	Emmelin and Feldberg, 1949; Collier and Chesher, 1956
U. parviflora Roxb.	leaves, hairs (ND)	Saxena et al., 1965
U. urens L.	leaves, hairs, stems, roots (120–180)19.58–29.37	Emmelin and Feldberg, 1947
FUNGI		
Agaricaceae		
Agaricus campestris (*Psalliota campestris*) (Fr.) Quel	mycelium (ND)	Heirman, 1939
Lactarius blennius Fr.	mycelium (ND)	Oury and Bacq, 1938
Hypocreaceae (Ascomycetes)		
Claviceps purpurea Talasne.	mycelium (ND)	Ewins, 1914
ANIMALS		
Rattus sp.	brain (0.564–5640)	Michelson and Zeimal, 1973
Rattus sp.	skeleton muscle (0.326–65200)	Lysov, 1982
Apis sp.	stinging cell (48.9)	Prosser, 1986

ND: There are no quantitative data

to be the cause of serious diseases of horses and workers engaged in railway building. Some other species of *Laportea* growing in New Guinea have effects so serious as to cause death. Australian aborigines used above-ground sprouts and fruits of this plant as a topical drug against rheumatism. The substance inducing a pain response may retain activity for 40 years when in dry plant material (Shaw *et al.*, 1955), as cited from Robertson and Macfarlane, 1957). The chemical composition of the secretion was studied by Rechter (1949 as cited from Robertson and Macfarlane, 1957). The secretion of stinging hairs was found to contain 0.01–0.025 μg of acetylcholine, 0.025–0.05 μg of histamine, and 0.001 ug of serotonin per hair (Robertson and Macfarlane, 1957).

A high concentration of acetylcholine was noted in breadfruit tree *Artocarpus integra* (Lin, 1955). It is unevenly distributed throughout the plant. In the stems, the concentration of the choline ester is low, about 55 nmole g^{-1} fresh weight. Leaf veins and seeds contain about 30–700 times more acetylcholine than the stems. It is noteworthy that acetylcholinesterase activity was not registered in tissues of breadfruit tree. This may account for the high concentration of acetylcholine in this plant (Lin, 1955). However, another species, *Artocarpus champeden*, contains relatively little cholinic ester in seeds (20 nmole g^{-1} fresh weight) and in leaves and stems.

Acetylcholine is often found in stinging hairs and nettles or princkles. Nettles are usually plants having stinging hairs. They are found mainly in four families: Urticaceae, Euphorbiaceae, Loasaceae, and Hydrophyllaceae. The nettle plant *Cnidoscolus texanus* (Muell. Arg) Small, belonging to family Euphorbiaceae and grown in the southwestern United States and Mexico, contains acetylcholine, histamine, and serotonin (Lookado and Pollard, 1991) in stinging trichomes. About 46 nmoles ACh g^{-1} of dry weight is found in the secondary pulvinus of *Macroptilium atropurpureum* cv. Siratro (Momonoki and Momonoki, 1992). Gas-liquid chromatography was used to study the changes in concentrations of acetylcholine in leaves of yard-long bean, cucumber, radish treated by heat shock and untreated radish. Under heat shock the amount of acetylcholine is 2–10 times higher in stems, leaves, nodes, and roots than in the plant grown in temperate zone without heating (Momonoki and Momonoki, 1991). In the work of Momonoki and Momonoki (1992), the changes in acetylcholine content in plant tissues of *Macroptilium atropurpureum* cv. Siratro following leaf wilt and leaf recovery after heat stress on detached and heat-stressed leaves to which the primary pulvinus was still attached were determined. The content of acetylcholine in the primary and secondary pulvini also changed dramatically within 3 min. after heat stress. The changes of acetylcholine content in plant tissues were then found to be correlated with leaf drop and leaf recovery. Acetylcholine is also found in reproductive cells such as pollen of willow *Salix caprea, Taraxacum officinale, Trifolium* sp., *Vicia faba* (Marquardt and Vogg, 1952) and in bee honey, which also contains pollen (Crane, 1955). The amount of acetylcholine in honey was near 2.5 mg per kg, and decreases through the third year to 75% approximately (Goldschmidt and Burkert, 1955).

The concentration of acetylcholine in different parts of the plants varies greatly. According to Hartmann and Kilbinger (1974b), the highest content of this substance was found in stems of some plant species. In leaves, the amount of acetylcholine was 3 times less and in seeds 70 times less than in stems. According to other authors (Miura and Shih, 1984a), the amount acetylcholine in leaves is 5–10 times higher that in root systems and 10–50 times that in stems. Acetylcholine is found not only in vegetative parts of plants (Miura and Shih, 1984a), but in seeds and green seedlings (Tretyn and Tretyn, 1988a). Comparison of the data obtained for cholinic ester-containing plants with the data for animal tissues indicates that the lower threshold of acetylcholine content in animal tissues is comparable with the values obtained for plants, and the upper value is 100 and even 1000 times higher. The concentration of this substance at the transmission of electric signal from motor nerve to muscle reaches 1 mM (Changeux and Revah, 1987), in synaptic vesicles of *Torpedo* electric organ 0.05–6.9 nmoles/mg protein (Whittaker, 1987), and in cell of human neuroblastoma only 1.9 nmoles/mg protein (Blusztajn *et al.*, 1987).

The localization of acetylcholine in a plant cell is poorly understood. In animals it is stored in secretory vesicles that are separately from the Golgi apparatus. Sometimes the vesicles are coated with the special protein clathrin. Similar vesicles are found in plant cells (Coleman *et al.*, 1988). Jaffe (1976) showed that the site of synthesis and primary localization of acetylcholine in plant cells is the membranes of the endoplasmic reticulum. Acetylcholine and butyrylcholine were also found in intact chloroplasts isolated from leaves of *Pisum sativum, Phaseolus aureus, Zea mays,* and *Urtica dioica* (Table 2). In some of the plant species studied, for instance false acacia *Robinia pseudoacacia,* the acetylcholine content has not been detected, whereas butyrylcholine was absent in maize *Zea mays* and common nettle *Urtica dioica* (Roshchina, 1989a, b). The difficulty in detecting cholinic esters could have several causes. First, acetylcholine and butyrylcholine are water-soluble and easily leach from plastids during the isolation procedure. The amount of endogenous cholinic esters may be small, and therefore it is necessary to concentrate extracts. As a result, cholinic derivatives appear to decompose. Besides, it should block the

Table 2. Cholinic esters in chloroplasts, nmol g^{-1} of fresh mass (Roshchina, 1989a, b)

Plant	Acetylcholine	Butyrylcholine
Pisum sativum L.	0.069–8.2	0–731
Phaseolus aureus Roxb.	0.1–49	3.8–100
Zea mays L.	0–2.3	0.0
Urtica dioica L.	2.0	0.0
Robinia pseudoacacia L.	0.0	260

enzymes hydrolyzing cholinic esters. All these factors may reduce the concentration of acetylcholine below the threshold of sensitivity of the detection method used.

The synthesis of acetylcholine is evidently stimulated by light and depends on photosynthesis (Tretyn, 1987; Tretyn et al., 1988). It should be noted that the acetylcholine content can undergo strong fluctuations and depends on the age of plant, its development phase, conditions of growth, and other factors. Therefore, in Table 2 a range is given within which the values varied.

1.1.2. Metabolism of Acetylcholine

The precursors of acetylcholine in plants, just as in animals, are acetyl-CoA and choline. The synthesis proceeds as follows (Fig. 1): Acetyl-CoA is synthesized from acetate by means of acetyl-CoA-synthetase or from pyruvate by means of the pyruvate dehydrogenase complex (Heise and Treede, 1987). Choline is formed in two ways: from the amino acid of serine and from phosphatidylcholine of membranes (Blusztajn et al., 1987). It is known that the pool of free choline in plants is very high (from 1000 in the dark to 3740 nmoles g^{-1} fresh mass in the light) and exceeds 100–1000 times the content of acetylcholine per se (Miura and Shih, 1984a). The source of free choline is serine and phosphatidylcholine from membranes and, probably, also acetylcholine, which release choline on hydrolysis.

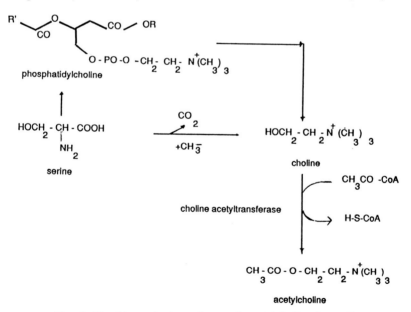

Fig. 1. The biosynthetic pathway of acetylcholine formation

Synthesis of acetylcholine per se proceeds with participation of the enzyme choline acetyltransferase (EC 2.3.1.6). Using (^{14}C)-acetyl-CoA as substrate, Barlow and Dixon (1973) first found the enzyme in leaves and single stinging hairs of nettle *Urtica dioica*. The choline acetyltransferase activity was also registered in extracts of pea *Pisum sativum* buds, cabbage *Brassica oleracea* var. *botrytis* and hypocotyls of beans (Biro, 1978) as well as in seeds of onion *Allium altaicum* (Hadacova et al., 1981). Smallman and Maneckjee (1981) carried out a comparative study of the acetylcholine synthesis. The rate of acetylcholine formation was found to be highest in *Urtica dioica* and was comparable with that in *Musca domestica*. As a whole, the activity of choline acetyltransferase and acetylcholine content in animal tissues are much higher than in plants. However, in some groups of plants, for example, in nettle, the activity is as high as in animal tissues. The presence of the enzyme in the blue-green bacterium *Oscillatoria agardhii* suggests that acetylcholine was probably synthesized in ancient plants.

Jaffe (1976) supposed that synthesis of acetylcholine takes place in vesicles of rough endoplasmic reticulum of root cells of *Phaseolus aureus*. He has isolated these vesicles and has shown the inclusion of label originated from ^{14}C-acetyl-CoA to acetylcholine. This author also demonstrated that red light 660 nm stimulates the synthesis of acetylcholine, unlike far red light (near 730 nm), which led him to conclude that phytochrome participates in this process. The breaking of vesicles by Triton X-100 increased 5-fold the rate of the reaction under red light, which demonstrates the localization of choline acetyltransferase within secretory vesicles. This enzyme has been isolated from some plant species (see 3.3).

Acetylcholine may also be involved in the biosynthesis of ethylene at the stage of 1-amino-cyclopropane-1-carbonic acid, thus inhibiting the formation of this gas (Jones and Stütte, 1988).

The hydrolysis of acetylcholine is accomplished by means of cholinesterases (see Chapter 3). Cholinesterase activity in plant cells was first found in agaric fungi (Oury and Bacq, 1938), then in characean algae *Nitella* (Dettbarn, 1962), lichens (Raineri and Modenesi, 1986), and higher plants (Riov and Jaffe, 1973b; Fluck and Jaffe, 1974b). Besides specialized enzymes, the ability to hydrolyze acetylcholine, although at lower rates (1000 times), is observed for other plant esterases such as pectinesterase, allylesterase (Fluck and Jaffe, 1976), and sinapinesterase (Tzagoloff, 1963b).

In plants, choline is a substrate for not only acetylcholine synthesis, but other cholinic esters as well. Figure 2 shows the formation of cholinic esters known for plants. Besides acetylcholine, propionylcholine (Miura and Shih, 1984b), butyrylcholine (Roshchina, 1991a, c), and sinapinecholine (Kefeli et al., 1977) have been identified. Sinapoylcholine (sinapine, sinapylcholine) is found in cotyledons of *Raphanus sativus* L. (Linscheid

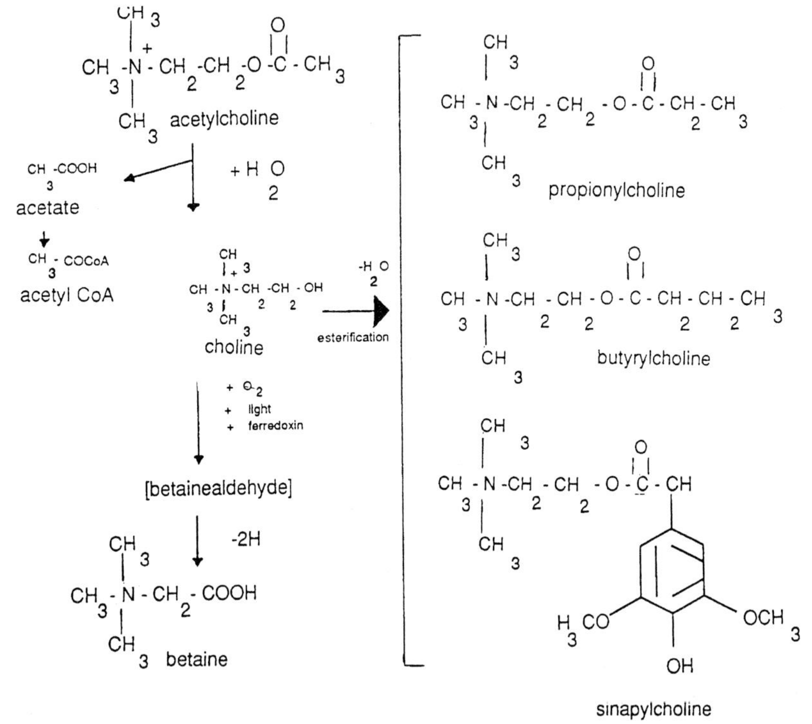

Fig. 2. Metabolism of acetylcholine in plants

et al., 1980) and feruloylcholine was isolated from *Cleome pungens* Willd. (Pagani and Romussi, 1969). The conditions and modes of production of these esters are not yet well studied. Propionylcholine (0.11–2.3 nmoles g^{-1} fresh mass) is found in leaves of *Codiaeum variegatum, Phaseolus aureus, Populus grandidentata, Betula pendula,* and *Plantago rugelli* (Miura and Shih, 1984b), while butyrylcholine (4.6–260 nmole g^{-1} fresh mass) is found in leaves of *Pisum sativum, Phaseolus aureus, Robinia pseudoacacia* (Roshchina, 1991a, c). Sinapinecholine was found only in plants belonging to the family Cruciferae (Brassicaceae), in particular in green tissues up to 56 mg% and in etiolated ones up to 10 mg% (Kutacek et al., 1981). It is formed from sinapic acid and choline. Sinapic acid is synthesized from cinnamic acid with participation of hydroxylating (hydroxylase of cinnamic acid and polyphenol oxidase) and methylating enzymes. Enzymes catalyzing synthesis of sinapinecholine are choline sinapinetransferase (Grawe and Strack, 1986) or choline acetyltransferase (Kutacek et al., 1981). Hydrolysis of this cholinic ester, as well as acetylcholine, can be due to cholinesterase (Kutacek et al., 1981). The study of partly purified protein extracts of pea, where the choline acetyltransferase and esterase activities are found, have shown that these preparations are able to synthesize cholinic esters of oxycinnamic acid (n-coumaroyl-, caffeyl-, feruloyl-, sinapoylcholines) and to hydrolyze sinapinecholine as well (Vackova et al., 1982a, b). In comparison with animals, plants synthesize excessive amounts of esters of choline and aromatic acids.

Besides the above-mentioned compounds, choline is the origin of betaine (or glycine betaine), which preserves plants against salt stress. For instance, choline and betaine are accumulated in *Trifolium alexandrinum* L. (Varsheney et al., 1988). Betaine is found in chloroplasts of some plants grown in arid regions, for instance barley, or halophytes as well as representatives of family Chenopodiaceae such as beet and spinach (Hanson and Grumet, 1985; Weigel et al., 1988). Moreover, betaine is formed in animal mitochondria (Zhang et al., 1992). Glycine betaine enhanced and stabilized oxygen release and ATP synthesis (Mamedov et al., 1991). Reactions occur with participation of enzyme pyridinenucleotide-dependent betaine aldehyde dehydrogenase with molecular mass ~120 kDa (Hanson and Grumet, 1985; Hanson, 1988). Activation of this enzyme is induced to the same extent by increase in salinity. Moreover, light and O_2 are necessary for betaine synthesis, as well as reduced ferredoxin. Choline also stimulated synthesis of some extracellular proteins in *Trichoderma* cells (Scheiber et al., 1986). Thus, utilization of acetylcholine in plants leads to the increase in the pool of choline, which easily transforms into betaine or esterifies into other esters. Choline chloride and allylcholine are rapidly incorporated into wheat protoplasts compared with the incorporation rate of benzylcholine bromide (Che et al., 1990). Earlier, choline was observed to stimulate oxygen liberation in photosynthesis by 10% (Hyeon et al., 1987). Choline (5–15 mole per m^{-3}) enhanced phospholipid and sterol levels prior to chilling (5°C) and maintained lipid levels throughout the chill-warm cycle, without any significant change in phospholipids (Guye, 1989). Moreover, it recovers the ethylene-forming enzyme activity when plants are transferred from chilling to warm conditions. Choline induced chill-tolerance in mung bean (*Vigna radiata* L. Wilcz.), via reduction of stomatal aperture prior to chilling, which decreases the development of symptoms of chilling-induced water stress in primary leaves at 5°C. Guye and coworkers (1987) showed that in nonchilled plants choline caused increases in concentration of chlorophylls, carotenoids and the carotenoid/chlorophyll ratio. Choline treatment prevents the photo-induced degradation of chlorophyll and carotenoids under chilling. This is due to enhanced concentrations of phospholipids and sterol formed under this treatment, which is favourable to isoprenoid metabolism as a whole and the pigments, in particular. Choline is oxidized by oxygenase to betaine aldehyde and then to betaine (Lerma et al., 1988). Acetic acid, perhaps, is easily metabolized to acetyl-CoA, actively participating in various syntheses, from organic acid in the Calvin cycle to terpenoids. The derivatives of choline can be biologically active. Among 17 artificial cholinic derivatives Vassilev (1987) found the substances having features of inhibitors, retardants, or growth stimulators.

1.1.3. Effect of External Factors on Metabolism of Acetylcholine

Light. Light may directly affect the level of acetylcholine in tissues (Jaffe, 1970; Hartmann and Kilbinger, 1974a, b; Miura and Shih, 1984a), thus regulating the activity of the enzymes responsible for its synthesis and degradation. The acetylcholine content in leaves of *Phaseolus aureus* exposed to continuous illumination with white light is twice that in leaves kept in the dark. In stems, it decreases abruptly (3-fold) and in roots it remains practically unchanged (Miura and Shih, 1984a). It is worth noting that in leaves the level of both acetylcholine and choline is higher than in roots and stems. This was found for many higher plants: sprouts of *Phaseolus vulgaris, Pisum sativum, Sinapis alba, Cucurbita pepo, Helianthus annuus, Spinacia oleracea, Amaranthus caudatus* (Hartmann and Kilbinger, 1974b), seeds and seedlings of *Avena sativa, A. fatua, Plantago lanceolata, Rumex obtusifolius, Triticum vulgare* (Tretyn et al., 1985), and moss *Funaria hygrometrica* × *Physcomitrium piriforme* (Hartmann and Kilbinger, 1974a). Etiolated sprouts of *Pisum sativum* and *Sinapis alba* (Hartmann and Kilbinger, 1974b) contained no acetylcholine. According to other authors, etiolated seedlings, for example, of *Avena sativa,* may contain small amounts of this compound (Tretyn et al., 1987a).

The necessity of light for acetylcholine synthesis has been demonstrated for many plants. Exposure to white light results in intensive accumulation of acetylcholine in secondary roots of *Phaseolus aureus* (Jaffe, 1970). Tretyn and coworkers (1987a) showed that the level of endogenous acetylcholine in etiolated plantlets of oats increases 2–3 times as they turn green, in both white and red light. According to Jaffe (1976), light stimulates synthesis of acetylcholine in membranes of endoplasmic reticulum. It was also found that the incorporation of labeled precursors in acetylcholine occurs only after illumination (Hartmann, 1979), indicating that synthesis of cholinic ester is associated with photosynthesis. This assumption seems to be very likely since acetylcholine was found in isolated chloroplasts of some plants (Roshchina, 1989a).

Light-stimulated synthesis of acetylcholine was not observed in all plants, presumably, owing to their taxonomic properties. For example, no marked difference was observed in the amount of acetylcholine in leaves of *Albizzia julibrissin* in the light and dark (Satter et al., 1972), whereas in Scots pine the level of acetylcholine in the dark is even higher than in the light (Kopcewicz et al., 1977).

There is evidence that accumulation of acetylcholine in tissues depends on the wavelength of light the plants are exposed to (Jaffe, 1970, 1972a, b; Hartmann, 1971; Hartmann and Kilbinger, 1974a; Kopcewicz et al., 1977; Tretyn et al., 1985). Far red light ($\lambda > 730$ nm) decreases the level of acetylcholine, probably by stimulating the acetylcholineesterase activity. Red light, on the contrary, enhances acetylcholine synthesis by stimulating the activity of choline acetyltransferase in the presence of choline and acetyl-CoA. According to some authors (Hartmann and Kilbinger, 1974a), red light is more efficient than white light. In white light, the acetylcholine level in callus of moss *Funaria hygrometrica* was only 27% (34–35 nmole g^{-1} fresh mass) of acetylcholine content in plants growing in red light, whereas exposure of callus consecutively to red and then to far red light inhibited sharply the synthesis of the cholinic ester (2.23 nmole g^{-1} fresh mass). In vesicles of crude endoplasmic reticulum from cells of roots acetylcholine synthesis in red light, increased 2–19 times compared to roots kept in the dark, whereas in far red light no changes were observed or the acetylcholine synthesis decreased 5 times (Jaffe, 1976). Acetylcholine is concentrated in illuminated parts of plants, for instance in cotyledons of tomato *Lycopersicon esculentum* seedlings (near 381 mmole g^{-1} of fresh weight) whereas the lowest amount upto 2×10^6 times in roots and ethiolated plants (Wisniewska and Tretyn, 1999). Under illumination the highest level of acetylcholine in *Cucumis sativus* and *Vigna unquiculata* was observed in the stem, than in leaves, but underground organs of *Raphanus sativus* contained more the mediator than the overground ones. Perhaps, in various plant species synthesis and decomposition of acetylcholine depend on the different location and activity of the enzymes choline acetyltransferase and cholinesterase. Different organs may percept light via different mechanisms.

Acetylcholine synthesis is believed to be controlled by phytochrome (Jaffe, 1970, 1972a, b; Hartmann and Kilbinger, 1974a), which occurs in two forms sensitive either to far red light or to red light. Using mutants of tomato varied in the phytochrome content, Wisniewska and Tretyn (1999) have shown that synthesis of acetylcholine is not exclusively under phytochrome rgeulation. In green plants the synthesis of acetylcholine probably takes place in chloroplasts and it's metabolism is linked with the light phase of photosynthesis (Roshchina, 1987a, b; 1989a). It is established that phytochrome participates in Emmerson effect during photosynthesis (Eichorn et al., 1987). From this premise, higher effectiveness of acetylcholine synthesis in red light due to the effect of both photosystems, whereas the decrease in the process under far red light is explained by low effectiveness of photosynthesis when only photosystem 1 functions. The stimulating effect of red light does not show up in all plants.

The amount of acetylcholine in *Albizzia, Mimosa pudica* in the red light and in the dark is approximately equal, 2–2.3 nmole g^{-1} (228–330 ng/g fresh mass of leaves (Satter *et al.*, 1972). The amount of acetylcholine accumulated in plants is also associated with photoperiodism. Indeed, in cells of *Lemna gibba* G 3 growing under short-day conditions, the amount of accumulated acetylcholine was greater than in plants growing under conditions of long day or continuous illumination (Hoshino and Oata, 1978). According to other authors (Ladeira *et al.*, 1982b), the acetylcholine level in plants growing under conditions of short day and continuous illumination, though different, has no direct effect on the flowering of plants of both short day (*Xanthium strumarium* and *Porophyllum lanceolatum*) and long day (*Sinapis alba*). Acetylcholine and cholinergic agents also regulate the photoperiodic flower induction of *Pharbitis nil* (Lukasiewicz-Rutkowska *et al.*, 1997).

Oxygen. The concentration of O_2 in tissues is probably dependent on the rate of cholinic ester synthesis. However, it should be noted that this process is poorly understood. In *Rumex obtusifolius* and *Cucumis anguria*, acetylcholine was found only in aerial parts of the plants: stems and leaves (Ladeira *et al.*, 1982a).

Temperature. The rate of acetylcholine synthesis depends on temperature, which affects primarily the activity of choline acetyltransferase and the activity of the enzyme cholinesterase hydrolyzing the cholinic ester. The optimal temperature range for choline acetyltransferase is between 25°C and 30°C (Barlow and Dixon, 1973) just as for cholinesterase (Riov and Jaffe, 1973b; Ernst and Hartmann, 1980; Kasturi and Vasantharajan, 1976). For cholinesterase isolated from chloroplasts, the optimal temperature is 28°C (Roshchina, unpublished data). Cholinesterases from many animals have a higher optimum of + 37 °C (Ellman *et al.*, 1961). As temperature increases, the rate of nonenzymic hydrolysis of acetylcholine rises abruptly and is comparable with the rate of enzymic degradation (Ernst and Hartmann, 1980). Under heat stress (30°C, 3 min.) in *Vigna unguiculata, Cucumis sativus,* and *Raphanus sativus,* the acetylcholine content decreased by 30–70% in leaf, stem, secondary pulvini, leaf blades, and petioles and increased by 160–200% in primary pulvini in comparison with control (22–23°C) (Momonoki and Momonoki, 1991). In the same conditions in *Macroptilium atropurpureum*, concentration of acetylcholine dramatically dropped in leaf, petiole, stem, root and secondary pulvini, whereas in primary pulvini it increased 2-fold from 41 to 70 nmol g^{-1} dry weight.

Effects of pH. The optimal pH value for acetylcholine hydrolysis by cholinesterases is 7.0–7.8 (Ernst and Hartmann, 1980). This value is close to that for cholinesterases from animal tissues. As pH changes from 7.6 to 8.0, the rate of nonenzymic hydrolysis of acetylcholine increases rapidly by several orders. Choline acetyltransferase is also effective at pH 8.0 (Barlow and Dixon, 1973). The optimal pH value for cholinesterases from chloroplasts compared to cholinesterases from animal tissues (8.0) (Ellmann *et al.*, 1961) was somewhat lower.

Ionic strength. Bivalent Mn^{2+} and Ca^{2+} (1–10 mM) decrease the activity of acetylcholinesterase 2–3 times and Mg^{2+} ion has no effect on it (Ernst and Hartmann, 1980).

Pesticides. Widely-used organophosphorus compounds such as chlorophos, phthalophos, and diisopropyl phosphofluoridate inhibit cholinesterase activity (Roshchina and Semenova, 1990). Organophosphate insecticides retard the germination of wheat seeds (Molyneux and McKinlay, 1989). The cholinesterase activity of the seedlings declined under this treatment, which according to Molyneux and McKinlay confirms the hypothesis of inhibition of cereal germination by insecticides as a result of the blocking of cholinesterase. Organophosphates depress plant growth and synthesis of chlorophyll (Wong and Chang, 1988). Besides these compounds, some retardants (derivatives of carbaminic acid or quaternary amines AMO-1618) also inhibit plant cholinesterases (Riov and Jaffe, 1973a, c). Substances synthesized by plants and having insecticidal qualities, for instance physostigmine from *Physostigma venenosum* and permethrin from *Chrysanthemum*, act by a similar mode (Fluck and Jaffe, 1974b; Hoshino, 1983).

1.2. CATECHOLAMINES

1.2.1 Occurrence in Plants

In 1956–1958, catecholamines were first found in studies of the mechanisms underlying the therapeutic effect of fruits of bananas *Musa* sp. (Waalkes *et al.*, 1958; Udenfriend *et al.*, 1959). To date, catecholamines have been found in 28 species of 18 families (Tables 3–6).

The basic methods for identification and quantitative determination of catecholamines in plants are anion-exchange, paper and thin layer chromatography (Waalkes *et al.*, 1958; Udenfriend *et al.*, 1959). In the last few years, the method of gas chromatography has been used in combination with mass-spectroscopy (Kamisaka and Shibata, 1982).

Table 3. The occurrence of dopamine in plants

Family and species	Part of plant and amount of dopamine ($\mu g\ g^{-1}$ of fresh mass)	Reference
Asteraceae (Compositae)	seedlings (ND)	Kamisaka, 1979;
Lactuca sativa L.		Kamisaka and Shibata, 1982
Cactaceae		
Carnegiea gigantea Britt. et Rose	above-ground part 3000–4000	Steelink et al., 1967
Lophophora williamsii Coult (Lemet SD)	above-ground part (ND)	Lundström, 1971a
Trichocereus pachanoi Br. et R.	above-ground part (ND)	Lundström, 1970
Chenopodiaceae		
Beta vulgaris L.	leaves (ND)	Gardner et al., 1967
Spinacia oleracea L.	leaves (ND)	Gewitz and Volker, 1961
Fabaceae		
Cytisus scoparius Link.	foliage, shoots (ND)	Tocher and Tocher, 1972
Mucuna pruriens DC (Holland)	culture of leaf tissue (ND)	Wichers et al., 1989, 1993
Lauraceae		
Persea L.	fruits 4–5	Udenfriend et al., 1959
Mimosaceae		
Entada pursaetha DC	seeds (ND)	Larsen et al., 1973
Monostromatadaceae		
Monostroma fuscum Wittr.	homogenate of cells (ND)	Tocher and Tocher, 1969
Musaceae		
Musa sp. L.	peel of fruit 700	Waalkes et al., 1958 Udenfriend et al., 1959
Musa sp. L.	pulp of fruit 8	Waalkes et al., 1958
Musa sp. L.	fruit 1–20	West, 1958
Nyctaginaceae		
Hermidium alipes (L.)	all parts of plant (ND)	Buelow and Gisvold, 1944
Papaveraceae		
Papaver bracteatum Lindl.	latex of fruits (ND)	Kutchan et al., 1986
Papaver somniferum L.	latex of fruits (ND)	Roberts, 1986
Piperaceae		
Piper amalago L.	leaves (ND)	Durand et al., 1962
Portulacaceae		
Portulaca grandiflora Hook	callus (ND)	Endress, 1977; Endress et al., 1984
Solanaceae		
Aconitum napellus L.	all parts 4,000–48,000 roots 4000–12,000 flower buds 125,000 flowers 58,000 flower heads 50,000 fruits 145,000	Faugeras at al., 1967

(Contd.)

Table 3. (*Contd.*)

Family and species	Part of plant and amount of dopamine ($\mu g\ g^{-1}$ of fresh mass)	Reference
Solanaceae		
Aconitum napellus L.	fruits 1000–14,500	Faugeras, 1967
A. paniculatum Lam.	fruits 100–3600	Faugeras, 1967
Solanum tuberosum L.	fruits (ND)	Bygdeman, 1960
Solanum tuberosum L.	whole cuttings 0.17 aerial parts 0.24 tubers 0.10 roots 0.19	Hourmant *et al.*, 1998
Verbenaceae		
Stachytarpheta jamaicensis Vahl.	leaves (ND)	Durand *et al.*, 1962

ND: Amount was not determined.

Table 3 presents the data on the level in plants of dopamine, which has been found so far in 20 species of 20 families, including green algae *Monostroma fuscum*. The greatest amount of dopamine is contained in banana peel; in the pulp it is 87 times less. Dopamine is found in all organs of *Aconitum napellus* (family Solanaceae), but especially in flowers and young fruits, whereas its concentration is lowest in roots (Faugeras *et al.*, 1967). In latex of poppy *Papaver bracteatum*, the content of dopamine is 0.1% (1 mg/ml) and it easily transforms into morphine and other alkaloids (Kutchan *et al.*, 1986). In plant culture of tissue from *Papaver somniferum* dopamine accumulated at about 1.0 mM g^{-1} of fresh mass after 5 days' exposure in liquid medium and after 20 days reached 4.0 mM g^{-1} of fresh mass (Anderson *et al.*, 1983). In seeds of *Entada pursaetha* DC (Mimosaceae), there are also glycosides of dopamine dopamine-3-o-glucoside and dopamine-4-o-glucoside (Larsen *et al.*, 1973). The occurrence of DOPA and dopamine is also demonstrated in the callus of *Portulaca grandiflora* (Endress, 1977). Quantity of dopamine rises sharply on wounding and other stresses. On the wounding of cactus *Carneginea gigantea* the defensive callus arises and its cortical tissue (pulpy cortex) contains dopamine as a main phenolic component (Steelink *et al.*, 1967). General concentration of dopamine on wounding can reach 1% of common pulp of callus. The concentration of the substance increases from 0.3–0.4% per g of fresh mass in unwounded parts to 0.58–0.64% per g of fresh mass in wounded ones.

Like dopamine, noradrenaline was first identified in banana fruits and was considered to be responsible for the enhancement of the therapeutic effect (Waalkes *et al.*, 1958; Wagner, 1988). It was found in 20 species of 10 families (Table 4). In leaves of some leguminous Fabaceae and

Table 4. The occurrence of noradrenaline in plants

Family and species	Part of plant and amount of noradrenaline ($\mu g\ g^{-1}$ of fresh mass)	Reference
Asteraceae (Compositae)		
Lactuca sativa L.	seedlings (ND)	Kamisaka, 1979; Kamisaka and Shibata, 1982
Cactaceae		
Trichocereus pachanoi Br. et R.	above-ground part of plant (ND)	Dobbins, 1968; Lundström, 1970
Crassulaceae		
Sedum pachyphyllum Rose	leaves 0.017–0.040	Kimbrough *et al.*, 1987
Crassulaceae		
Sedum pachyphyllum Rose	leaves 0.017–0.040	Kimbrough *et al.*, 1987
S. morganianum E. Walth	leaves 0.012–0.0301	Kimbrough *et al.*, 1987
Fabaceae		
Mucuna pruriens DC (Holland)	culture of leaf tissue (ND)	Wichers *et al.*, 1989, 1993
Phaseolus aureus Roxb.	leaves 0.49–1234	Roshchina, 1991a, c
P. multiflorus Lam.	pulvini 0.6, leaves 0.6	Applewhite, 1973
Pisum sativum L.	tendrils 1.8, stems 0.8, leaves 1.0	Applewhite, 1973
Pisum sativum L.	leaves 0.027–42.7	Roshchina and Mukhin, 1985a
Robinia pseudoacacia L.	leaves 0–6760	Roshchina, 1991a
Samanea saman Merr.	secondary pulvini 8.3, leaves 0.5, osts 5.0	Applewhite, 1973
Mimosaceae		
Albizzia julibrissin Durazz.	secondary pulvini 4.6, leaves 2.8, osts 3.4	Applewhite, 1973
Mimosa pudica L.	primary pulvini 3.5, petioles 0.6	Applewhite, 1973
Musaceae		
Musa sp. L.	fruit peel 1–20	West, 1958
Musa sp. L.	fruit (peel) 122, fruit (pulp) 2.0	Udenfriend *et al.*, 1959
Papaveraceae		
Corydalis pallida Yatabe	cultivated cells of stem callus (ND)	Iwasa *et al.*, 1993; 1995
Passifloraceae		
Passiflora quadrangularis L.	tendrils 0.3, leaves (traces)	Applewhite, 1973
Portulacaceae		
Portulaca grandiflora Hook	callus (ND)	Endress *et al.*, 1984
P. oleracea L.	above-ground part 2500	Feng *et al.*, 1961
Solanaceae		
Aconitum napellus L.	dormant buds (ND)	Faugeras, 1967; Faugeras *et al.*, 1967
A. paniculatum Lam.	dormant buds (ND)	Faugeras, 1967; Faugeras *et al.*, 1967

(Contd.)

Table 4. (*Contd.*)

Family and species	Part of plant and amount of noradopamine (μg g^{-1} of fresh mass)	Reference
Solanum tuberosum L.	fruit (ND)	Udenfriend *et al.*, 1959
Solanum tuberosum L.	fruits (pulp) 0.006–0.045 fruits (peel) 0.038–0.160	Bygdeman, 1960
Urticaceae		
Urtica dioica L.	leaves 591	Roshchina and Mukhin, 1985a

ND: Amount was not determined.

Urticaceae its level is approximately 4000 nmole g^{-1} fresh mass. In plants that respond to mechanical stimuli with motor activity, such as *Albizzia julibrissian* and *Mimosa pudica*, noradrenaline accumulates in motor organs and axes of leaves (Applewhite, 1973). The accumulation of noradrenaline in leaves of some plants may be related to stress. Besides, the increase in noradrenaline level in banana peel from 5.1–6.2 μg g^{-1} in unmatured and yellow matured bananas to 15.2 μg g^{-1} of fresh mass in yellow-black supermatured ones as well in pulp from 2.5 μg g^{-1} yellow matured to 10.1 μg g^{-1} of fresh mass in supermatured is associated with oxidative processes during ripening (Foy and Parratt, 1960). At the ripe or overripe stage, the concentration of noradrenaline is 2–4 times higher that in unripe fruits.

As distinct from other catecholamines, adrenaline was found in only six species belonging to five families of Dicotyledoneae (Table 5). Especially large amounts of adrenaline occur in leaves of *Zea mays*. Catecholamines are found primarily in leaves, above-ground parts of plants, and

Table 5. Occurrence of adrenaline in plants

Family and species	Part of plant and amount of adrenaline (μg g^{-1} of fresh mass)	Reference
Asteraceae (Compositae)		
Lactuca sativa L.	seedlings (ND)	Kamisaka, 1979
Fabaceae		
Pisum sativum L.	leaves 0.0–34.2	Roshchina, 1991a
Phaseolus aureus Roxb.	leaves 0.22–284	Roshchina, 1991a
Gramineae		
Zea mays L.	leaves 3833	Roshchina, 1991a
Musaceae		
Musa sp. L.	leaves (ND)	Askar *et al.*, 1972
Portulacaceae		
Portulaca grandiflora Hook	callus (ND)	Endress *et al.*, 1984

ND: Amount was not determined.

latex of fruits. The concentrations of catecholamines in tissues of plants and animal are of the same order of magnitude. Under stress, the catecholamine level increases sharply.

Subcellular localization of biogenic amines in plants is not yet well studied. Dopamine was found in vacuoles of poppy *Papaver bracteatum* and *P. somniferum* (Kutchan et al., 1986). Noradrenaline and adrenaline also occur in vacuoles, but they have also been identified in isolated chloroplasts of some plants of families Fabaceae and Urticaceae (Table 6). In animals catecholamines are stored mainly in secretory vesicles, sometimes clathrine-coated. They also are found in fractions of mitochondria (Zieher and de Robertis, 1964; de Robertis, 1967).

Table 6. Catecholamines in isolated chloroplasts (nmoles g^{-1} of fresh leaf mass) (Roshchina, 1989a)

Plant	Noradrenaline	Adrenaline
Phaseolus aureus Roxb.	5	700
Pisum sativum L.	120	87
Robinia pseudoacacia L.	0–4000	0
Urtica dioica L.	3120	3930

Light period stimulates the accumulation of noradrenaline in tissues of *Sedum morganianum* and *Sedum pachyphyllum* (Kimbrough et al., 1987), but opposite results were received for *Portulaca grandiflora* when this process occurred in the dark (Endress et al., 1984).

1.2.2. Metabolism in Plants

Biosynthesis. Monoamines dopamine, noradrenaline, and adrenaline are products of plant common nitrogen metabolism. Their biosynthesis in animals and plants includes decarboxylation and hydroxylation of corresponding amino acids (Smith, 1981). A group of phenylethylamines, involving catecholamines, are formed from phenylalanine (Fig. 3) by two modes. The first, tyrosine conversion, is via DOPA formation. Phenylalanine is first hydroxylated and then transformed into tyrosine and then in to dihydroxyphenylalanine (DOPA). These processes are catalyzed by phenylalanine hydroxylase or phenylalanine monoxidase (I) and tyrosine hydroxylase or tyrosine-3-monoxidase (II). Dopamine, an immediate precursor of noradrenaline and adrenaline (Fig. 3), arises from DOPA through decarboxylation by means of the enzyme decarboxylase dioxyphenylalanine and decarboxylase of aromatic amino acids (EC 4.1.1.26), which was found for some plants, such as *Cytisus scoparius* (Tochter and Tochter, 1972), *Lophophora williamsii* (Lundström, 1971a, c), *Monostroma fuscum* (Tochter and Tochter, 1969), and *Trichocereus pachanoi* (Lundström, 1970).

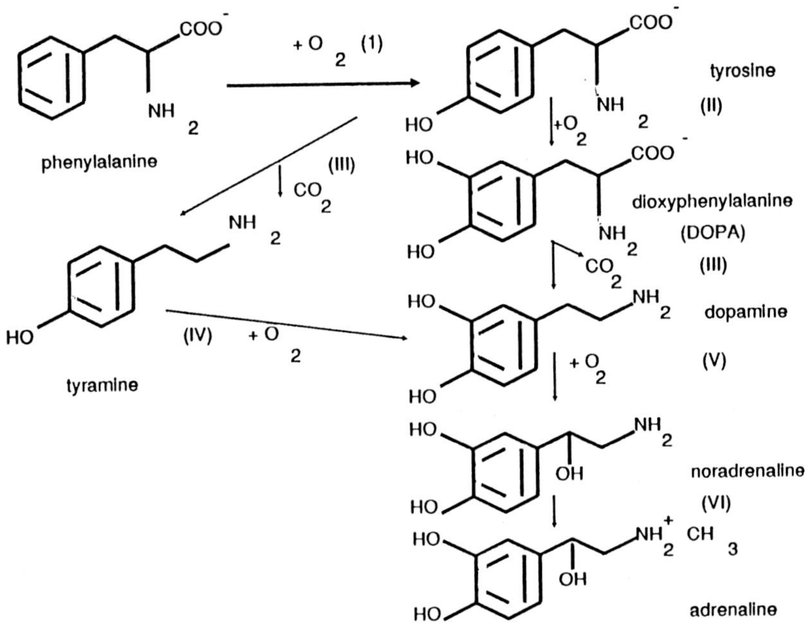

Fig. 3. Formation of dopamine, noradrenaline and adrenaline from phenylalanine. I, phenylalanine hydroxylase; II, tyrosine hydroxylase; III, decarboxylase of aromatic acids; IV, tyramine hydroxylase; V, dopamine-β-hydroxylase; VI, phenylethanolamine-N-methyltransferase

Another pathway of tyrosine transformation is also possible. By decarboxylation, it is transformed into tyramine and then by hydroxylation with the participation of tyramine hydroxylase into dopamine. Dopamine is further oxidized to noradrenaline by copper-containing enzyme β-hydroxylase 3,4-dioxyphenylethylamine (the enzyme was first found in banana by Smith and Kirshner (1960)). Then, under the influence of transmethylase of phenylethanolamines, the formation of adrenaline takes place.

The site of catecholamine synthesis in the plant cell remains unclear. It is believed that, as in animals, these substances occur in cisternae of the Golgi apparatus.

Synthesis of catecholamines depends on external factors. In betacyanin-forming callus of the red-flowering (var JR) *Portulaca grandiflora* Hook, containing dopamine, adrenaline, noradrenaline, and epinine (Endress et al., 1984), the production of DOPA, dopamine, and betacyanin is much higher in callus kept in the dark, than in illuminated callus. The other catecholamines are only accumulated in the dark. The synthesis of dopamine via DOPA can be demonstrated by the use of inhibitors. Stage tyrosine → dopamine is inhibited by hydroxylamine, inhibitor of tyrosine decarboxylase, whereas methyl-L-DOPA, inhibitor DOPA decarboxylase, inhibits the stage DOPA → dopamine. Hydroxylamine stimulates the dopamine production from tyramine in the light, whereas the radioactive label is not determined in noradrenaline and in betacyanin. Accumulation of noradrenaline in *Portulaca grandiflora* increases in the dark. Utilization of dopamine is inhibited by light.

Catabolism. It is supposed that catabolism of catecholamines in plants is similar to that in animals. There are two catabolic processes for the compounds:

1. Oxidative deaminations according to the total equation

$$R\text{–}CH_2\text{–}NH_2 + H_2O + O_2 \xrightarrow{} R\text{–}COOH + NH_3 + H_2O_2$$

Catecholamine Monoamine oxidase Vanillic acid

2. o-Methylation

$$\text{Catecholamine–}NH_2 + CH_3 \xrightarrow{\text{Catecholamine-o-transferase}} \text{Methylcatecholamine–}NH_2$$

During catabolism, catecholamines are oxidized with participation of copper-containing enzymes aminooxidases, including pyridoxal phosphate or (and) flavin in reaction center, deaminated, and then methylated by catechol-o-methyltransferases, to physiologically inactive products. Unlike in animals, diamine oxidases of plants play a major role in oxidative deamination of catecholamines, since monoamine oxidases use other substrates. Aminooxidases, both monoamine- and diamine oxidases, are localized in mitochondria of animals. As for catecholamine-o-methyltransferases, they are present in all animal tissues, and especially active in nervous cells. A detailed diagram of catabolism of catecholamines is shown in Fig. 4. In animal tissues, catecholamines are converted to metanephrine, normetanephrine, vanillic aldehyde, and dehydroxymandelic and vanillic acids. In plants, this pathway is also possible because the last three compounds are ordinary products of their metabolism.

The metabolic processes, however, may involve not only deactivation of catecholamines, but also the formation of toxic products, alkaloids, which are often taxon-specific. These substances are essentially formed from amines and aldehydes in the Mannich reaction

$$\begin{array}{c} \text{O} \\ | \quad \| \\ -\text{C}-+\text{C}-\text{H}+\text{H}-\text{N}-\text{R}' \\ | \quad | \quad | \\ \text{R} \quad \quad \text{R}'' \end{array} \longrightarrow \begin{array}{c} \text{H} \\ | \quad | \\ -\text{C}-\text{C}-\text{N}-\text{R}' \\ | \quad | \quad | \\ \text{R} \quad \text{R}'' \end{array}$$

in which catecholamines and the products of their metabolism aldehydes dioxybenzaldehyde, vanillic aldehyde, etc. participate. It is in this way that morphine, papaverine, norcoclaurine, reticuline and their derivatives, and colchicine arise from catecholamines (Figs. 5–7). The formation

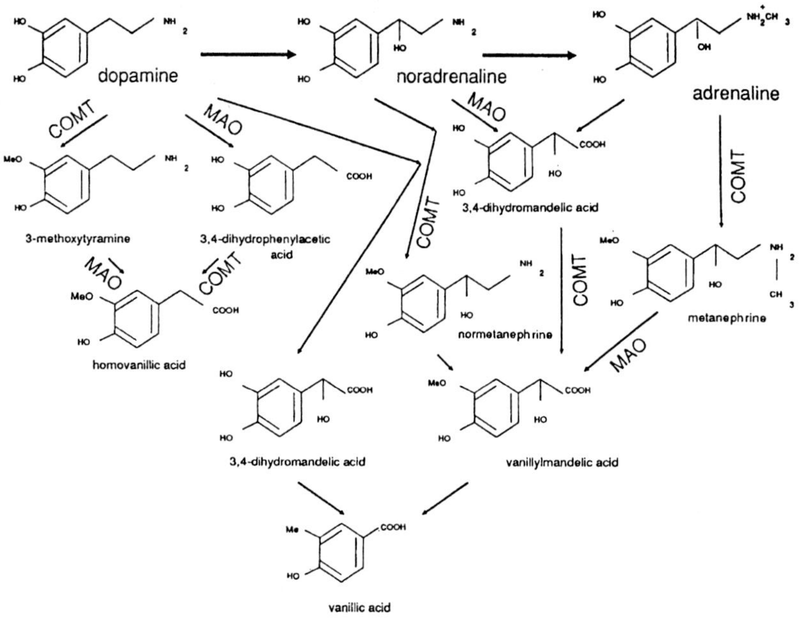

Fig. 4. Catabolism of catecholamines until vanillic acid. MAO, monoamine oxidase; COMT, catecholamine-o-methyltransferase

of two first alkaloids from product of deamination of DOPA (DOPA-aldehyde) and dopamine is demonstrated in laticifers of poppy *Papaver somniferum* and *P. bracteatum* (Kutchan et al., 1986; Roberts, 1986). Dopamine condenses with acetaldehyde to give isoquinoline alkaloid salsolonol in cultured cells of *Corydalis pallida* (Papaveraceae) (Isawa et al., 1993). Alkaloids derived from dopamine have been known since Spencer and Gear (1962) discovered the incorporation of radioactive dopamine-^{14}C in C3-hydrastine in *Hydrastis canadensis* (Fig. 7). In a paper by Leete and Murrill (1964), the analogous label of dopamine was included to C16 morphine of *Papaver somniferum* and chelidonine of *Chelidonium majus* through the Mannich reaction as dopamine condensation with 3,4-dihydroxyphenyl pyruvate.

Another group of alkaloids is formed by cyclization and condensation of rings of catecholamines give rise to the synthesis of alkaloids such as berberine, in particular in *Hydrastis canadensis* (Monkovic and Spencer, 1965a, b). This was shown for the families of Berberidaceae, Ranunculaceae, Menispermaceae, Rutaceae, etc. (Monkovic and Spencer, 1965a, b). The main radioactive label of 1-^{14}C-dopamine is incorporated into berberine, whereas DL-2-^{14}C-noradrenaline is primarily incorporated into berberastine (Fig. 5). Berberine is produced in lesser degree than another isoquinoline alkaloid, cryptopine (Anderson et al., 1983). In such

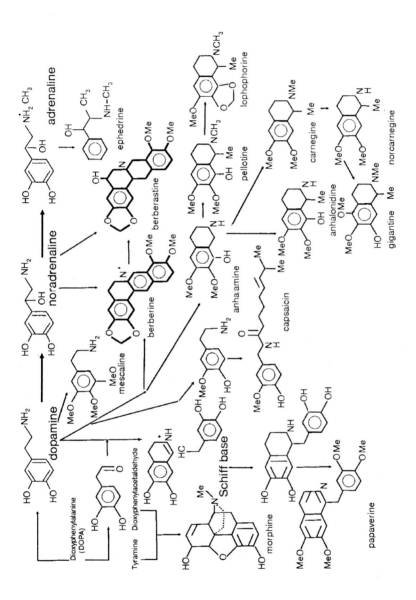

Fig. 5. Formation of alkaloids from catecholamines

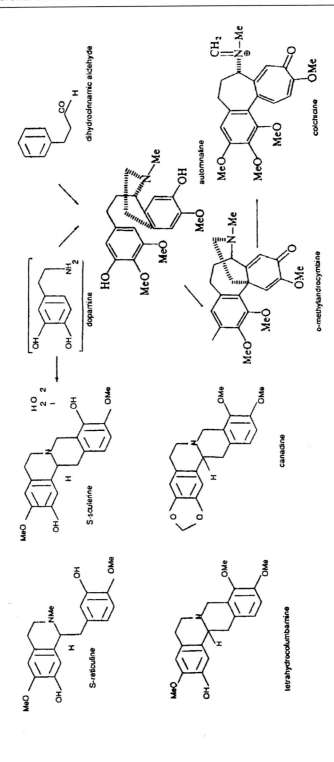

Fig. 6. Formation of alkaloids from dopamine and dihydrocinnamic aldehyde

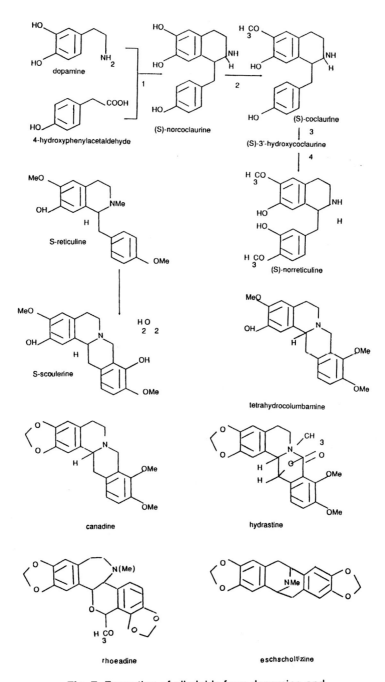

Fig. 7. Formation of alkaloids from dopamine and 4-hydroxyphenylacetaldehyde

plants as *Annona reticulata* and *Berberis stolonifera,* intermediate products in the pathway of berberine synthesis are benzylisoquinoline alkaloids norcoclaurine and reticuline, which are formed by condensation of dopamine and 4-OH-phenylacetaldehyde or 3,4-dihydroxyphenylacetaldehyde (Rueffer, 1988). This process occurs with participation of enzyme S-norcoclaurine synthase (Fig. 7).

It is proposed that condensation of dopamine with 3,4-dihydroxyphenylacetaldehyde leads to the formation of norlaudanosoline (Löffler *et al.,* 1987; Schütte, 1989). The morphine skeleton originates from the condensation of dopamine with 4-hydroxyphenylacetaldehyde. Then norcoclaurine and coclaurine are formed. 1-^{14}C-norcoclaurine and 1-^{14}C-coclaurine fed to *Papaver somniferum* afforded highly enriched thebaine and morphine specifically labelled at the C-9 position. H_2O_2 is formed (Schütte, 1989) in the following reaction: (S)-reticuline→(S)-Scoulerine (see Frenzel *et al.,* 1988). Then tetrahydrocolumbamine, canadine, and berberine are formed. In the reaction of the conversions of alkaloids reticuline and berberine, the bridge enzyme operates highly stereospecifically. The berberine bridge enzyme from *Berberis beaniana* cell cultures has Mr = 52 kDa and participates in the conversions of reticuline and the derivatives into tetrahydroprotoberberinic alkaloids in the presence of O_2 (Steffens *et al.,* 1985). In this reaction H_2O_2 is formed as well.

Successive methylation of dopamine in Mexican cactuses *Lophophora williamsii* and *Trichocereus pachanoi* (fam. Cactaceae) results in a formation of alkaloids mescaline (Fodor, 1980), anhalamine, pellotine, and lophophorine (Lundström, 1971a; Paul, 1973), which induce hallucinations in humans. Mescaline binds with adrenoreceptor in post-synaptic membrane and so prevents the action of noradrenaline and adrenaline. Biosynthesis of alkaloids mescaline, anhalamine, and anhalonidine (Fig. 5) is derived from dopamine throughout, with the following intermediates (Lundström and Agurell, 1971, 1972):

>   + 3,4,5-trihydroxyphenethylamine
> |
> **dopamine**
> |
>   +3-methoxy-4-hydroxyphenethylamine-
>   ⇓
> 3,4-dihydroxy-5-methoxyphenethylamine
>   ⇓
>   anhalamine (R = H) or anhalonidine (R = Me)
>   ⇓
>   mescaline

By methylation of adrenaline agonist of this compound, alkaloid ephedrine is formed in plants of genus *Ephedra* (Fodor, 1980).

In cactus *Carnegiea gigantea*, catecholamine precursors of alkaloids carnegine, norcarnegine, and gigantine are found (Augurell et al., 1972), while in cactus *Anhalonium lewinii* alkaloid anhalamine and its derivatives are observed (Hoffmann, 1969). One of the metabolic ways in which alkaloids may arise is from the interaction of dopamine with amino acids. For instance, condensation of dopamine with lysine leads to the formation of securinin and phyllandin (Talapatra et al., 1974). Dopamine interacts with ornithine, and alkaloids norsecurinin and 4-methoxynorsecurinin are formed (Hegnauer, 1964; Talapatra et al., 1974).

Alkaloids capsaicin, d-tubocurarine, and perhaps atropine are formed from products of the catecholamine catabolism vanillin and vanillic acid as precursors. d-Tubocurarine and atropine are blockers of nicotinic and muscarinic cholinoreceptors of animals (see Chapter 3) and have antimediatory action on post-synaptic membranes. Among alkaloids reported to be formed from dopamine originally are hordenine, lycopine, crinine, tazettine, homolycorine, galanthamine, isocraugsodine, rhoeadine, hydrastine, and eschscholtzine (Fig. 7).

Besides the above-mentioned metabolic processes, catecholamines are oxidized by oxygen of air, forming oxidized products—red pigments aminochromes and brown pigments melanins—which are polymers of indole (Heacock and Powell, 1973). This oxidation can be either nonenzymatic (spontaneous in solutions and enhanced by the presence of bases, salts Na_2HPO_4 and $NaHCO_3$, oxidants, metal ions Fe^{2+}, Fe^{3+}, Cu^{2+}, Mn^{2+}, etc.) or (and) enzymatic (in the presence of polyphenoloxidases, etc). The formation of semiquinone and quinone is predicted to cause cycling of catecholamines to aminochromes, as shown in Fig. 8. The electron transfer to the air oxygen appears to induce the formation of hydrogen peroxide, and the mechanism of oxidation per se is connected with the arising of the superoxide radical, active oxygen $O_2^{\cdot-}$. The blocking of oxidation of dopamine by superoxide dismutase confirms this possibility (Heacock and Powell, 1973). In the air, aminochrome is transformed into melanins. Enzymatic oxidation of catecholamines to melanins by polyphenoloxidase has been demonstrated in banana fruits (Thomas and Nair, 1971).

Another mode is oxidation to dark pigments melanins, found in both plant and animals. This process may involve participation of polyphenoloxidase in the first step and then spontaneous oxidation, as shown in the following diagram adapted from Mason (1948) by Palmer (1963), and Harrison et al. (1967):

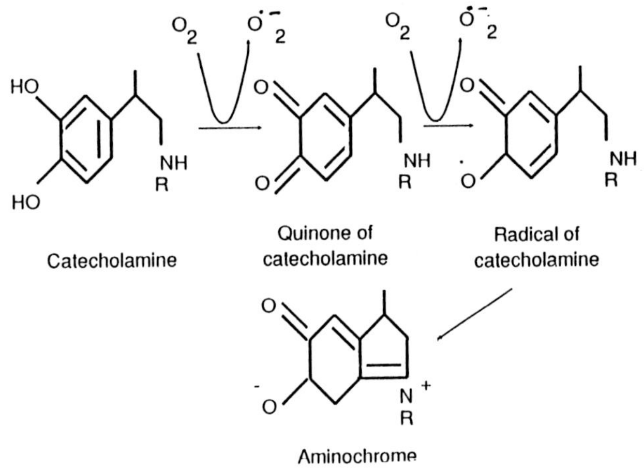

Fig. 8. The non-enzymatic oxidation of catecholamines by oxygen

$$\text{Dopamine} \xrightarrow[\text{polyphenoloxidase}]{\text{fast} \\ +1/2 O_2} \text{Dopamine quinone} \xrightarrow{\text{fast}} 2,3\text{-}$$

$$\text{Dihydroindole-5,6—quinone} \xrightarrow{\text{slow}} \text{5,6-Dihydroxyindole} \xrightarrow{+1/2 O_2}$$

$$\text{Indole-5,6-quinone} \longrightarrow \text{Melanin}$$

1.3. SEROTONIN

1.3.1. Occurrence in Plants

Erspamer (1940, cited in Erspamer, 1961) and Rapport and coworkers (1948) discovered serotonin in animals. Ten years later, this mediator was found in plant fruits (Bowden *et al.*, 1954; Collier and Chesher, 1956). To date, it has been found in 42 species of 20 families (Table 7).

Serotonin (5-oxytryptamine) in plants is usually determined by paper and thin-layer chromatography (Regula, 1970, 1981) and also by high performance gas-liquid chromatography (Grosse *et al.*, 1983; Lembeck and Skofitsch, 1984).

As shown in Table 7, the amount of serotonin varies with the taxonomic position of the plant and is especially high in representatives of families Juglandaceae, Fabaceae, and Lygophyllaceae. In comparison with animals (Table 7 below), its quantities in plants are the same (average) or lower than in mammals, whereas in some species of mollusks *Octopus* it correlates with values observed in fruits and seeds.

Besides free serotonin, conjugated serotonins such as N-feruloylserotonin, N-(p-coumaroyl) serotonin, and N-(p-coumaroyl) serotonin mono-β-D-glucopyranoside were isolated from safflower *Carthamus tinctorius* L. seed (Sakamura *et al.*, 1980)

Serotonin is not equally distributed along the plant. It is accumulated mainly in reproductive organs of some plants and *Urtica* stinging hairs (Table 7). A particularly large amount of serotonin (0.2%) was found in seeds of West African leguminous plant *Griffonia simplicifolia*. Analysis of Table 7 shows that fruits and seeds of may plants contain 100 times as much serotonin as vegetative organs, and sometimes more.

Table 7. The occurrence of serotonin in plants

Family and species	Part of plant and amount of serotonin ($\mu g\ g^{-1}$ of fresh mass)	Reference
PLANTS		
Araceae		
Symplocarpus foetidus L.	leaves (ND)	Bulard and Leopold, 1958
Bromeliaceae		
Ananas comosus Merril.	fruit (ND)	Bruce, 1960
Ananas comosus Merril.	fruit 1.5	West, 1960
Ananas comosus Merril.	fruit 19.0, juice of fruits 13–22*	Foy and Parratt, 1961
Ananas comosus Merril.	juice of fruits 2.9–25	Foy and Parratt, 1960
Caricaceae		
Carica papaya L.	fruit (ND)	Foy and Parratt, 1960
Carica papaya L.	fruit 1.1–2.1	Erspamer, 1961
Crassulaceae		
Sedum pachyphyllum Rose	leaves 0.02–0.03	Kimbrough *et al.*, 1987
S. morganianum E. Walth.	leaves (ND) (0.065 $\mu g\ \mu g^{-1}$ of protein)	Kimbrough *et al.*, 1987
Cucurbitaceae		
Citrullus vulgaris Schrader	fruit (ND)	Dannenburg and Liverman, 1957
Momordica charantia L.	fruit (ND)	Dhalla *et al.*, 1961
Elaeagnaceae		
Elaeagnus umbellata Thunb.	leaves (ND)	Regula, 1972
Shepherdia argentea (Pursh) Nutt	leaves (ND)	Regula and Devide, 1979
S. canadensis (Pursh) Nutt	leaves (ND)	Ayer and Brown, 1970
Euphorbiaceae		
Cnidoscolus oligandrus (Mull. Arg) Pax and Hoffm.	prickles (ND), leaves and stems (ND)	Cordeiro *et al.*, 1983

(Contd.)

Table 7. (*Contd.*)

Family and species	Part of plant and amount of serotonin ($\mu g\ g^{-1}$ of fresh mass)	Reference
Cnidoscolus phyllacanthus L.	prickles (ND)	Cordeiro *et al.*, 1983
Cnidoscolus texanus (Muell. Arg)	prickles (ND)	Lookado and Pollard, 1991
Fabaceae		
Griffonia simplicifolia	leaves 0.0017–0.007, seeds 200,000	Fellows and Bell, 1970
Mucuna pruriens DC.	stinging hairs of pods (fruit) 150	Bowden *et al.*, 1954
Phaseolus multiflorus Lam.	leaves 0.6–1.0	Applewhite, 1973
Piptadenia peregrina Benth.	leaves (ND)	Fellows and Bell, 1971
Pisum sativum L.	leaves, stems 0.9–1.0	Applewhite, 1973
Samanea saman Merr.	stems 0.1–4.0	Applewhite, 1973
Juglandaceae		
Juglans ailanthifolia Carr.	embryo of fruit 95	Regula *et al.*, 1988, 1989
J. mandshurica Maxim.	embryo of fruit 251	Regula, 1985
J. nigra L.	embryo of fruit 180	Regula, 1986
J. nigra L.	seeds (ND)	Lembeck and Skofitsch, 1984
J. regia L.	fruit 178–337	Kirberger and Braun, 1961; Bergmann *et al.*, 1970; Grosse *et al.*, 1983; Regula, 1985, 1986
Lauraceae		
Persea sp. Mill	fruit 10	Udenfriend *et al.*, 1959
Loasaceae		
Loasa vulcanica ed Andre	leaves (ND)	Regula, 1981
Lygophyllaceae		
Peganum harmala L.	culture of leaf tissue 18,200	Nettleship and Slaytor, 1974; Sasse *et al.*, 1982
Peganum harmala L.	root culture	Berlin *et al.*, 1987
Peganum harmala L.	hairy root cultures	Berlin *et al.*, 1992, 1993
Malvaceae		
Gossypium hirsutum L.	fruit (ND)	Bulard and Leopold, 1958
Mimosaceae		
Albizzia julibrissin Durazz.	leaves 2.7	Applewhite, 1973
Mimosa pudica L.	leaves (ND), primary pulvini (ND)	Applewhite, 1973
Musaceae		
Musa sapientum L.	fruit (peel) 1–41	Marshall, 1959

(*Contd.*)

Table 7. (*Contd.*)

Family and species	Part of plant and amount of serotonin ($\mu g\ g^{-1}$ of fresh mass)	Reference
Musaceae		
Musa sapientum L.	fruit (peel) 40–150	Waalkes *et al.*, 1958
Musa sapientum L.	fruit (pulp) 19–28	Udenfriend *et al.*, 1959
Nitrariaceae		
Nitraria schoberi L.	all parts (ND)	Üstünes *et al.*, 1991
Passifloraceae		
Passiflora foetida L.	fruit 1.4–3.5	Foy and Parratt, 1960, 1961
Passiflora foetida L.	leaves (ND)	Erspamer, 1961
Passiflora quadrangularis L.	leaves (ND), tendrils 1.0	Applewhite, 1973
Rosaceae	red fruit 10.0, blue-red	Udenfriend *et al.*, 1959
Prunus domestica L.	fruit 8.0	
Solanaceae	stems, leaves 0.0–0.5,	West, 1958, 1959a, b
Lycopersicon esculentum L. (Mill.)	fruit 3.75	
Lycopersicon esculentum L. (Mill.)	fruit 1.5–12.0	Udenfriend *et al.*, 1959
Solanum melongena L.	fruit 2.0	Udenfriend *et al.*, 1959
Urticaceae		
Blumenbachia contorta Hook. Fil.	stems, leaves 3.1	Regula, 1987
Girardinia heterophylla Gandich. (Decne.)	stinging trichomes 0.15	Saxena *et al.*, 1966
Laportea moroides Wedd.	stinging trichomes 1.0**	Robertson and Macfarlane, 1957
Urtica cubensis Klotzsch. ex. Herd.	shoots with leaves 0.42	Regula and Devide, 1980
U. dioica L.	stinging trichomes 3–5	Collier and Chesher, 1956
U. dioica L.	stinging trichomes 3.5**	Regula and Devide, 1980
U. ferox Forst.f	shoots with leaves 0.33	Regula and Devide, 1980
U. membranacea Poir. ex Savigny.	shoots with leaves 0.26	Regula and Devide, 1980
U. pilulifera (L.) Aschers	shoots with leaves (ND)	Regula, 1974
U. thunbergiana Sieb. et Zucc.	shoots with leaves 0.31	Regula and Devide, 1980
	FUNGI	
Agaricaceae	mycelium (ND)	Wieland and Motzel, 1953
Amanita mappa Batsch.		
Panaeolus campanulatus Fr. ex L.	mycelium (ND)	Tyler, 1958

(*Contd.*)

Table 7. (*Contd.*)

Family and species	Part of plant and amount of serotonin ($\mu g\ g^{-1}$ of fresh mass)	Reference
ANIMALS		
Mammalia	brain 0.75	Erspamer, 1961
Canis familiaris (Dog)		
Rattus sp. (rat)	brain 0.21–0.96	Erspamer, 1961
Oryctolagus (rabbit)	brain 0.45–0.57	Erspamer, 1961
Mollusca	salivary glands	Erspamer, 1961
Octopus	68–500	

**μg per 1000 of stinging hairs; * –μg per 1 ml of juss;
ND: Amount was not determined.

In particular, roots and callus formed from cotyledons of walnut *Juglans regia* lack serotonin. It is also absent in leaves, stems, and roots of adult vegetating plant, concentrating (0.4-0.6 $\mu g\ g^{-1}$ fresh mass) in embryos (Bergmann et al., 1970). The amount of serotonin contained in fruits is comparable with that found in animal tissues (Erspamer, 1961). The fruits of some plants show a high level of serotonin (Table 8). Its distribution in tissues is not uniform. The greatest amount serotonin is found in the outer skin. In the inner skin and in the pulp, the amount is considerably less. As fruits ripen, the serotonin level decreases (for example, in *Ananas*) or rises sharply (in the skin and pulp of tomato *Lycopersicon*). In other parts of the plant, the amount of serotonin is substantially less than in fruits, especially in leaves (<0.08 $\mu g\ g^{-1}$ fresh weight).

Perhaps, serotonin plays a role in plant reproduction as suggested from seasonal dynamics of the substance. In the reproductive period

Table 8. The content of serotonin in maturing fruits ($\mu g\ g^{-1}$ of fresh mass)

Plant species	Part of fruit	Immatured	Matured	Superma- tured	Reference
Ananas comosus L.	pulp	50–60	19	—	Foy and Parrat, 1961
Lycopersicon esculentum L. (Mill.)	pulp	0.18	3.75	2.9	West, 1958
Musa sapientum L.	pulp	24	36	35	Udenfriend et al., 1959
Musa sapientum L.	outer peel	74	96	161	Udenfriend et al., 1959
Musa sapientum L.	inner peel	13	38	170	Udenfriend et al., 1959

(March) in legume *Griffonia simplicifolia,* leaves accumulate it up to 1.2–1.3 µmole g^{-1} fresh mass, whereas in vegetative period (November and December) they accumulate only up to 0.3 µmole g^{-1} fresh mass (Fellows and Bell, 1970). An analogous event is observed in the temperate species *Juglans regia,* where the amount of serotonin is lowest in spring and sharply increases, especially in fruits, near autumn, the period of reproduction. The highest quantity is concentrated in the outer part of the peel. The inner part of the peel and the pulp, as a rule, contain smaller amount of the amine. Light regime does not influence the accumulation of 5-hydroxytryptamine (serotonin) in *Sedum morganianum* (Crassulaceae); it does not change in white light (300–700 nm) and in the dark during 12 hours (Reynolds *et al.*, 1985). The accumulation is about 65.36 ng mg^{-1} of protein, whereas indole acetic acid is lower in the light (by about 25%). Serotonin in *Sedum morganianum* and *Sedum pachyphyllum* accumulates more in light period than in the dark (Kimbrough *et al.*, 1987).

A special role is characteristic of serotonin in stinging hairs of some taxas, for instance common nettle *Urtica dioica* (Emmelin and Feldberg, 1947, 1949) or beans *Mucuna pruriens* (Bowden *et al.*, 1954). It preserves these plants from being eaten by herbivorous animals.

Little research has been done on subcellular localization of catecholamines and serotonin. In animals up to 40% of serotonin is in microsomes and 49–60% in mitochondria of nervous endings, whereas in nuclei there is only 14% (Zieher and de Robertis, 1964). In neurons it is localized in clathrinic vesicles. As for plants, serotonin is present here in vacuoles of nettle (Regula, 1981). It is found in cells of stinging hairs in genus *Urtica* (Collier and Chesher, 1956; Regula and Devide, 1980). It also accumulates in protein bodies of cotyledons of developing embryos in which there are still no vacuoles (Grosse, 1982; Regula *et al.*, 1989). In particular, serotonin is accumulated in protein bodies of seeds of *Juglans regia,* which lack vacuoles (Grosse, 1984). It appears to be a means for the detoxication of ammonium because as the amount of ammonium in seeds decreases from 2.1 to 1 mmoles the concentration of serotonin increases from 0.2 to 0.7 mmoles/g of fresh mass.

1.3.2. Metabolism in Plants

Synthesis. The metabolism of tryptophan, which is unable to accumulate in the cell, is supposed to exist as a mode of detoxication of ammonium. Serotonin is synthesized in plants from tryptophan formed by the shikimate pathway and localized in plastids. This process has two pathways, either via 5-hydroxytryptophan or via tryptamine formation, as shown in Fig. 9. Grosse and coworkers (1983) showed by means of liquid gas chromatography and [^{3}H]-labelled tryptophan and [^{14}C]-labelled tryptamine that the first step of serotonin biosynthesis is decarboxylation

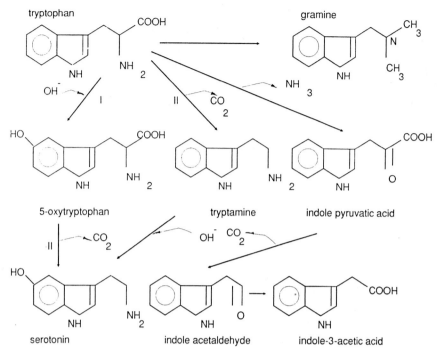

Fig. 9. Formation of serotonin and indole acetic acid from tryptophan

of tryptophan, which transforms in plants into tryptamine by the enzyme tryptophan decarboxylase (EC 4.1.1.27) or decarboxylase of aromatic amino acids (EC 4.1.1.26/27). Then tryptamine is transformed into serotonin by hydroxylation with participation of tryptamine-5-hydroxylase or L-tryptophan-5-hydroxylase (EC 1.14.16.4). The enzyme was found in leaves, stalks, petals, roots, and seeds of *Griffonia simplicifolia*. The enzyme functions in the presence of oxygen; in the presence of nitrogen, its activity is reduced by 90%.

In a plant cell, another pathway of transformation of tryptophan to serotonin, which is typical of animals, is also possible. Hydroxylation of tryptophan leads to formation of 5-oxytryptophan in the presence of tryptophan-5-hydroxylase (EC1.14.16.4). At the next stage, 5-oxytryptophan is decarboxylated by decarboxylase of aromatic acids to yield serotonin (Fellows and Bell, 1971; Grosse et al., 1983). The increased production of serotonin by suspension and root cultures of *Peganum harmala* is transformed with a tryptophan decarboxylase cDNA clone from *Catharanthus roseus* (Rugenhagen et al., 1993).

The biosynthesis of serotonin from endogenous tryptamine began immediately, whereas when tryptophan was added into the reaction medium a 12-hour lag phase was observed (Grosse, 1984). The latter phenomenon can be explained by the enzymic limit for the reaction, such as tryptophan carboxylase deficit.

Like serotonin, phytohormone indole acetic acid is formed from tryptophan through indolyl pyruvate and indolyl acetaldehyde. Grosse (1982) proposed that metabolism of tryptophan leads to the formation of serotonin and indole acetic acid as a means of detoxication of ammonium. Besides, tryptophan serves as a precursor of the alkaloid gramine, which is found in barley. It is toxic for many herbivorous animals and can be a strong allelochemical, i.e., a natural growth regulator in competition in phytocenosis (Andreo et al., 1984).

The localization of the serotonin synthesis in the plant cell is not exactly established. In animals, this biogenic amine is synthesized in the Golgi apparatus.

Catabolism. During catabolism, serotonin is oxidized and deaminated by aminooxidases such as diamine oxidase and then methylated by catechol-o-methyltransferases to physiologically inactive products (Fig. 10). Unlike in plants, in animals the first step of serotoninic metabolism is realized by the enzyme monoamine oxidase, which is concentrated mainly in mitochondria (Zieher and de Robertis, 1963). In animal tissues, serotonin is catabolized to 5-hydroxyindolylacetic acid, which is a final product of deactivation of the biogenic amines. This compound is found in

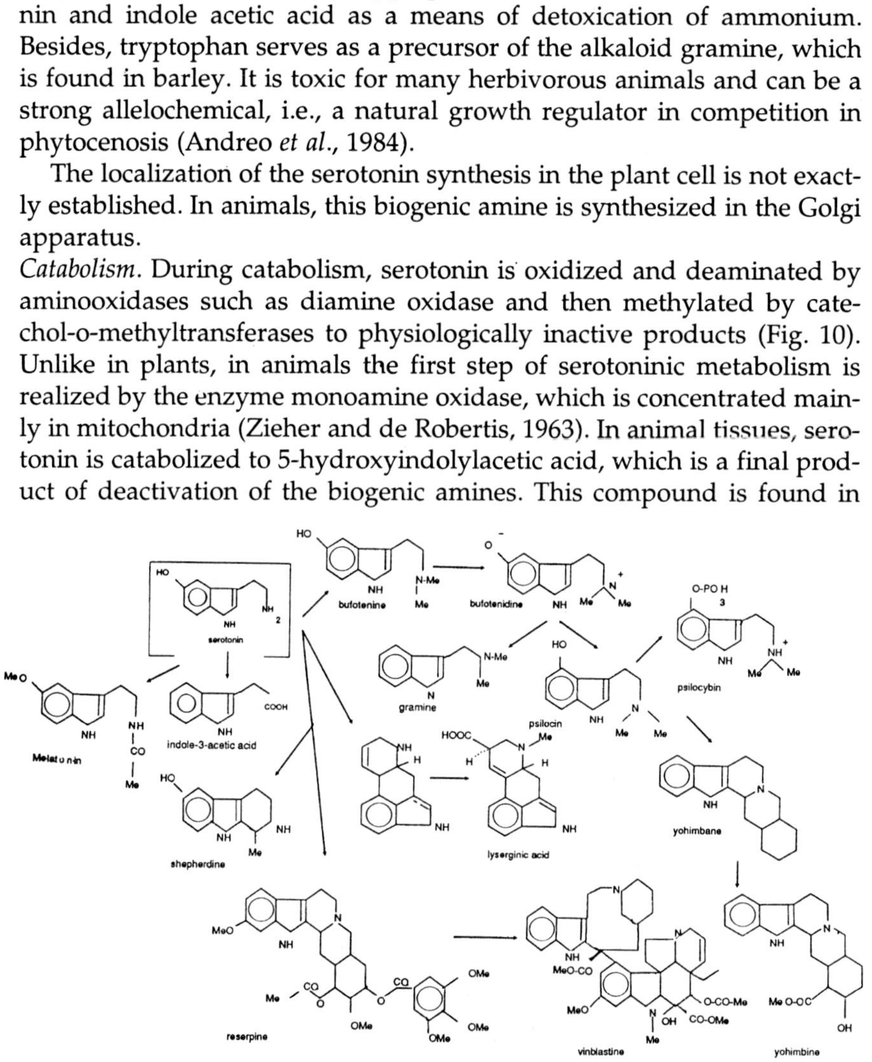

Fig. 10. Metabolism of serotonin in plant

plants as well (Umrath and Thaler, 1981). There is a possibility that the simple dehydroxylation of 5-hydroxyindolylacetic acid leads to the formation of indole acetic acid.

The deactivating catabolism of serotonin in plants has not yet been understood in detail. Acetylation of serotonin to N-acetylserotonin, catalyzed by hydroxytryptamine acetyltransferase, and then its methylation to 5-methoxy N-acetyl tryptamine, named melatonin, occurs as a biosynthetic pathway of flowering regulation (O'Neill and van Tassel, 1994). Melatonin is a trigger of flowering. Methylated derivatives of serotonin and the alkaloids bufotenine, bufotenidine, psilocin, and psilocybin were found in amphibians and plants (Atta-ur-Rahman and Basha, 1999). For example, bufotenine occurs in toadstool *Amanita mappa*, in seeds of plants belonging to genus *Piptadenia: P. peregrina, P. colubrina, P. macrocarpa* (up to 9.4 $\mu g\ g^{-1}$ fresh mass) (Fellows and Bell, 1971). Serotonin is a precursor of bufotenine and is found in the South American and Caribbean tree *Piptadenia peregrina*, which is rich in indoleamines. Fellows and Bell (1971) showed that, in this species, 80% 2-^{14}C-serotonin is incorporated into bufotenine. Psilocin and psilocybin are formed directly from serotonin in fungi *Psilocybe aztecorum*. All these alkaloids induced human hallucinations. There is a hypothesis that endogenous formation of the alkaloids in humans may cause psihic disorders (Metzler, 1978). Serotonin is a precursor of such toxic drugs as amides of D-lysergenic acid. In *Ipomoea tricolor* Cav., there are amides of D-lysergic acid (0.035%) and of D-isolysergic acid (isoergin) (0.005%), as well as lysergol (0.005%) (Hoffmann, 1969). In *Rivea corymbosa* (L.) Hall, the same substances are found, respectively 0.0065, 0.0020, and 0.0005% (Hoffmann, 1969). Biosynthesis of serotonin and related β-carboline alkaloids also occurs in hairy root cultures of *Peganum harmala* (Berlin et al., 1993).

Besides the above-mentioned pathways of metabolism condensation of the isopentyl group with the indole nucleus of serotonin may give rise to synthesis of alkaloids of the reserpine group reserpine, yohimbane and yohimbine, vinblastine. Reserpine from the Indian species *Rauwolfia serpentina* is used against snake bites. It decreases the blood pressure, liberating tissues from excess of serotonin, dopamine, and noradrenaline. Yohimbane and yohimbine are found in bark of the African tree *Corynanthe yohimbe* and were first used to increase sexual potency. Yohimbine is also used in scientific study as adrenoblocker. Vinblastine is isolated from peri-winkle *Vinca rosea* and *Catharanthus roseus* and practically used as an antiviral and antitumour agent. In *Shepherdia canadensis* (L.) Nutt. (Eleagnaceae), serotonin is found to be converted to alkaloid shepherdine (Ayer and Browne, 1970). Among alkaloids that are proposed to originate from serotonin are also gramine, girinimbine, borreline, and paspalicine.

1.4. HISTAMINE

1.4.1. Occurrence in Plants

Significant amounts of histamine can occur in plants. This amine was first found in ergot fungi *Claviceps purpurea* (Barger and Dale, 1910; Kutcher, 1910). Later it was observed in higher plants (Werle and Raub, 1948). Now histamine is known to be present in 49 plant species belonging to 28 families, from basidiomycetes to angiosperms (Table 9). Representatives of families Chenopodiaceae and Urticaceae are studied better than others. Leaves contain 2–3 µmoles g^{-1} fresh mass of histamine, whereas flowers contain up to 10 µmoles g^{-1} fresh mass (Werle, 1955). Quantitative data are known for various plant organs: flower of *Spinacia oleracea*, stinging trichomes of *Laportea moroides*, *Girardinia heterophylla*, *Jatropha urens*.

Table 9. The presence of histamine in plants

Family and species	Part of plant and amount of histamine ($\mu g\ g^{-1}$ of fresh mass)	Reference
PLANTS		
Araceae		
Hydrosme rivieri (Durieu) Engl.	spadix (ND)	Smith and Meeuse, 1966
Asclepiadaceae		
Calotropis gigantea Dryand.	latex (ND)	Saha and Kasinathan, 1963
Asteraceae (Compositae)		
Helianthus annuus L.	seeds (ND)	Appel and Werle, 1948; Haartmann et al., 1966;
Helianthus annuus L.	roots (3.5)	Korobova et al., 1988
Lepidium sativum L.	seedlings (ND)	Haartmann et al., 1966
Brassicaceae (Cruciferae)		
Capsella bursa pastoris (L.) Moench	leaves (ND)	Jurisson, 1971; Hill, 1984
Silybum marianum (L.) Gaerth	leaves (ND)	Volinskii et al., 1981.
Sinapis alba L.	seedlings (ND)	Haartmann et al., 1966
Chenopodiaceae		
Beta vulgaris var. *rapa* L.	leaves 5.2	Werle and Raub, 1948, Werle, 1955
Beta trigyna (L.)	leaves 4.2	Werle and Raub, 1948, Werle, 1955
Chenopodium album L.	leaves and fruits 6.0	Werle and Raub, 1948, Werle, 1955

(Contd.)

Table 9. (*Contd.*)

Family and species	Part of plant and amount of histamine ($\mu g\ g^{-1}$ of fresh mass)	Reference
Chenopodium bonus-Henricus (L.)	leaves, roots 45–52	Werle and Raub, 1948; Werle, 1955
Chenopodium quinoa Willd.	fruits 20.8	Werle and Raub, 1948; Werle, 1955
Kochia childsii (Hort.)	leaves, fruit, stems 5.2–10.5	Werle and Raub, 1948; Werle, 1955
Salsola kali L.	leaves, fruits, stems 13.4–14.2	Werle and Raub, 1948; Werle, 1955
Spinacia oleracea L.	leaves, seeds, flowers 12.6–292	Werle and Raub, 1948; Werle, 1955
Droseraceae		
Drosera sp. L.	leaves 1.2–6.7	Werle and Raub, 1948; Werle, 1955
Euglenaceae	homogenate of cells	Klein, 1975
Euglena sp. L.	(ND)	
Euphorbiaceae	Prickles 0.126–0.753	Cordeiro *et al.*, 1983
Cnidoscolus oligandrus (Muell. Arg) Pax and Hoffm.	leaves 0.001–0.01 stems 0.637	
Cnidoscolus phyllacanthus	all parts (ND)	Mors *et al.*, 1959
Cnidoscolus phyllacanthus	prickles (ND)	Cordeiro *et al.*, 1983
Cnidoscolus texanus (Muell. Arg)	prickles (ND)	Lookado and Pollard, 1991
Jatropha urens (*Cnidoscolus urens*)	leaves and prickles (ND)	Cordeiro *et al.*, 1983
Jatropha urens (*Cnidoscolus urens*)	leaves and prickels 620–1250	Villalobos *et al.*, 1974
Fabaceae		
Trifolium pratense L.	leaves 3–4.5	Werle and Raub, 1948; Werle, 1955
T. repens L.	leaves and seedlings 4–13.1	Fowler, 1962; Haartmann *et al.*, 1966
Geraniaceae		
Erodium cicutarium L. Her.	hairs (ND)	Eijk, 1957
Gramineae		
Agrostis alba L.	pollen (ND)	Marquardt and Vogg, 1952
Alopecurus pratensis L.	pollen (ND)	Marquardt and Vogg, 1952
Avena sativa L.	seedlings (ND)	Haartmann *et al.*, 1966
Bromus erectus L.	pollen (ND)	Marquardt and Vogg, 1952
Cynosurus cristatus L.	pollen (ND)	Marquardt and Vogg, 1952

(*Contd.*)

Table 9. (*Contd.*)

Family and species	Part of plant and amount of histamine (μg g^{-1} of fresh mass)	Reference
Corylus avellana L.	pollen (ND)	Marquardt and Vogg, 1952
Dactylis glomerata L.	leaves (ND)	Fowler, 1962
Dactylis glomerata L.	pollen (ND)	Marquardt and Vogg, 1952
Gelsemium sempervirens L.	pollen (ND)	Marquardt and Vogg, 1952
Lolilum perenne L.	leaves (ND)	Fowler, 1962
Lolilum perenne L.	pollen (ND)	Marquardt and Vogg, 1952
Phleum pratense L.	pollen (ND)	Marquardt and Vogg, 1952
Poa pratensis L.	pollen (ND)	Marquardt and Vogg, 1952
Secale cereale L.	pollen (ND)	Marquardt and Vogg, 1952
Labiatae		
Lamium album L.	leaves 20–20.2	Werle and Raub, 1948; Kwasniewski, 1959
Salvia sp. L.	leaves 2.9	Werle and Raub, 1948; Werle, 1955
Lentibulariaceae		
Pinguicula sp. L.	leaves 2–10	Werle and Raub, 1948; Werle, 1955
Loranthaceae		
Viscum album L.	all plant (ND)	Sajner and Veris, 1958a
Malvaceae		
Gossypium hirsutum L.	leaves 2–23	Lloyd and Nicholis, 1964, 1965
Mimosaceae		
Mimosa sp. L.	leaves 2.7–13.3	Werle and Raub, 1948
Musaceae		
Musa sapientum L.	fruits 1.0	Askar *et al.*, 1972
Nepenthaceae		
Nepenthes sp. L.	leaves 1.7–13.8	Werle and Raub, 1948; Werle, 1955
Oleaceae		
Syringa vulgaris L.	pollen (ND)	Marquardt and Vogg, 1952
Papaveraceae		
Chelidonium majus L.	leaves 5.7–107	Werle and Raub, 1948; Werle, 1955; Kwasniewski, 1958
Plantaginaceae		
Plantago lanceolata L.	leaves 0.7–6.2	Werle and Raub, 1948; Werle, 1955

(*Contd.*)

Table 9. (Contd.)

Family and species	Part of plant and amount of histamine ($\mu g\ g^{-1}$ of fresh mass)	Reference
Primulaceae		
Cyclamen sp. L.	leaves, roots and flowers 0.6–6.2	Werle and Raub, 1948; Werle, 1955
Ranunculaceae		
Delphinium sp. L.	flowers 2.9	Werle and Raub, 1948; Werle, 1955
Rosaceae		
Crataegus sp. L.	flowers (ND)	Sajner and Veris, 1958b
Rutaceae		
Citrus vulgaris (L.)	leaves 17.3	Werle and Raub, 1948
Sarraceniaceae;		
Sarracenia sp. L.	leaves 2.6–10.2	Werle and Raub, 1948; Werle, 1955
Solanaceae		
Lycopersicon sp. L.	fruits and seedlings (ND)	Holtz and Janisch, 1937; Haartmann et al., 1966
Tiliaceae		
Tilia cordata Miller	pollen (ND)	Marquardt and Vogg, 1952
Tilia platyphyllos Scop.	pollen (ND)	Marquardt and Vogg, 1952
Urticaceae		
Girardinia heterophylla Tandich	stinging hairs 3.75	Saxena et al., 1966
Laportea sp. L.	leaves, fruits 2.2–50	Werle and Raub, 1948
L. moroides Wedd.	stinging hairs 25–50	Robertson and Macfarlane, 1957
Urera sp. Gaudich.	all plant 30–3473	Werle and Raub, 1948
Urtica dioica L.	leaves 11.3–43.6	Werle and Raub, 1948
Urtica dioica L.	stinging hairs (ND)	Emmelin and Feldberg, 1949
U. parviflora Roxb.	leaves (ND)	Werle, 1955
U. urens L.	leaves 20–143.5	Werle and Raub, 1948; Saxena et al., 1965; Vialli et al., 1973
FUNGI		
Basidiomycetae		
Phallus impudicus L. ex. Fr.	mycelium (ND)	List and Reinhard, 1962
Hypocreaceae (Ascomycetes)		
Claviceps purpurea Talasne	mycelium (ND)	Barger and Dale, 1910; Kutcher, 1910

(Contd.)

Table 9. (*Contd.*)

Family and species	Part of plant and amount of histamine (μg g^{-1} of fresh mass)	Reference
ANIMALS		
Invertebrata	body 100	Ackermann and List, 1957
Geodia gigas (giant siliceous gigas)		
Vertebrata	liver 100	Vaisfeld and Kassil, 1981
Equus caballus (horse)		
Canis familiaris (dog)	liver 8	Vaisfeld and Kassil, 1981
Mammalia	milk 0.5	Vaisfeld and Kassil, 1981

For stinging hairs, the amount of histamine is given as μg per 1000 stinging hairs.
ND: Amount was not determined.

An especially large amount is observed in species of family Urticaceae, and that could be one of the taxonomic characteristics. The Brazilian stinging shrub *Jatropha urens* (family Euphorbiaceae) contains 1250 μg histamine per 1000 hairs. Histamine of stinging hairs frightens off animals by inducing burns, pain, and allergic reactions. According to Paton (1958) the average amount of histamine in plants is near 1.34 μg g^{-1} of fresh mass, which is enough for a strong pain sense. In animal cells, the amount depends on species and organ of the same species, reaching 110 μg g^{-1} fresh mass in horse but only 0.5 μg g^{-1} fresh mass in milk of mammalia (Table 9). Histamine also occurs in a large amount in giant sponge *Geodia gigas* (Ackermann and List, 1957; Hill, 1984). It is found in pollen, mainly of wind-pollinated species *Agrostis alba, Alopecurus pratensis, Bromus erectus, Corylus avellana, Cynosurus cristatus, Dactylis glomerata, Gelsemium sempervirens, Lolium perenne, Phleum pratense, Poa pratensis, Secale cereale,* and *Zea mays,* as well as of some insect-pollinated species, *Syringa vulgaris, Tilia platyphyllos,* and *T. cordata* (Marquardt and Vogg, 1952). Under stress reactions a sharp increase in histamine is observed in plants, as in animals. In plants the rise in histamine is found during drought, for instance for sunflower (Korobova *et al.*, 1988). Histamine is also found in pollen and may be an agent of allergic reactions in humans (Stanley and Linskens, 1974). Histamine derivatives are found in plants, e.g., N-acetylhistamine in *Spinacia* sp. (Appel and Werle, 1959), N, N-dimethylhistamine in mushroom *Coprinus comatus* (Appel and Werle, 1959; Hill, 1984) and seeds of *Casimiroa edulis* (Major and Dursch, 1958; Hill, 1984), and feruloylhistamine in *Ephedra* roots (Hikino *et al.*, 1983).

In animal cells, histamine is located in nuclei and mitochondria as well as in microsomes (Vaisfeld and Kassil, 1981). The localization of this amine in plant cell is presently unknown, perhaps invaculoes.

1.4.2. Metabolism in Plants

Synthesis. In animal tissues, histamine is formed directly by intracellular decarboxylation of histidine by histidine decarboxylase (Fig. 11). It is

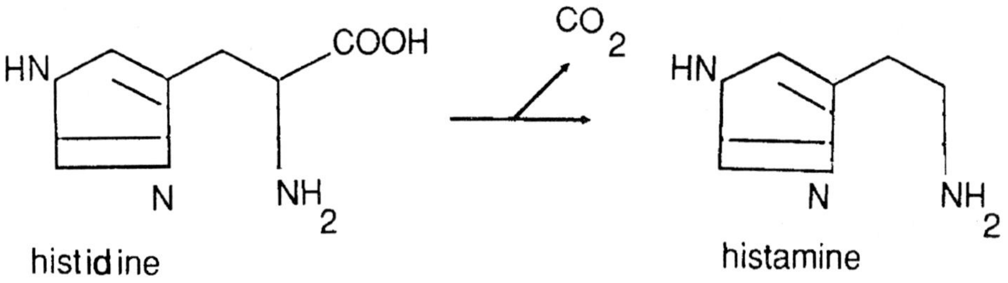

Fig. 11. Biosynthesis of histamine from histidine

likely that synthesis of the enzyme in plants, where its activity was observed, occurs by the same pathway (Werle and Pechmann, 1949). Thus, histamine is formed by very simple one-step reactions, unlike catecholamines and serotonin. Histidine decarboxylase in animals is found in mitochondrial and microsomal fractions. There is no information on its location in plant cells.

Catabolism. Catabolism of histamine in animals and bacteria occurs by oxidative deamination in which monoamine oxidase and diamine oxidase are involved (Vaisfeld and Kassil, 1981). A similar process is supposed for plants as well (Fig. 12), since these enzymes are present in plants (Werle and Pechmann, 1949; Mann, 1955). Monoamine oxidase is usually present in mitochondria, whereas diamine oxidase of the cytoplasm is contained in vesicles originating from the elements of Golgi apparatus and endoplasmic reticulum. Monoamine oxidase is involved in catabolism of primary, secondary, and quaternary amines, whereas diamine oxidase inactivates primarily histamine and diamines putrescine, spermidine, and cadaverine. Oxidative deamination of histamine leads to formation of imidazole acetic acid. In plants, as in animals and bacteria, catabolism of histamine occurs also via methylation or acetylation in the presence of histamine-N-methyltransferase or histamine-N-acetyltransferase. In the histamine molecule there are two reactive groups: imino group (-NH-) of nucleus or amino group ($-NH_2$) in branch chain. These groups participate in the formation of histamine derivatives (Figs. 12, 13). Methylation of nitrogen in imidazole ring of histamine and following its oxidation to methylimidazole acetic acid is also one of the main pathways of

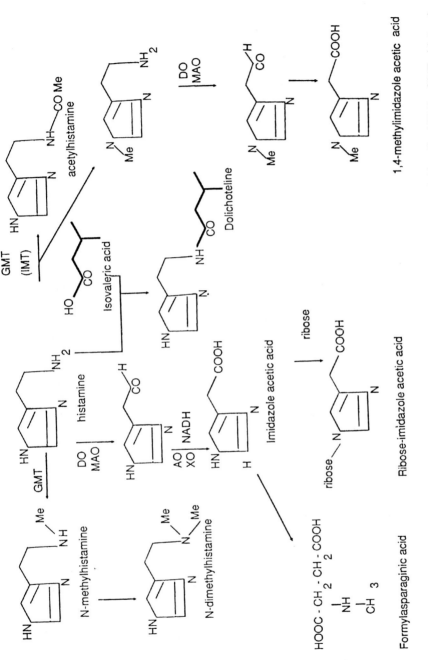

Fig. 12. Metabolism of histamine (after Rosenberg and Stohs, 1976; Vaisfeld and Kassil, 1981). GMT, histamine-N-methyltransferase; DO, diamine oxidase; AO, aldehyde oxidase; MAO, monoamine oxidase; XO, xanthine oxidase

histamine catabolism in many tissues and organs. Imino nitrogen of the ring can undergo methylation forming 3-methylhistamine or N,N-dimethylhistamine or conjugates with ribose forming riboside and histamine dinucleotide. Amino group in lateral chain can be methylated to N-methylhistamine and N',N'-dimethylhistamine, or bind with aliphatic acids (N-acetylhistamine), heterocyclic acids (N-imidazolepropionylhistamine), or amino acids (β-alanylhistamine). The third mode of inactivation consists in acetylation of amino group in lateral chain of histamine to acetylhistamine. Methylated and acetylated derivatives of histamine are found in plants as well. These are N-methylhistamine and N'-methylhistamine. N^α, N^α-dimethylhistamine has been isolated from them and it is supposed to serve as an active matter (Major and Dursch, 1958). This component was found earlier only in spongy *Geodia gigas*. N-N-dimethylhistamine is observed also in mushroom *Agaricus campestris*, whereas N-acetylhistamine is observed in spinach (Appel and Werle, 1959).

Histamine is a precursor of the imidazole alkaloid dolichotheline (Rosenberg and Stohs, 1976), which is formed by condensation of the biogenic amine and isovaleric acid or isovaleryl-CoA in (cactus) *Dolichothele* (Fig 12). Dolichotheline was discovered in cactus *Dolichothele sphaerica* (Dietrich) Britton and Rose, indigenous to southern Texas and northern Mexico, by Rosenberg and Paul (1970).

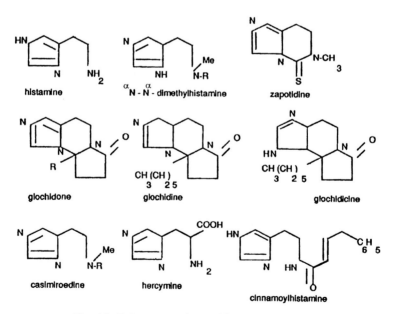

Fig. 13. Substances derived from histamine

Histamine is a precursor of many other alkaloids (Fig. 13) such as glochidone, an alkaloid from leaves of *Glochidion venulatum* (Euphorbiaceae) (Ahmad and Zaman, 1973). Leaves of tropical (Asia, Africa, New Guinea) genus *Glochidion* contain histaminic derivatives. Among them are alkaloids 1-glochidine, glochidicine, N^α-4'- oxodecanoylhistamine, and N^α-cinnamoylhistamine (Johns and Lamberton, 1966, 1967). In the bark of *G. philipicum* (Cav.) C.B. Rob, only glochidicine and N-cinnamoylhistamine are present. Other imidazole alkaloids were also isolated from plants, in particular N-(4'-oxodecanoyl) histamine from *Glochidion phillippicum* and *G. multiloculare* and dimedone from *G. multiloculare* (Talapatra et al., 1974). Seeds of *Casimiroa edulis* Llave et Lex grown in Mexico and Central America have been used as hypnotic and sedative drugs in native medicine. Seeds of *Casimiroa edulis* Lal Llav et Lejarza contain the alkaloid casimiroedine, originated from histamine (Aebi, 1956). Alkaloid cinnamoylhistamine is found in leaves and stems of the tree *Argyrodendron peralatum* (F.M. Bail) H.L. Edlin ex JH Boas (Johns et al., 1969), which belongs to Sterculiaceae and grows in the tropical rain forest in northern Queensland. The antitumoral activity of the component has been proposed. Cinnamoylhistamine is also found in *Acacia polystacha* (Lukes, 1930), up to 0.4% in leaves of *Acacia argentea* Maidand, in bark of *A. polystacha* A. Cunn (Fitzgerald, 1964), and in *Acacia spirorbis* Labill. (Poupat and Sevenet, 1975). Alkaloid zapotidine from *Casimiroa edulis* (Rutaceae) also originates from histamine (Mechoulam et al., 1961).

CONCLUSION

Thus acetylcholine and biogenic amines are synthesized by plants. They may be involved in cell metabolism. Acetylcholine is a precursor of choline required for the formation of other cholinic esters and phosphatidylcholine of membrane. Catecholamines are the precursors of vanillic acid and alkaloids, serotonin is a precursor of alkaloids and 5-hydroxyindolylacetic acid, histamine is a precursor of alkaloids.

Response of Plants to Acetylcholine and Biogenic Amines

2.1. GROWTH, DEVELOPMENT AND MOVEMENTS OF PLANTS

2.1.1. Growth

The physiological role of acetylcholine in plants was first investigated by Jaffe (1970). He found that this substance acts on the plant growth reactions controlled by phytochrome. The effect of acetylcholine on growth of plants is demonstrated in Table 10. Most often it stimulates the elongation of coleoptiles and hypocotyls of seedlings but has no effect on

Table 10. The effects of acetylcholine on the growth reactions

Plant	Organ	Effect	Reference
Avena sativa	coleoptiles	stimulation of elongation	Evans, 1972
Cicer arietinum	etiolated seedlings	no effect	Gupta, 1983
Cucumis sativus	hypocotyls (segments)	stimulation of elongation	Verbek and Vendrig, 1977
Glycine max	hypocotyls (intact)	stimulation of elongation	Mukherjee, 1980
Lens culinaris	root tips	inhibition of growth	Penel *et al.*, 1976
Phaseolus aureus	plants grown in the light	inhibition in formation of secondary roots	Jaffe, 1970; 1972a, b
Pisum sativum	plants grown in the light	inhibition in formation of secondary roots	Kasturi, 1978
Pisum sativum	etiolated seedlings	no action on growth	Kasturi, 1978
Sinapis alba	etiolated seedlings	no action on growth	Kasemir and Mohr, 1972
Triticum vulgare	seedlings	stimulation of growth	Dekhuijzen, 1973
Triticum vulgare	coleoptiles (segments)	stimulation of growth	Lawson *et al.*, 1978
Tropaeolum majus	stems	no action on elongation	Reed and Bonner, 1971
Vigna sesquipedalis	hypocotyls	depression of growth in red light	Hoshino, 1983
Vigna sesquipedalis	epicotyls	stimulation of growth	Hoshino, 1983

elongation of etiolated seedlings. In three cases an inhibition of the growth of root tips and secondary roots or the formation of secondary roots was noted. The formation of secondary roots in *Phaseolus aureus* and *Pisum sativum* was observed only in plants grown in the light. In the concentration range 0.1–500 μM, acetylcholine accelerated the growth of apical segments of coleoptiles of wheat *Triticum vulgare* by 20% (Lawson et al., 1978). Exposure to red light at 660 nm caused an elongation of the segments by 62% compared to control. The addition of acetylcholine 30 min. after switching on of the light decreased the rate of the process by 50%. It should be noted that marked changes in growth processes were induced only by high concentrations of acetylcholine ($> 10^{-3}$ M). The effect was no more than 20–30% of the control and mainly within the standard deviations. Unlike in higher plants, in moss *Athyrium filix-femina*, low concentrations of acetylcholine (10^{-6}–10^{-5} M) sharply stimulated the elongation of protonema (Bahre, 1975).

It has been shown (Tretyn and Kendrick, 1991) that acetylcholine (1 μM) stimulates unfolding of leaves in etiolated seedlings of wheat *Triticum aestivum* cv. Arminda, and its action is similar to that of red light, which regulates the activity of phytochrome.

Little research has been done on the effect of catecholamines on plant growth. It was shown (Martin et al., 1987) that dopamine at concentrations exceeding 5 mg/l (3.2×10^{-5} M) suppresses the growth of the callus tissue of banana *Musa acuminata*, and at a concentration of 50 mg/l (3.2×10^{-4} M) completely blocks it. In another set of experiments, noradrenaline and adrenaline enhanced synergistically the gibberellin-stimulated elongation of salad seedlings (Kamisaka and Shibata, 1982). Catecholamines also stimulate growth of plant tissue culture of potato (Protacio et al., 1992; Dai et al., 1993).

Serotonin is an active regulator of growth. As early as 1958 it was assumed that this compound, which is similar in structure to indole acetic acid, acts as a hormone since it induces the growth hook of oats coleoptiles (Niaussat et al., 1958). The auxin-like effect of serotonin was also observed on hypocotyls of *Lupinus albus* (Umrath and Thaler, 1980), whereas the product of its deamination hydroxyindolylacetic acid appeared to be ineffective in this reaction (Umrath and Thaler, 1981). It has been established by Regula and coworkers (1989) that serotonin, even to a greater extent than indolylacetic acid, stimulates the formation of roots in the leaf tissue culture of hybrid poplar *Populus tremuloides* × *P. tremula*. Moreover, serotonin inhibits the formation of rhizosphere on roots of poplar and at the same time stimulates the growth of secondary roots (Regula et al., 1988). In barley seeds, serotonin induces an increase in the mitotic index and accelerates growth of root, root mass, and cotyledons (Csaba and Pal, 1982). It has also been found that the biomediator inhibits the formation and growth of gallic tumors on potato disks (Regula et al., 1989). Now serotonin is considered a common growth signal (Lin, 1993).

The action of histamine on plant growth is not yet well studied. In animals it is a stimulator of the synthesis of proteins, nucleic acids, and lipids in some cases, such as at the development of embryos, regeneration of tissues, and restoration of wounded tissues (Vaisfeld and Kassil, 1981).

2.1.2. Germination of Seeds, Pollen, and Spores

Acetylcholine has essentially a stimulating effect on germination of seeds, pollen of plants, and fungi spore tube (Table 11). Hartmann and Gupta (1989) deny the general stimulating effect of acetylcholine on germination of seeds because the effect is not strong and becomes manifest only at high concentrations of acetylcholine (~10^{-3} M). However, there is evidence in the literature indicating a substantial increase in the number of germinating seeds under the influence of acetylcholine (Kostir et al., 1965). Thus, at concentration of 10^{-3} M, acetylcholine and its analog carbamylcholine stimulated the germination of seeds of *Rumex obtusifolius* in the light by 25–50% (Tretyn et al., 1985, 1988). Holm and Miller (1972) reported that acetylcholine stimulates germination of seeds of *Echinochloa crusgalli* three times, of *Chenopodium album* four times, and of *Setaria viridis*

Table 11. Action of acetylcholine on the germination of seeds, pollen and spores

Plant	Organ	Effect	Reference
Agropyron repens	seeds	stimulation of germination	Holm and Miller, 1972
Arachis hypogaea	pollen	stimulation of pollen tube elongation	Chhabra and Malik, 1978
Brassica kaber	seeds	stimulation of germination	Holm and Miller, 1972
Chenopodium album	seeds	stimulation of germination	Holm and Miller, 1972
Echinochloa crusgalli	seeds	stimulation of germination	Holm and Miller, 1972
Equisetum arvense	vegetative microspores	stimulation of germination	Roshchina and Melnikova, 2000
Hippeastrum hybridum	pollen	stimulation of germination at low viability	Roshchina and Melnikova, 1998a, b
Lathyrus sativus	pollen	inhibition of pollen tube elongation	Gharyal (cit. Hartmann and Gupta, 1989)
Raphanus sativus	seeds	weak effect	Roshchina, 1992
Rumex obtusifolius	seeds	stimulation of germination	Tretyn et al., 1988
Setaria viridis	seeds	stimulation of germination	Holm and Miller, 1972
Trichoderma viride	spores	stimulation of germination	Gressel et al., 1971

eight times compared to control. Seeds of *Brassica* and *Agropyron repens* did not germinate at all without acetylcholine treatment. Thus, the role of acetylcholine in germination of seeds may be more significant than was proposed previously. On the other hand, according to experimental evidence on germination of seeds of radish *Raphanus sativus*, acetylcholine, dopamine, histamine, and adrenaline have weak or unverifiable effects, whereas noradrenaline and serotonin significantly stimulated the process, by 40 – 50% (Fig. 14). The stimulating effects on seeds of *R. sativus* are better seen on material with low rate of emergence (Roshchina, 1991a, 1992, 1994). According to Gharyal (cited from Hartmann and Gupta, 1989), acetylcholine also had no effect on pollen of *Pisum sativum*, *Cajanus cajan*, and *Lathyrus*. In pollen of *L. sativus*, acetylcholine even inhibited the growth of pollen tube. From other papers is known that the influence of acetylcholine on pollen of *Arachis hypogaea* (Chhabra and Malik, 1978), or *Hippeastrum hybridum* (Roshchina and Melnikova, 1998a, b) depended on the concentration. Viability of the tested material was also significant, because germination of poorly viable pollen grains was stimulated by acetylcholine, unlike those that are easily germinated (Roshchina and Melnikova, 1998a; Roshchina et al., 1998b; Roshchina et al., 1999b).

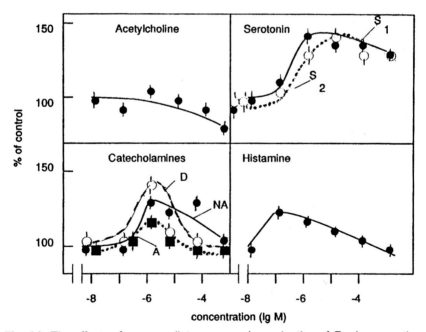

Fig. 14. The effects of neuromediators on seed germination of *Raphanus sativus* (after Roshchina, 1992, 1994). S_1 and S_2, serotonin creatine phosphate and serotonin chloride; D, dopamine; NA, noradrenaline, A, adrenaline.

Noradrenaline at 10^{-6}–10^{-4} M usually inhibited the pollen germination of *H. hybridum* (Roshchina and Melnikova, 1998b), while dopamine and serotonin stimulated the process at concentrations of 10^{-6}–10^{-4} M (Fig. 15) (Roshchina *et al.*, 1998a, b).

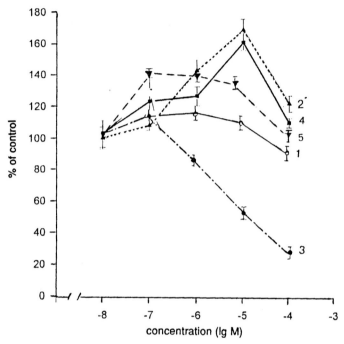

Fig. 15. The effects of neurotransmitters on pollen germination of *Hippeastrum hybridum* during 24 hours (after Roshchina and Melnikova, 1998a, b; Roshchina *et al.*, 1998a). 1, acetylcholine; 2, dopamine; 3, noradrenaline; 4, serotonin; 5, histamine.

In literature there is evidence that histamine has an effect on germination of seeds. At a concentration of 24.5 mol m^{-3}, it reduces the number of germinating seeds by 50% compared to control and at higher concentrations suppresses the process completely (Reynolds, 1989). Probably, like other derivatives of histidine, histamine is a regulator of seed germination. Recently the stimulatory effects of histamine (Fig. 15) on pollen germination of *H. hybridum* were established (Roshchina and Melnikova, 1998a, b). The effect was stronger on pollen with low rate of emergence (stored for half a year).

2.1.3. Development of Plants

In 1972, Kandeler found that under continuous illumination acetylcholine prevents flowering of long-day plant *Lemna gibba* but accelerates flowering of short-day plant *Lemna perpusilla* (Kandeler, 1972). These results were confirmed by Oata and Hoshino (1974) in experiments on *L. gibba*. Other authors (Ladeira *et al.*, 1982b) did not observe the effect of acetylchoine on the development of both short-day plants *Xanthium strumarium* and *Porophyllum lanceolatum* and long-day plant *Sinapis alba*.

As distinct from acetylcholine, catecholamines are more effective in regulating the processes of development and morphogenesis. In some cases, they suppress the flowering in *L. gibba* G-3 (Oata, 1974), in others they can stimulate the flowering of the same plant (Oata and Nakashima, 1978) and *Lemna paucicostata* (Khurana *et al.*, 1987). Norepinephrine (noradrenaline) and its metabolites are considered a flower-inducing component of water extracts from *L. paucicostata* (Takimoto, 1992). According to Hourmant *et al.* (1998), catecholamines are involved in morphogenesis *in vitro* of potato (*Solanum tuberosum* L.) plants. Exogenous dopamine (100–500 μmoles/L) in cultural medium causes a stimulation of growth, especially of aerial parts—leaves and stems—approximately of 20%, whereas its effect on tubers and roots was smaller, and even slightly toxic for roots, at concentration 500 μmoles/L. The level of endogenous dopamine was also higher in aerial parts. When inhibitor of spermidine biosynthesis, methylglyoxal-bis(guanylhydrazone), was added to the cultural medium, it inhibited the growth of aerial parts while it stimulated tuber formation and totally blocked rhizogenesis.

Serotonin also demonstrates its influence on morphogenesis. It is perhaps a precursor of melatonin (O'Neill and van Tassel, 1994), which triggers flowering.

2.1.4. Influence of Environmental Factors on the Activity of Neurotransmitters

The action of environmental factors on the activity of neurotransmitters was studied, mainly for acetylcholine. Light is not only a powerful regulator of acetylcholine synthesis in plants, it also plays an important role in the effects induced by this cholinic ester. Kasturi (1978) showed that the acetylcholine-produced inhibition of formation of secondary roots in *Pisum sativum* depends on exposure of seedlings to light. The strong stimulation (about 50% of the control) of seed germination of *Rumex obtusifolius* by acetylcholine was also observed mainly in the light (Tretyn *et al.*, 1985).

In some cases, the effects induced by acetylcholine, such as the inhibition of formation of secondary roots in *Phaseouls aureus* (Jaffe, 1970, 1972a), stimulation of sporulation in fungi *Trichoderma viride* (Gressel *et al.*, 1971), stimulation of germination of seeds of six different plant types (Holm and Miller, 1972), and inhibition of growth of hypocotyls and stimulation of growth of epicotyls of *Vigna sesquipedalis* (Hoshino, 1983), are similar to those induced by illumination with red light at $\lambda < 680$ nm. The acetylcholine-stimulated formation of conidia and mycelium of *T. viride* grown in the dark was similar to the effect produced by blue light (Gressel *et al.*, 1971). These facts underlie the discussion about the action mechanism of light of different qualitative composition (Fluck and Jaffe, 1976; Hartmann and Gupta, 1989). It was assumed that the mechanism of action of red light is mediated through acetylcholine (Fluck and Jaffe, 1976; Jaffe, 1976). The similarity of the effects produced by red light and acetylcholine was also observed in special experiments, which were analyzed in the review of Hartmann and Gupta (1989). However, it was impossible to resolve this question unambiguously.

Some authors proposed that the effect of acetylcholine depends on pH of the medium. In special experiments it was shown (Dekhuijzen, 1973), for example, that acetylcholine at concentrations of 10^{-3}–10^{-2} M stimulates the growth of seedlings of *Triticum vulgare* only at pH 7.5.

It should be noted that the use of exogenous acetylcholine and catecholamines in experiments on whole plant or its intact organs presents some difficulties. First of all, this refers to acetylcholine, 90% of which can be digested by cholinesterase localized in the plasmalemma and cell wall of surface cells (Hartmann, 1978). The effects of high concentrations ($> 10^{-3}$ M) of the substances are difficult to interpret since they can induce nonspecific reactions, acting as retardants or desensitizing the surface receptors, which blocks the responses of plants. Animal cholinoreceptors are offen desensitized by higher concentrations of acetylcholine, if the cholinic ester is not broken by cholinesterase (Changeux *et al.*, 1984). A similar phenomenon is observed when the receptors are inhibited by pesticides, which can strengthen the effects induced by acetylcholine due to the blockage of the enzyme hydrolyzing the mediator. For instance, stimulation of seed germination in *Agropyron repens* L. by exogenous acetylcholine is enhanced three-fold by physostigmine, blocker of cholinesterase (Holm and Miller, 1972). On the other hand, the cholinesterase inhibitors themselves can act negatively. In particular, physostigmine inhibits the elongation of pollen tube in *Crinum asiatum* (Martin, 1972). However, neostigmine has no similar effect on the pollen tubes of *Lathyrus latifolius* (Fluck and Jaffe, 1974b). Flowering of *Lemna* also did not react to inhibitors of cholinesterase. Therefore, it is necessary to keep in mind individual sensitivity of plants to pesticides in experiments with mediators.

2.1.5. Mechanism of Action on Growth and Development

The mechanism of action of a substance on growth consists either in activating the proteins involved in growth, including the enzymes of phytohormone synthesis, or in affecting directly the genetic apparatus, i.e., DNA and RNA synthesis as well as the processes of transcription and translation, repression, or activation of separate genes. In animals acetylcholine and catecholamines promote mitosis (Morgan et al., 1984). Special experiments with biomediators on plants have not been carried out.

Interest in the problem of biomediators arose in view of the investigations of the mechanisms of photomorphogenesis in the 1970s (Jaffe, 1970; Hartmann, 1971). Jaffe was able to show that the effect of acetylcholine on growth reactions is similar to that of red light at < 680 nm (Jaffe, 1970; 1972a, b). This finding led him to a hypothesis that phytochrome is involved in growth processes induced by acetylcholine (Fluck and Jaffe, 1976). However, other groups failed to reproduce all the experiments. Moreover, the effects of acetylcholine have not been observed for many other phytochrome-dependent reactions (Kasemir and Mohr, 1972; Saunders and McClure, 1973). In the review of Hartmann and Gupta (1989), this hypothesis was subject to a sharp criticism. However, evidence accumulated over 20 years since the first works of Jaffe is not sufficient to give a better insight into the problem. Therefore, there is no reason to reject this hypothesis completely (Chapter 3). There is no information on what growth-regulating proteins other than phytochrome are activated by acetylcholine. Growth effect of acetylcholine is possibly connected with activation or depression of cholinesterase (see Chapter 3). In some cases, the effect of cholinesterase inhibitors neostigmine and physostigmine on growth was similar to that of acetylcholine itself (Jaffe, 1970; Hoshino, 1983). Moreover, some artificial retardants AMO-1618 also inhibit cholinesterase (Riov and Jaffe, 1973c). When the mechanism of action of biomediators is being considered, their possible effect on the genetic apparatus of the cell should be taken into account. It is believed that in animals acetylcholine and noradrenaline may directly affect the RNA synthesis in nuclei (Golikov et al., 1985). Arkhipova and coworkers (1988) and Tretyak and Arkhipova (1992) showed that catecholamines and serotonin stimulate RNA synthesis in tissues of brain and liver and in nuclei isolated from the cells of these organs. Serotonin is also capable of stimulating DNA synthesis in fibroblasts (Seuwen et al., 1988).

Some workers relate the mechanism of action of the biomediators to their ability to interact with phytohormones. Some phytohormones and biomediators are synthesized by a similar mechanism from the same precursor. For instance, serotonin and indolylacetic acid are derived, in the final analysis, from tryptophan (Chapter 1). In other cases, phytohormones and mediators can have different precursors, but they are synthesized simultaneously and show a similar activity. For instance, in fruits and callus of banana *Musa acuminata colla* A.A.A., along with cytokinins, dopamine is found (Martin et al., 1987), which at a concentration of

>5 mg/l inhibits the growth of soybean callus and at 50 mg/l blocks the process completely, acting in the same way as the cytokinin zeatin. The third possible relationship between the biomediators and phytohormones may be the synergism of action of these substances. Catecholamines adrenaline, noradrenaline, dopamine, and 3,4-dihydroxymandelic acid accelerate the elongation of hypocotyls of lettuce *Latuca sativa* gibberellin (Kamisaka, 1979). The derivatives of catecholamines metanephrine, normetanephrine, 3-methoxy-4-hydromandelic acid, or their precursor DOPA (dioxyphenylalanine) did not cause this effect. The catecholamine-induced reactions were inhibited by trans-cinnamic acid, which competed for the binding site (receptor) not only with catecholamines but with their structural analog the dihydroconiferyl alcohol. It should be noted that there is a similarity in chemical structure between catecholamines and metabolites such as cinnamic acid, coniferyl alcohol, and sinapic acid on the one hand, and serotonin and indolylacetic acid and gramine on the other.

The interactions between biomediators and hormones may be competitive. An example of such interactions is the acetylcholine-induced inhibition of release of the hormone ethylene by etiolated seedlings of *Phaseolus vulgaris* (Parups, 1976) and leaf disks of soybean *Glycine max* (Jones and Stutte, 1984, 1986, 1988). At low concentrations, ethylene stimulates the elongation of cells of moss *Athyrium filix-femina*, and addition of acetylcholine (10^{-5} M) inhibits the process by 1.5–2 times (Bahre, 1975, 1977). Acetylcholine (0.1–500 μM) decreases the stimulation and growth of wheat seedlings pretreated with indolylacetic acid (Lawson et al., 1978). According to Hoshino (1979), treatment of *Lemna gibba* G-3 by indolylacetic acid reduces the amount of acetylcholine by 30%, whereas the antagonist of acetylcholine atropine (>10^{-6} M) levels off the effect of auxin on the growth and reduces the number of flowers on plants compared to experiments in which only indolylacetic acid was used. On the basis of these findings, Hoshino proposed that auxin and cholinic ester are bound to the receptor at the same site on the membrane, or independently but near each other.

The observed effects are of interest not only in view of intra- and intercellular effects of biomediators in the organism itself. It should be noted that acetylcholine, histamine, and serotonin occur widely in plant secretions (Roshchina and Roshchina, 1989, 1993) and as exogenous products can exert allelopathic action in phytocenosis (Roshchina, 1994, 1999a). Clearly, this phenomenon may be of great ecological significance.

2.1.6. Movements and Motility Reactions of Plants

Mediators are known to be involved in motor responses of animals controlled by the cholinergic and adrenergic systems. Contractile elements of plant cells are also sensitive to mediators. According to Oniani (1975), the movement of the cytoplasm in cells of *Nitella* slows down at a concentration of acetylcholine in the medium of 10^{-7}–10^{-6} M, and at higher concentrations it stops. This phenomenon is reversible, and if the cells are washed free of the agent the velocity of movement returns to the initial level. Adrenaline and noradrenaline (10^{-7}–10^{-6} M) decelerate the cyclosis in cells of charanian algae *Nitella* (Oniani, 1975; Oniani et al., 1977a, b). Higher concentrations of the substances given rise to irreversible changes inducing death of cells. The correlation between these phenomena and the effect of biomediators on contractile proteins was confirmed by experiments of Toriyama (1975, 1978), who studied the contractile properties of protoplasts of single cells from roots of *Phaseouls aureus* and showed that they are sensitive to exogenous acetylcholine. Acetylcholine-induced contraction of protoplasts was inhibited by its antagonist atropine. Toriyama (1978) showed that protoplasts of cells from the pericycle of root tips are also capable of contracting under the influence of acetylcholine. It is possible that this phenomenon is connected with the phytochrome function. This author supposed contractile proteins to be incorporated into plasmatic membrane of cells sensitive to acetylcholine. Contractile movements of organelles, for instance chloroplasts (Ohnishi, 1964; Schönbohm and Meyer-Wegener, 1989), may also be regulated with neuromediators, as seen from the use of colchicine and cytochalasin B (blockers of self-assembling tubulin and of actin microfilaments, respectively), on the neuromediator-regulated photophosphorylation in isolated chloroplasts of *Pisum sativum* (Roshchina, 1990c). On animal cells, serotonin, dopamine, or noradrenaline is associated with an actin-like component (Small and Murtman, 1985).

A few plants respond with rapid motion to various external stimuli, including chemical substances. These plants were used to study the effect of exogenous mediators on motor responses of plants. Some authors believe that in plants such as *Albizzia julibrissin*, *Mimosa pudica*, and *Samanea saman* a rapid closing of leaves with turgor changes is related to the transmission of the electric signal in the form of the action potential (Toriyama and Jaffe, 1972) that arises in animals with participation of neuromediators. However, the concentration of endogenous acetylcholine in leaves and pulvini (motor organs) of these plants appeared to be approximately the same (Satter et al., 1972). Exogenous acetylcholine at concentrations (10^{-5}–10^{-7} M) exceeding its amount in pulvini did not induce movement of leaves. However, Kumaravel and coworkers (1979) favor the hypothesis that acetylcholine is involved in movements of leaves of *Samanea saman*, although the amount of acetylcholine in closed leaves is higher than in open ones.

The function of acetylcholine and acetylcholinesterase is associated with the motility of leaves in plants sensitive to mechanical irritation. The enzyme activity was found only in motile organs, the primary and secondary pulvini (Momonoki and Momonoki, 1993a, b). Other functions of acetylcholine are considered factors of geotropism (Momonoki and Momonoki, 1993a, b). It seems more likely that noradrenaline is involved in plant movement, since its amount in pulvini and axes is higher than in leaf lamina (Applewhite, 1973).

The meadow-grown plant *Tragopogon arvense* demonstrates a biological clock: its flower is opened at 9 hours morning (nasty movements). Recently the neurotransmitter influence on the flower opening was studied (Roshchina, unpublished data). When cutted shoots of the species are put in the vessel with a solution tested (neurotransmitters acetylcholine or catecholamines or antitransmitter drugs atropine and neostigmine 10^{-6}–10^{-4} M) at 18 hours, all flowers were closed, and from 6 hour of the following day the observation lasted upto 9 hour morning when all flowers were opened. Acetylcholine alone and in lesser degree in the mixture with atropine induced the flower opening earlier (at 7.30 or 8.00), than in the control at 9.00. It shows a possible acetylcholine participation in regulation of the nasty movements.

The swelling of protoplasts of wheat seedlings can be induced by acetylcholine and other hormonal substances via receptors of plasmatic membranes where the function is common with animal K^+/Na^+ channels and sensitive to GTP Ca^{2+} channel (Bossen et al., 1991). It is also proposed that such movements are concerned with contractile elements of protoplasm.

Zholkevich and coworkers (1979, 1989) and Korolev and Zholkevich (1990) found that acetylcholine (5×10^{-5}–10^{-3} M) significantly increases (by 40–50% of control) the pressure in roots of *Helianthus annuus* and exudation of xylem sap by seedlings of *Zea mays*. Noradrenaline, adrenaline, and serotonin also stimulate the xylem sap exudation of roots in these plants (Zholkevich and Chugunova, 1995; Zholkevich et al., 1998). The authors relate the findings obtained to the effect of the mediator on contractile systems of conducting (transforming) vascular elements of plants, especially on contractile proteins.

It was generally thought that acetylcholine may be effective only inside the organism. However, there is reason to believe that it participates in the interactions between the organisms. In favor of this view are the following facts. Acetylcholine is also known to act as inhibitor of chemotaxis in bacteria. This was found in *Pseudomonas fluorescens* (Chet et al., 1973) and may be considered to result from the action of compounds excreted by some species of plants and, probably, microorganisms. The effect of acetylcholine may be very strong. At a concentration of 1.5×10^{-3} M it inhibits the motility of photosynthesizing bacteria *Rhodospirillum rubrum* and *Thiospirillum jenense* (Faust and Doetsch, 1971). Acetylcholine may act to maintain the symbiosis between rhizobia and soybean plant (Fluck and Jaffe, 1976), since it regulates the motility of *Rhizobium* in a soil solution. The enzyme acetylcholinesterase, which hydrolyzes acetylcholine, was found in small nodules. However, there is no evidence that the enzyme belongs to *Rhizobium* per se or to plant cells (Fluck and Jaffe, 1976). Another example is washing out of acetylcholine from leaf hairs where it is present in great amounts (Emmelin and Feldberg, 1947, 1949). In the rain, in the case of dew and fog, the soil on which nettle grows may be enriched with cholinic esters (Tukey, 1966), which influence coexisting organisms.

In the 1970s it was found that adrenaline and noradrenaline (10^{-7} – 10^{-6} M) retard the cyclosis in Charophyta algae *Nitella* (Oniani *et al.*, 1974, 1977a, b). In some soybean plants such as *Albizzia julibrissin*, *Mimosa pudica*, and *Samanea saman*, the amount of noradrenaline in motor organs pulvini is higher than in laminae (Applewhite, 1973), which is considered to be associated with the ability of leaves to close quickly under the influence of mechanical irritation. Catecholamines and serotonin are efficient in the induction of chemotaxis in some bacteria, indicating, presumably, the effect of these substances on contractile proteins (Sastry and Sadavongvivad, 1979).

2.2. MEMBRANE PROCESSES

The key mechanism by which neuromediators act in animal cells consists in changing membrane permeability, membrane potential, and associated processes (McQueen, 1987). As hydrophilic substances, neuromediators do not penetrate the plasmalemma, but act outside.

2.2.1. Ion Permeability and Membrane Potential of Whole Cells

Jaffe (1970) was the first to show that acetylcholine is capable of increasing the ionic permeability of plant membranes and inducing changes in membrane potential. He found that under the action of acetylcholine or red light the protons exit from root cells of *Phaseouls aureus* to the external medium. As a result, there arises a positive bioelectric potential. This result was confirmed by the experiments of Tanada (1972), who showed that root tips of etiolated seedlings of *Phaseouls aureus* adhere to the negatively charged glass surface in red light (Tanada's effect). A similar effect was induced by acetylcholine. It is likely that acetylcholine and red light give rise to a more positive membrane potential than the resting potential. The depolarizing effect of acetylcholine was also demonstrated in other experiments.

In Table 12 are presented data showing the effects of acetylcholine on the membrane potential and ionic permeability of plants. The difference in potential between the apical part or tip of the etiolated hypocotyl of *Phaseouls vulgaris* and the site of cutting is usually approximately −30 mV (Hartmann, 1975). Blue light λ 430 nm increases the hyperpolarization to −37 mV, and acetylcholine at a concentration of 10^{-9}–10^{-3} M may suppress the effect of the blue light (Hartmann, 1977) and induce a weak polarization. Hartmann (1977) also showed that the K^+ ion uptake from the external medium increases when etiolated bean hypocotyls are incubated with acetylcholine (10^{-6} M) in the dark. The changes in absorption occur during 30–60 min. of exposure. Using a microelectrode technique, Yurin and coworkers (1979a, b) showed that acetylcholine at a concentration of 10^{-5}–10^{-3} M induces a reversible depolarization of cells of *Nitella flexilis*. At acetylcholine concentration of 10^{-5} M the potassium permeability of the membranes from these algae increases and the membrane potential falls from −136 to −123 mV. Slezak (1984) demonstrated that under the influence of acetylcholine the permeability of membranes to K^+ roots of *Pisum sativum* increases, and the membrane potential falls. Like red light, acetylcholine (10^{-4} M) accelerates the Ca^{2+} uptake by coleoptiles of oat *Avena*, which is accompanied by secretion of protons (Tretyn, 1987).

Table 12. Changes in membrane potential and ionic permeability induced by acetylcholine in plants

Plant	Organ	Membrane potential (MP)	Ion permeability	Reference
Avena sp.	Coleoptiles		Ca^{+2} and H^+ permeability increases	Tretyn, 1987
Nitella	Cells	MP decreases from 136 to 123 mV (depolarization)	K^+ permeability increases	Yurin et al., 1979a, b, 1990
Phaseolus aureus	Roots	MP decreases (depolarization)	H^+ permeability increases	Jaffe, 1970
Phaseouls vulgaris	Etiolated hypocotyls	MP decreases (slight depolarization on blue light)	K^+ permeability increases	Hartmann 1977
Pisum sativum	Roots	MP decreases (depolarization)	K^+ permeability increases	Slezak, 1984, 1987
Pisum sativum	Chloroplasts	MP decreases	Na^+/K^+ permeability increases	Roshchina and Mukhin, 1985a; Roshchina, 1987a

The mediator (10^{-6} M) also induced the Ca^{2+}-dependent swelling of protoplasts from leaf mesophyll of etiolated seedlings of wheat *Triticum aestivum* (Tretyn et al., 1990a, b). Antagonists of acetylcholine (see Chapter 3, Section 3.1), muscarine, and nicotine mimicked the neuromediator effect both in the Ca^{2+}- and in the K^+/Na^+-dependent protoplast swelling and unrolling of etiolated wheat seedlings induced with acetylcholine, whereas its antagonists atropine and d-tubocurarine nullified the responses (Tretyn et al., 1990b; Tretyn and Kendrick, 1990). Acetylcholine (1 μM) and Bay K-8644, inducing Ca^{2+}-channel opening, stimulated unrolling of etiolated wheat seedlings, while nifedipine, Ca^{2+}-channel closing agent, as well as trifluoperazine, the serotonin antagonist and calmodulin inhibitor, reduced this effect (Tretyn and Kendrick, 1990). Verapamil, Ca^{2+}-channel blocker, inhibited the phytochrome-induced swelling of the etiolated wheat mesophyll protoplasts (Bossen et al., 1988). Acetylcholine usually hyperpolarizes animal outer hair cells and increases Ca^{2+} concentration (Doi and Ohmori, 1993). It can be seen in Table 12 that most often acetylcholine induces depolarization—a decrease in membrane resting potential and a change of the potassium permeability. Effects of exogenous acetylcholine, K^+, or Ca^{2+} on detached and heat-stressed leaves to which the primary pulvinus was still attached were determined (Momonoki and Momonoki, 1992). The application of acetylcholine together with K^+ or Ca^{2+} salts remarkably enhanced leaf recovery from wilting in detached and heat-stressed primary pulvinus leaves. The results suggest that acetylcholine may control ion or hormone fluxes regulating the opening of ion channel in pulvini.

Exogenous acetylcholine (1 mM) induced 80% closure of stomata within 5 min., whereas butyrylcholine and propionylcholine at the same concentrations induced only 10% and 15% (Madhavan et al., 1995). The opening of stomata is K^+-regulated, and K^+ ions (10 mM), on the contrary, induced further 10% increase in the opening of the stomata. Thus, acetylcholine can regulate the aperture of the stomata via an effect on the K^+ exchange (Madhavan et al., 1995). Moreover, the presence of acetylcholinesterase in the guard cells has been shown. Hence, this confirms the occurrence of cholinergic regulatory system in stomata (Madhavan et al., 1995).

Like acetylcholine, biogenic amines are capable of changing the membrane potential. Adrenaline has a marked effect on bioelectric processes in the conducting system of squash (Salleo et al., 1977). Adrenaline, noradrenaline, and serotonin are able to shift the membrane potential of cells of green algae *Nitella syncarpa* (Oniani et al., 1973). The authors relate these changes to Ca^{2+} permeability of plasmalemma. In beet *Beta* cells, serotonin depresses the active Na^+ transport across plasmalemma (Pickles and Sutcliffe, 1995). We compared the influence of neuromediators on the permeability of plasmalemma of root beet disks for ions (Table 13). Na$^+$ efflux was decreased by acetylcholine, dopamine, serotonin and histamine, whereas Ca^{2+} efflux—only by serotonin. Ca^{2+} output was also stimulated by acetylcholine. Antagonists of acetylcholine (see Chapter 3, section 3.1.1) α-bungarotoxin, and atropine as well as antagonist of serotonin (see Chapter 3, section 3.2.1) trifluoroperazine strongly decreased Na$^+$ efflux. Although atropine stimulated K$^+$ efflux, trifluoroperazine, as calmodulin blocker, also decreased Ca^{2+} efflux, but stimulated K$^+$ efflux. Antagonist of noradrenaline (see Chapter 3, section 3.2.1) yohimbine stimulated Mg^{2+} efflux. Besides the effects on plasmalemma permeability, the red pigment betacyanin efflux was also observed in variants with histamine, serotonin, acetylcholine, and trifluoroperazine. This showed on the additional action on the tonoplast membrane permeability.

The changes in permeability of the plasmalemma induced by exogenous acetylcholine are reminiscent of the phenomena proceeding in synapses of animal cells, i.e., when coming in contact with the postsynaptic membrane, acetylcholine increases the permeability to Na$^+$ and K$^+$ ions, and biogenic amines for Ca^{2+}, thus generating the action potential that propagates through the organism. As distinct from animals, the transmission of the signal from cell to cell in plants proceeds via continuous cytoplasmic strands rather than through synaptic contacts. Consequently, in plant cells, acetylcholine and other mediators are involved in the transmission of signal inside the cell but not between cells via intercellular

Table 13. Effects of neuromediators and their antagonists on ion efflux from and ion uptake by root disks of red beet *Beta vulgaris* var. *rubra* measured by atomic absorbance spectrophotometer Perkin-Elmer (Roshchina, unpublished data).

Added substance (10^{-6} M)	Ions, μg/L				Betacyanin efflux (A540 nm)
	Na$^+$	K$^+$	Ca^{2+}	Mg^{2+}	
Control (without additions—only water)	3.47±0.03	5.30±0.01	6.51±0.03	5.80±0.02	No effect
Acetylcholine	0.84±0.08	5.8±0.01	9.13±0.7	5.30±0.01	Stimulates
Dopamine	0.95±0.08*	4.8±0.5	3.35±0.02	5.8±0.01	No effect
Noradrenaline	3.38±0.02	5.30±0.5	6.61±0.05	4.8±0.5	No effect
Serotonin	1.03±0.10	4.80±0.05	1.60±0.07	5.30±0.5	Stimulates
Histamine	0.77±0.005	4.30±0.04	3.95±0.3	4.80±0.05	Stimulates
α-Bungarotoxin	1.12±0.30	5.80±0.01	2.35±0.02	4.30±0.04	No effect
Atropine	0.81±0.1	25.8±0.5	5.38±0.02	5.80±0.01	No effect
Yohimbine	2.43±0.01	5.81±0.01	6.26±0.01	25.8±0.5	No effect
Trifluoroperazine	0.62±0.1	12.45±0.90	1.88±0.50	5.81±0.01	Stimulates

The collection of the medium after 1 hour of exposure of discs (1 g/ml)

contacts. Since mediators are also present in unicellular organisms (Janakidevi et al., 1966a, b; Levitzki, 1988) it was assumed that neuromediators participate in signal transmission from the plasmalemma to organelles and between the organelles.

2.2.2. Ion Permeability and Membrane Potential of Organelles

To check the hypothesis that mediators are involved in transmission of intracellular information, the effect of acetylcholine and biogenic amines on individual organelles was studied. The first studies in this line were done on mitochondria isolated from different organs of animals (Vdovichenko and Demin, 1965; Vdovichenko, 1966). The investigators studied the swelling of mitochondria treated with acetylcholine ($\sim 10^{-3}$ M) and found an inhibition of swelling. A similar phenomenon was observed in chloroplasts of *Pisum sativum* subjected to the same treatment (Roshchina and Mukhin, 1985b): acetylcholine at a concentration of $10^{-4} - 10^{-3}$ M inhibited swelling of the chloroplasts in isotonic medium in the light to a greater extent than in the dark. The swelling is known to be due to entering of water inside the organelle, where the concentration of ions is higher. Since high concentrations of acetylcholine (10^{-3} M) inhibit the process it could be assumed that the cholinic ester increases the permeability of chloroplasts to ions. In favor of this assumption are the changes in membrane potential of plastids, which can be judged from an increase in the amplitude of the light-induced changes in absorption at 520 nm. $\Delta 520$ nm characterizes the energized state of reaction centers of photosystems in the light. In the experiments performed, an increase in acetylcholine content in the medium resulted in a decrease of $\Delta 520$ nm by $\sim 40\%$ (Fig. 16). This finding can be interpreted to indicate a decrease of the membrane potential under the influence of the cholinic ester. Analysis of the concentration-dependence curve for acetylcholine indicates that its effect on $\Delta 520$ becomes manifest when the amount of the agent in the medium exceeds 10^{-7} M.

In order to determine which ions are responsible for the fall of the acetylcholine-induced membrane potential of plastids, special experiments with intact chloroplasts from *Pisum sativum* were designed (Roshchina, 1987a; Roshchina and Mukhin, 1987a, b). The addition of acetylcholine to the medium gave rise to the efflux of Na^+ and K^+ from the plastids (Fig. 17). In this process, the concentration of ions in the medium increased. The concentration-dependent curves reached a plateau at a acetylcholine concentration of $10^{-6}-10^{-5}$ M. This may suggest that the binding sites of acetylcholine in chloroplast membranes have limited dimensions. The reaction was the same whether or not the plastids were illuminated. The concentration of Na^+ in the ambient solution increased 15 times due to

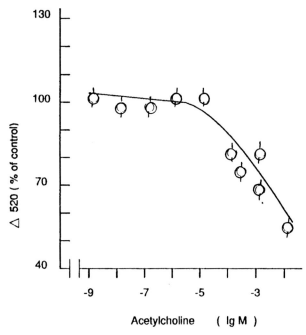

Fig. 16. Concentration curve of the acetylcholine action on light-induced absorption at 520 nm in pea chloroplasts (Roshchina and Mukhin, 1985a)

output of ions from plastids, whereas that of K^+ increased two times compared to the initial value. The efflux of ions was, most probably, passive (with respect to concentration gradient), since the initial amount of ions in the external medium was negligible. When acetylcholine was not added to the solution, no spontaneous efflux of Na^+ and K^+ from chloroplasts was observed during 40 min. exposure, which exceeded 10 times the duration of the experiment (4 min.). Therefore, the observed effects of acetylcholine cannot be explained by the damage of the envelope of chloroplasts on isolation. Analysis of the amount of the same ions inside the chloroplasts carried out after destruction of plastids indicated a clear correlation between the output of sodium and potassium into the external medium and the rest of these ions inside the plastids. The amount of Na^+ ions in chloroplasts of *Pisum sativum* is usually higher than in the cytoplasm (Robinson and Downton, 1984). Therefore, the experiments can be considered to model the situation that may occur in the cell *in vivo*.

The efflux of Na^+ and K^+ from intact chloroplasts is accompanied by the H^+ uptake by intact chloroplasts and thylakoids in the dark (Roshchina and Mukhin, 1987a, b). pH gradient was increasing only at low concentrations (10^{-9}–10^{-7} M) of acetylcholine in the medium (Roshchina, 1991c).

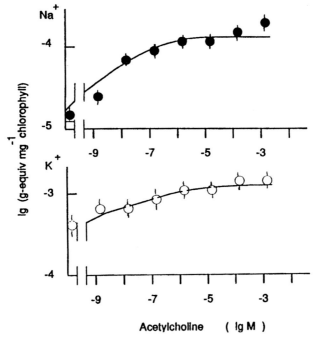

Fig. 17. Concentration curves of the acetylcholine action on the Na⁺ and K⁺ effluxes from intact pea chloroplasts (Roshchina, 1987, 1989a, b; Roshchina and Mukhin, 1987a)

Acetylcholine may regulate the movement of molecules across the membrane by interacting with the lipid fraction of membranes (Hartmann et al., 1980). It has been established that at low concentrations (10^{-8} M) acetylcholine inhibits the incorporation of radioactive phosphorus into phospholipids that control the transport of substances in animal cells.

As distinct from acetylcholine, noradrenaline, adrenaline, and serotonin had no effect on K⁺ and Na⁺ input from chloroplasts. However, they stimulated the release of Ca^{2+} and Mg^{2+} ions (Fig. 18) (Roshchina, 1989a, b). The effect of the biogenic amines began with low concentrations (10^{-9} M) and strengthened at concentrations up to 10^{-7} M, after which a saturation occurred.

2.3. ENERGETIC AND METABOLIC REACTIONS

Since growth responses of plants are the sum of many processes the effect of any chemical substance on growth may be mediated by stimulating or inhibiting the metabolism of substances and energy.

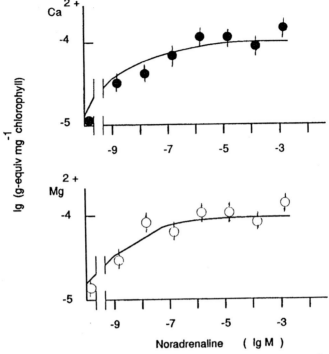

Fig. 18. Concentration curves of the noradrenaline action on Ca^{2+} and Mg^{2+} effluxes from intact pea chloroplasts (Roshchina, 1989a, b)

2.3.1. Electron Transport and Coupled Phosphorylation

White and Pike (1974) showed that very small amounts of acetylcholine (10 nM) induce a 70% fall in the endogenous ATP level in stems of *Phaseouls vulgaris*. Analysis of the effect of acetylcholine on the ATP content in etiolated bean plumules shows that the mediator (10^{-11}–10^{-3} M) in both darkness and red light reduces the level of adenosine nucleotide triphosphate (Kirshner *et al.*, 1975). Light plays a great role in manifestation of the effect of acetylcholine. Red light per se significantly increases the ATP level in bean plumules, whereas acetylcholine reduces it by half. In the dark, on the contrary, the reduction of the ATP level by acetylcholine was insignificant (Kirshner *et al.*, 1975). In experiments with mitochondira from roots of *Phaseouls aureus*, Yunghan and Jaffe (1972) also observed a fall in the ATP level and an increase in the amount of ADP and inorganic phosphate induced by acetylcholine. However, other authors failed to observe any changes in the ATP level in bean roots (Burcky and Kauss, 1974).

The observations of Yunghans and Jaffe (1972) indicated that acetylcholine stimulates the uptake of oxygen by whole tissues and isolated mitochondria of secondary roots of *P. aureus* but does not practically affect the P/O ratio and utilization of inorganic phosphate. In rat liver, the mediator stimulated the substrate-induced phosphorylation and in isolated mitochondria it induced oxidative phosphorylation (Doliba, 1986; Kondrashova and Doliba, 1989), whereas in mitochondria from roots of *P. aureus* (Yunghans and Jaffe, 1972) acetylcholine had no effect on the synthesis of ATP. The experiments of Roshchina and Mukhin (1987a, b) with isolated chloroplasts demonstrated that acetylcholine had no changes in the rate of the photosynthetic electron transport (photoreduction of $NADP^+$, 2,6-dichlorphenolindophenol, cytochrome f, and plastocyanin). However, the coupled ATP synthesis in chloroplasts may be regulated by acetylcholine (Roshchina and Mukhin, 1985a, 1987a; Roshchina, 1987a, b). As can be seen in Fig. 19, at low concentrations ($<10^{-5}$ M) this cholinic ester increases the photophosphorylation in garden pea chloroplasts twofold compared to control, whereas at higher concentrations it inhibits the synthesis. This effect is evident on illumination with 68 Wt m^{-2} and temperatures higher than 10°C. A pattern can be observed: the stronger the illumination and temperature the more efficient are low concentrations of acetylcholine (Roshchina, unpublished results). The effect of acetylcholine on photophosphorylation cannot be explained by an increase of $\Delta\mu H$ under the influence of light since the biomediator stimulates the uptake of protons in the dark (Roshchina and Mukhin, 1987b). It was assumed that the stimulation of ATP synthesis in chloroplasts catecholamines and histamine stimulated coupled photophosphorylation (Fig. 20). Serotonin was inefficient in this reaction. The stimulation of ATP synthesis was maximal (two-fold) at low (10^{-8} M) concentrations of catecholamines. In the same concentrations acetylcholine also enhanced the rate of photophosphorylation. When dopamine was used in experiments in the presence of oxygen, synthesis of ATP was not changed greatly, whereas after the evacuation of air the effect of stimulation achieved the level observed for other catecholamines. Differences can be explained by the formation of oxidized derivatives of dopamine in the presence of oxygen (Udenfriend *et al.*, 1959), and due to this event the concentration of active matter may be decreased.

2.3.2. Redox Reactions

Dopamine, adrenaline, and noradrenaline, as compounds having double links and hydroxylic groups, can participate in redox reactions. The simplest reaction of the catecholamine autooxidation occurs, perhaps, through the formation of peroxides at the electron transfer from mediator to molecular oxygen. It is enhanced significantly when metal ions such as Fe^{2+},

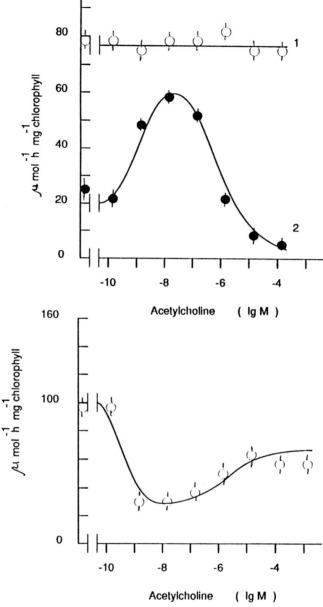

Fig. 19. Action of acetylcholine on the photochemical reactions of isolated pea chloroplasts (after Roshchina and Mukhin, 1987a; Roshchina, 1989a). Above, concentration curves of acetylcholine action on NADP⁺ photoreduction (1) and coupled photophosphorylation (2). Below, concentration curve of acetylcholine action on Mg^{2+}-dependent light-induced ATPase in pea chloroplasts.

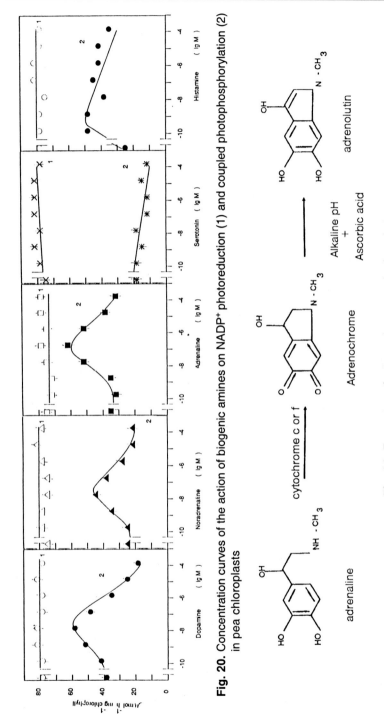

Fig. 20. Concentration curves of the action of biogenic amines on NADP⁺ photoreduction (1) and coupled photophosphorylation (2) in pea chloroplasts

Fig. 21. Formation of adrenochrome and adrenolutin

Fe^{3+}, and Cu^{2+} form complexes with catecholamines (K–Me) and react with oxygen (hydrogen peroxide), according to the following equation of Fenton (Sutton and Winterbourn, 1989):

$$K-Me^{n+} + H_2O_2 \rightarrow K-Me^{n+1} + {}^{\bullet}OH + OH^-$$

In medium lacking oxygen, copper and iron form a complex with catecholamines in the ratio 1:2, in particular with adrenaline, which easily transforms to adrenochrome in the air (Heacock and Powell, 1973), and then in base medium and in the presence of ascorbate to adrenolutine (Fig. 21).

Adrenochrome is a toxic product for animal tissue and in some cases is known as a cause of various diseases, when an organism accumulates many catecholamines under stressor factors (Nalbandyan, 1986; Jewett et al., 1989). It is formed in the presence of ions Fe^{3+} and Cu^{2+} and, moreover, at an even higher rate (2–3 times) when iron-containing proteins ferritin, methemoglobin, or oxidized cytochrome C are in the medium (Green et al., 1956; Walaas et al., 1963).

Roshchina (1989a) studied the interaction of catecholamines with isolated electron carriers ferredoxin, ferredoxin-$NADP^+$-reductase, and plastocyanin from pea, as well as cytochrome f (C_{553}) from *Chlorella*. There were no redox reactions between catecholamines and ferredoxin or ferredoxin-$NADP^+$-reductase. However, in high concentrations (>10^{-5} M) catecholamines reduce oxidized cytochrome f and plastocyanin (Figs. 22, 23). In absorbance spectra of a mixture of catecholamine and cytochrome f, a maximum 480 nm is clearly seen, which appears to belong to aminochrome, in particular to adrenochrome in Fig. 22. If reactions with electron carriers take place after the evacuation of air containing oxygen, aminochrome is not formed in variant with cytochrome f. The cytochrome f-dopamine system usually has a red color peculiar to the reduced form of the electron carrier when the reaction occurs in anaerobic conditions. If oxygen is present, the color becomes blood-red and in absorbance spectra a maximum 480 nm characteristic for aminochrome arises. When dopamine (10^{-5} M) reacts with plastocyanin, it not only accepts an electron (the reduction of carrier), but also binds with copper, as chelators do. This is confirmed by a yellow coloration as well as disappearance of blue color of plastocyanin (which is reduced to colorless protein), as is usually observed. There is perhaps a phenomenon similar to the reduction of copper-containing protein neurocuprein from brain (Gasparov et al., 1979) or blood protein cerulloplasmin (Walaas et al., 1963) by catecholamines. Thus, adrenaline, noradrenaline, and dopamine are capable of altering into oxidation-reduction reactions with the electron carriers localized between the photosystems.

The capacity of catecholamines to act as electron donors in Electron Transport Chain (ETC) has been checked on isolated chloroplasts

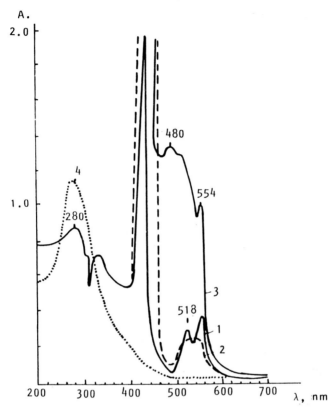

Fig. 22. Interaction of adrenaline with isolated proteins-electron carriers from *Chlorella* cytochrome f (C_{553}). 1, Reduced cytochrome f 3.3×10^{-5} M; 2, cytochrome f oxidized by ferricyanide; 3, 2+ adrenaline 10^{-5} M; 4, adrenaline 10^{-5} M (after Roshchina, 1989a).

(Roshchina, 1989a, b). The electron transport from water to $NADP^+$ blocked by diurone was restored weakly by dopamine, noradrenaline, and adrenaline (Fig. 24). However, in the presence of the redox pair catecholamine-sodium ascorbate the restoration was greater. The two components—dopamine and ascorbate—accumulated in chloroplasts may serve as a potential natural electron donor. In favor of this is also the ability of catecholamines, serotonin, and histamine to inhibit the fluorescence of phycoerythrin (De Lange and Glazer, 1989). Oxygen formed by chloroplasts is also capable of oxidizing adrenaline to the red product adrenochrome (Asada et al., 1974) and this oxidation occurs at the level of photosystem 2 (Lysenko et al., 1987). On the other hand, the exogenous dopamine, as an oxidant, may be involved in the mechanism of water decomposition on illumination of grana of spinach chloroplasts (Gewitz and Volker, 1961).

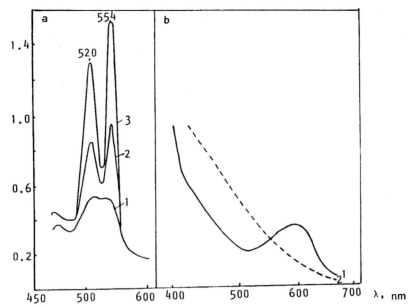

Fig. 23. Interaction of dopamine with the electron carriers (after Roshchina, 1990b). (a) The absorbance spectra of dopamine-cytochrome f system from *Chlorella*. 1, cytochrome f 2.5×10^{-5} M; 2, 1+ dopamine 10^{-5} M; 3, dopamine 2×10^{-5} M; (b) The absorbance spectra of dopamine-plastocyanin system from pea leaves. 1, Plastocyanin 3×10^{-5} M, 2, 1+ dopamine 10^{-5} M.

The participation of acetylcholine, serotonin, and histamine in plant redox reactions in pollen grains is also proposed in light of the following: (1) catecholamines are found in the excretions of the cells (Roshchina and Melnikova, 1999) and may initiate free radical reactions with the substances; (2) free radical reactions occur on the pollen surface enriched in pigments (phenols, carotenoids, azulenes) and forming superoxide anion radicals and peroxides as active oxygen species interact, in turn, with these neurotransmitters (Roshchina and Melnikova, 1998a).

2.3.3. Fluorescence of Individual Cellular Components and Intact Cells

Fluorescence of individual cellular components and intact plant cells as an indicator of the redox state and energy exchange is also sensitive to neuromediators. De Lange and Glazer (1989) show the quenching of the phycoerythrin fluorescence with adrenaline (5×10^{-7}–10^{-3} M) by 62–100%, with noradrenaline (3.5×10^{-7}–10^{-6} M) by 63–83 %, with dopamine (3×10^{-7} to 6×10^{-7} M) by 41–62%, with serotonin (10^{-6}–10^{-5} M) by 75–88%, and with histamine (10^{-6}–10^{-5} M) by 19–41%.

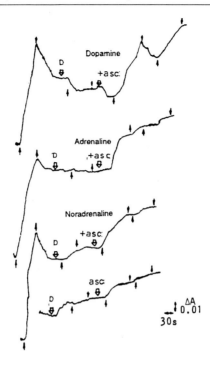

Fig. 24. Kinetics of the NADP⁺ photoreduction by isolated chloroplasts of pea in the presence of redox pair catecholamine/sodium ascorbate (from Roshchina, 1990b). Arrows ↓ and ↑, the switch on and off of the actinic red light; ⇓, the addition of the substance to the reaction medium; D, diuron 10^{-6} M; DP, dopamine 8×10^{-4} M/sodium ascorbate 10^{-4} M; Ad/Asc, adrenaline 10^{-4} M/sodium ascorbate 10^{-4} M.

The effect of treatment with transmitters on the autofluorescence (excited with ultraviolet light 360–380 nm) of intact pollen grains and the pistil stigma has also been studied (Roshchina and Melnikova, 1998a, b). Pollen fluorescence was unaffected by acetylcholine and histamine (Roshchina and Melnikova, 1998a), whereas noradrenaline decreases the emission intensity (Roshchina and Melnikova, 1998b). The inhibitory effect of noradrenaline (10^{-6}–10^{-4} M) from 20 to 50% was connected with superoxide radical formation. Significant influence of acetylcholine and histamine on the fluorescence of pistil stigma of *Hippeastrum hybridum* was observed (Fig. 25). The fluorescence spectrum of the pistil stigma surface (Fig. 25) also shows strong changes in the presence of own pollen. It is known that pollen may contain acetylcholine and histamine (Marquardt and Vogg, 1952). When acetylcholine or histamine was added to the pistil stigma, fluorescence increased approximately two-fold, as compared to the control, just as after pollen addition. Although the spec-

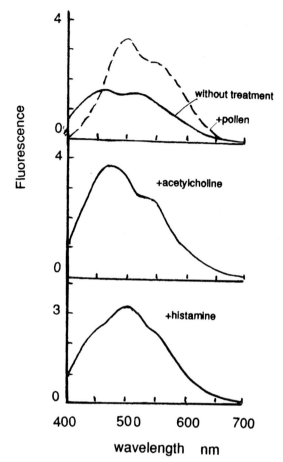

Fig. 25. The changes in the autofluorescence spectra of the *Hippeastrum hybridum* pistil stigma without and after the addition of the pollen and neurotransmitters 10^{-5} M (after Roshchina and Melnikova, 1998a, b)

tra had different shapes in the presence of acetylcholine or histamine, a maximum of 475 nm was clearly seen after the addition of acetylcholine, and there was a shoulder at 520 nm. In the presence of histamine, peaks at 500 nm and shoulders at 450 and 550 nm were seen. Thus fertilization is sensitive to addition of intact pollen and neurotransmitters (which also occurred in the pollen grains).

Fluorescence of vegetative microspores of *Equisetum arvense* was also changed under the treatment of neurotransmitters (Fig. 26). In the spectra, components fluorescing in blue (450–480 nm) and yellow (500–550 nm) spectral region are seen. Microspores began developing when blue-fluorescing grains began to be red-fluorescing, mainly because of the

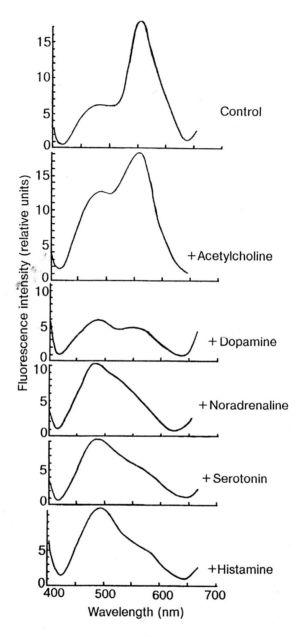

Fig. 26. Effects of neurotransmitters 10^{-5} M on the autofluorescence of the vegetative microspores of *Equisetum arvense* after 1 day of growth (Roshchina, unpublished data)

increased chlorophyll amount (Table 14). On the basis of the more intensive fluorescence at 680 nm, under acetylcholine at 10^{-8}–10^{-6} M an increase 2-8 times was found in the number of red-fluorescing grains in comparison to the control. This means that neuromediator stimulates the first stages of microspore development.

Table 14. The effect of acetylcholine on the fluorescence of vegetative microspores of *Equisetum arvense* (+, yes; ++, many; +++, highest amount; –, absence).

	Blue-fluorescing microspores				Red-fluorescing microspores			
Time	0 h	1 h	24 h	48 h	0 h	1 h	24 h	48 h
Control	+++	+	+	–	–	+	++	+++
Acetylcholine 10^{-8} M	+	+	+	–	–	++	+++	+++
10^{-7} M	+	++	+	+	–	++	+++	+++
10^{-6} M	+	+	+	+	–	++	++	++
10^{-5} M	+	+	+	+	–	+	+	+

2.3.4. Effects on Metabolic Processes

Acetylcholine and biogenic amines may influence the metabolism of plant cells by regulating the activity of enzymes and/or acting as substrates in some reactions. At concentrations of 10^{-4}–10^{-3} M, acetylcholine stimulates and at higher concentrations it inhibits the peroxidase activity of plant tissues, the inhibition reaching 50% (Penel et al., 1976). In *Hordeum vulgare* these mediators inhibit the activity of phenylalanine-ammoniumlyase, a basic enzyme of phenol synthesis (Saunders and McClure, 1973). Roshchina (1989a) described the stimulating effect of acetylcholine, adrenaline, and noradrenaline on the activity of the photo-induced Mg^{2+}-dependent ATPase of garden pea chloroplasts (section 2.3). Hartmann et al. (1980) have reported that acetylcholine inhibits ^{32}P incorporation into phospholipids of etiolated bean hypocotyls. Acetylcholine is supposed to be antioxidant when it is released from the animal cell (Kawashima et al., 1990) or when the same event occurs at the interaction of pollen and pistil in plants (Roshchina and Melnikova, 1998b). High concentrations of acetylcholine (>10^{-3} M) inhibited the ethylene emission by the soya leaves disks (Jones and Stütte, 1986). This phenomenon could be explained by the formation of the ethylene excess during the acetylcholine splitting either due to free radical mechanism or Hoffmann's splitting reaction as it was shown *in vitro* (Kurchii and Kurchii, 2000). The ability of histamine to suppress the lysozyme activity of latex of *Asclepias syriaca* was observed by Lynn (1989). Regulatory action of catecholamines on the acetylcholinesterase activity is shown on isolated preparation of the enzyme (see Chapter 3, section 3.1.2). Catecholamines also stimulate ethylene production in potato cell suspension cultures (Dai et al., 1993). In model experiments on the interaction between acetylcholinesterase and dopamine molecules during autooxidation of the amine, modification of the enzymic activity has been shown (Klegeris et al., 1995). The catalytic activity is supposed to depress by the products of the autooxidation of dopamine-quinones and hydrogen peroxide that are generated, as well as superoxide radical.

CONCLUSION

According to data described in this chapter, sensitive reactions to neurotransmitters are plant growth and morphogenesis, membrane potential, water and ion exchange, nastic movements and motility of unicellular plants, photophosphorulation, and ethylene release. Hence, non-synaptic functions of neurotransmitters could be (1) regulation of growth and morphogenetic processes; (2) regulation of energetic and metabolic processes; (3) protective role as oxidants and antioxidants; (4) chemosignaling role in intercellular and intracellular contracts (see Chapter 4); (5) regulation of motile reactions; and (6) regulation of ion and water exchange.

The Regulatory Systems with Participation of Neuromediators

Complex regulatory systems are known to function in animal tissues, such as cholinergic, adrenergic, dopaminergic, serotoninergic, and histaminergic. Each system includes a mediator of low molecular weight, enzymes of its synthesis and catabolism, and a receptor sensitive to the mediator. They are involved in operative reception and transmission of external excitation along the organism and realization of fast reaction in a response. The signal is directly transmitted by diffusing acetylcholine and biogenic amines in special contacts—synapses that contain two cells distinguished by narrow synaptic chink (North, 1986, 1994).

Reader can also see modern fundamental books, related to the problem as a whole (Urich, 1994; Stone, 1996; Cooper et al., 1996; Baumgarten and Göthert, 1997; Eglen, 1997; Nederkoorn et al., 1997; Barrantes, 1998).

The mediator is synthesized in the Golgi apparatus of presynaptic cell and is stored in secretory vesicles from which it is excreted into synaptic chink under the excitation; then it is diffused to the plasmatic membrane of postsynaptic cell—the signal detector or sensor. Binding with the receptor of plasmatic membranes, the mediator transmits the information. In this case the excited receptor undergoes conformational changes that alter ion permeability of this membrane by opening ion channels or switching on the system of synthesis of secondary intracellular messengers which regulate physiological processes.

In both cases, the cell responds quickly to the irritation. The sequence of the regulation with participation of secondary messengers is shown in the following diagram:

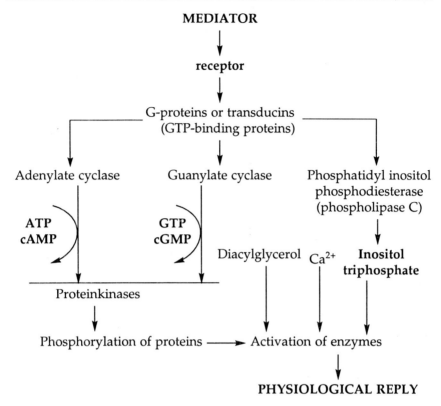

Unlike animals, plants have no nervous system or synapses, and therefore they lack a specialized transmission of nerve impulse. However, they contain compounds that are known as mediators in animals and can induce visible physiological changes in plant cells (see Chapter 2). An analogous phenomenon is characteristic for low-organized organisms lacking nervous system and embryos (Buznikov, 1967, 1989, 1990). This became the basis for the supposition that components of cholinergic, adrenergic, dopaminergic, serotoninergic, and histaminergic systems are present in any living cell (Buznikov, 1987). Below we shall consider the known data for plants.

The main features of the presence of receptors for mediators and hormones in animals and plants can be postulated as follows (Rozen, 1986; Firn, 1987; Barnard, 1992; Vizi and Lendvai, 1999):

1. The receptor needs to be highly selective and specific to the ligand and must differ on the basis of the ligand structure. The effect should be induced by low concentrations (10^{-10}–10^{-7} M) of agents or their agonists and be selectively sensitive just to their compounds.

2. The kinetics of a ligand binding must be described by a curve that has a plateau showing that the number of receptor molecules in the membrane or cell is limited.

3. Antagonists of the agent-effector, when they are binding with the receptor, have to prevent the appearance of the reaction induced by the effector.

The characteristics of a membrane receptor to any substance is usually studied in dose-effect experiments in which the membrane response is analyzed in the presence of its agonists and antagonists (Mikhelson and Zeimal, 1970, 1973). When special receptors are present in the membranes, their molecules (R) can bind with a mediator (M) or its agonist (Ma) and/or antagonist (A), forming a complex according to the following equations:

$$(R) + (M) \leftrightarrow (RM) \rightarrow \text{response}$$
$$(R) + (Ma) \leftrightarrow (RMa) \rightarrow \text{response}$$
$$(R) + (A) \leftrightarrow (RA) \rightarrow \text{response}$$

The affinity of a substance to a possible membrane receptor can be analyzed by the estimation of constants of the agent ligand, binding with receptor (K_B) and dissociation of the receptor-ligand complex (K_D) when the system is in equilibrium $K_B = K_D$ (Mikhelson and Zeimal, 1973) according to the following formula:

$$K_B = \frac{[L][R]}{[LR]}$$

where K_B is equivalent to the concentration of the ligand (L) at which half of the binding sites are occupied (when (R) = (LR)) and is also equal to the concentration, giving a half-maximum response. The formula is equivalent to the Michaelis-Menten equation of the enzyme kinetics.

The main difference between receptor binding and non-receptor binding is in certain obligatory physiological responses, even if the concentration curves of the ligand binding have a saturation (Firn, 1987). Firn (1987) proposed to introduce the term "physiological limit" in studying response relating to the receptor. This includes fast response to the hormonal or mediatory effector, its agonist and antagonist, and estimation of the constants of their binding (constants of dissociation of the receptor-compound complex).

3.1. CHOLINERGIC SYSTEM OF REGULATION

The cholinergic system of regulation in animals includes four components: acetylcholine, cholinoreceptor, enzyme of synthesis of acetylcholine choline acetyltransferase, and enzyme of its hydrolysis cholinesterase.

3.1.1. Cholinoreception

3.1.1.1. Main Features of Cholinoreception

Receptor systems participate in many forms of mediation within and outside the cell, including cell proliferation, adhesion, migration, secretion, and differentiation (Laufenburger, 1991). Anticholinoreceptor drugs affect growth and metamorphosis of animals (Marchi et al., 1996). The cholinoreceptor of animals is a glycoproteid having affinity to acetylcholine and certain pharmacological agents that can block the physiological action of acetylcholine or, on the contrary, imitate its effects (Changeux et al., 1984). For the study of the characteristics of isolated or membrane-bound receptor, the pharmacological method or method of radioactively labeled ligands is used, or a combination of the two methods. The pharmacological method records the concentration curves for the acetylcholine effects and also for its agonists (imitators) and antagonists (Fig. 27) when some reactions of the studied organ, tissue, or organelle conducted by cholinoreceptors are analyzed. In animal physiology, these reactions consist of changes in membrane potential, action potential, and ion exchange, for instance Na^+, K^+, Ca^{2+}.

From the sensitivity to a certain agonist of acetylcholine, specifically nicotine or muscarine, one distinguishes nicotinic or muscarinic cholinoreceptors (Changeux et al., 1984; North, 1986). These types of cholinoreceptor are blocked by different antagonists of acetylcholine: nicotinic by d-tubocurarine and α-bungarotoxin and muscarinic by atropine and quinuclidinyl benzylate. Within each type there are separate groups (subtypes) of the same receptors that differ from the above-mentioned ones in their affinity to radioligands and localization. They are identified from their different physiological and biochemical effects. The intermediate type of cholinoreceptor has been found in some mollusks. It is blocked by the acetylcholine antagonist tetraethylammonium, which acts on K^+ channels (Prosser, 1986).

The method of radioactive ligands usually consists of isolation of receptors or the enriching of the preparation by them, followed by study of kinetic characteristics of their binding with ligands. L-(^{3}H)-quinuclidinyl benzylate is used for the study of muscarinic receptor (Sillard et al., 1985) and (^{125}I) α-bungarotoxin for the study of nicotinic receptor (Changeux and Revah, 1987).

The acetylcholine receptor was first isolated in the early 1970s from electric ray *Torpedo* by the use of polypeptide neurotoxins from snakes (Changeux et al., 1984; Changeux and Revah, 1987). This confirmed the hypothesis of Nachmanson (Nachmanson and Neumann, 1975), who in 1953 predicted the existence of the protein-receptor. When the receptor interacts with acetylcholine it undergoes conformational changes that lead

86 Neurotransmitters in Plant Life

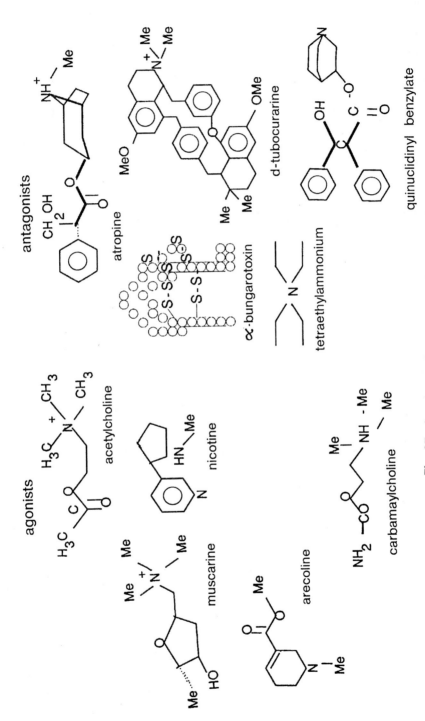

Fig. 27. Agonists and antagonists of acetylcholine

to the formation of a transmembrane ion channel. Electric organs of marbled (crampfish) electric ray *Torpedo marmorata* or Pacific electric ray *T. californica* are richest in acetylcholine receptors: 100 mg of receptor per 1 kg of tissue. In skeleton muscles of mammalia the amount of cholinoreceptor is approximately 600-fold less.

Nicotinic cholinoreceptor from animal tissue (Changeux and Revah, 1987) is a complex protein with molecular mass 250–300 kDa and consists of 5–6 subunits: α, β, γ, and δ-subunits with molecular mass approximately 40, 48, 58, and 66.5 kDa that form an ion channel (Fig. 28). Under excitation of nerve cells, α-subunits of cholinoreceptor bind with acetylcholine, conformational changes of the protein occur, and the channel is open. Unlike nicotinic receptor, muscarinic cholinoreceptor only directs the ion channel work but is not itself included in this channel.

Cholinoreceptor has two active centers of acetylcholine binding. The first is negatively charged anionic center, including carboxylic groups of glutamate or aspartate. This part of the receptor interacts with positively charged cationic "head" of the acetylcholine molecule (Golikov et al., 1985; Changeux et al., 1987).

The action of any receptor is considered a two-step process: binding of ligand and initiation of the action signal. The second step can be completed in two different ways. Receptors acts as ionophore, opening an ion channel that leads to impulse uptake of Na^+ by cell and efflux of K^+ from it. In this case membrane potential decreases sharply. When depolarization of membrane achieves a certain value, the action potential arises that induces the spontaneous opening of potential-dependent ion channels. Thus, the action potential spreads in the form of electric impulse along

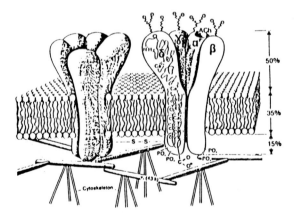

Fig. 28. Localization of nicotinic cholinoreceptor in animal cell membrane. Separate subunits of the receptor form sodium channel (Oberthür et al., 1987).

the membrane. The duration of the impulses of the order of milliseconds. Ions Na^+, K^+, Ca^{2+}, and Mg^{2+} as regulatory and secondary messengers influence the effectiveness of the metabolic exchange. The energy conversions, ATP formation, and ATP utilization are controlled by these ions.

The other mode of the receptor action consists in the triggering of mechanism of secondary messenger synthesis. For instance, adenylate cyclase is activated, catalyzing the formation of the secondary messenger cAMP on the internal side of plasmalemma, which in turn induces a cascade of reactions within the cell (see section 3.2). The rate of the reaction in this case is lower than ionophore mechanism and is from 0.1 to some seconds.

Besides the opening of ion channels the acetylcholine binding with muscarinic cholinoreceptor is accompanied by the increase in concentration of cyclic nucleotides cAMP and cGMP within the cell (Golikov et al., 1985) or (and) diacylglycerol and inositol triphosphate being formed at the hydrolysis of phosphatidylinositol phosphate on the other (Berridge, 1984; Berridge and Irvine, 1984) within the cell. Interaction of acetylcholine with nicotinic cholinoreceptor leads to only the opening of ion channels.

3.1.1.2. Cholinoreception in Plants

Sensitivity to agonists and antagonists. Earlier study of cholinoreception outside the animal kingdom has been done on photosynthesizing bacteria. Faust and Doetsch (1971) showed that motility of bacteria *Rhodospirillum rubrum* and *Thiospirillum jenense* are completely blocked by antagonist of acetylcholine atropine.

The same antagonist of acetylcholine increases the adenosine triphosphate (ATP) level in bean buds (Kirshner et al., 1975), activity of choline kinase (Hartmann and Schleicher, 1977), diminishes the stimulating effect of acetylcholine on the growth of *Vigna* seedlings (Hoshino, 1983), counteracts the acetylcholine-stimulated shrinking of the cells of bean pericycle (Toriyama, 1978), nullifies flowering stimulated by the indole-3-acetic acid in *Lemna gibba* G_3 (Hoshino, 1979), and inhibits the incorporation of the radioisotope of phosphorus (^{32}P) into the phospholipids of bean hypocotyls (Hartmann et al., 1980).

Jaffe and Thoma (1973) first searched for cholinoreceptor in plants by the use of pharmacological methods. They showed that d-tubocurarine, antagonist of acetylcholine, inhibits the ^{14}C-acetate uptake by the roots of beans *Phaseolus aureus* L. stimulated by red light (< 700 nm). d-Tubocurarine weakly inhibited the root pressure in sunflower *Helianthus annuus*, which is usually stimulated by acetylcholine (Zholkevich et al., 1979; Zholkevich, 1981). The influence of acetylcholine and its antagonist atropine on the ethylene release from soya leaves has been studied by Jones and Stutte (1986). Acetylcholine (11 mM) inhibited the ethylene emission by disks of soya leaves about two-fold in comparison with control, whereas atropine (2.75 mM) stimulated this process three-fold .

Agonist of acetylcholine nicotine was also studied. It induced morphogenesis of tobacco roots (Peters *et al.*, 1974) and stimulated root growth of maize (Rizvi and Rizvi, 1987).

New data on the action of agonists and antagonists of acetylcholine have appeared since the 1980s (Table 15). Wheat protoplast swelling was stimulated by acetylcholine and its agonist carbamoylcholine (Tretyn *et al.*, 1990a, b) and sunflower root exudation was stimulated by both

Table 15. Plant reactions sensitive to acetylcholine, its agonists and antagonists

Reaction sensitive to acetylcholine	Effect of Agonist	Effect of Antagonist	Reference
Motility of photosynthesizing bacteria	acetylcholine	atropine	Faust and Doetsch, 1971
Uptake of ^{14}C-acetate	acetylcholine	d-tubocurarine	Jaffe and Thoma, 1973
Root pressure	acetylcholine	d-tubocurarine	Zholkevich *et al.*, 1979; Zholkevich, 1981
Release of ethylene	acetylcholine	atropine	Jones and Stutte, 1986, 1988
Growth of embryos	nicotine		Rizvi and Rizvi, 1987
Na$^+$ efflux from intact chloroplasts	acetylcholine, arecoline, muscarine carbamoylcholine	atropine, d-tubocurarine, α-bungarotoxin	Roshchina, 1987a, 1989b Roshchina and Mukhin, 1978b
K$^+$ efflux from intact chloroplasts	acetylcholine arecoline, muscarine	tetraethylammonium, atropine, d-tubocurarine	Roshchina, 1987a, 1989a, b Roshchina and Mukhin, 1987a
Photophosphorylation in chloroplasts	acetylcholine carbamoylcholine, muscarine, arecoline	atropine, d-tubocurarine, α-bungarotoxin, quinuclidinyl benzylate	Roshchina, 1987a, 1989a, b, 1991a, c Roshchina and Mukhin, 1987a
Unrolling of leaves	acetylcholine, muscarine, nicotine	atropine, d-tubocurarine	Tretyn and Kendrick, 1990
Pistil stigma autofluorescence	acetylcholine	atropine, d-tubocurarine	Roshchina and Melnikova, 1998a, b
Fertilization after pistil treatment	acetylcholine	atropine, d-tubocurarine	Roshchina and Melnikova, 1998a, b
Water uptake by dry pollen	acetylcholine	atropine, d-tubocurarine	Roshchina, unpublished data

acetylcholine and nicotine (see Zholkevich et al., 1990). Tretyn (1990) showed that effects of acetylcholine on the wheat protoplasts are not seen at the presence of atropine and d-turbocurarine. Ion changes in isolated chloroplasts and photophosphorylation were also sensitive to agonists and antagonists of acetylcholine (Roshchina, 1987a; 1989a, b; 1990a; Roshchina and Mukhin, 1987a, b).

Researchers have concentrated special interest on fertilization after the treatment of pistil with antagonists of acetylcholine (atropine and d-tubocurarine) blocked seed maturation (Roshchina and Melnikova, 1998a, b). Participation of the cholinergic system may be very important for the breeding mechanism (see below). Acetylcholine (10^{-5}) stimulated dry pollen swelling in water two-fold, and this process was blocked after preliminary treatment with the same concentration of atropine or d-tubocurarine (Roshchina, unpublished data).

As is seen in Table 15, the most effective reactions to low concentration (10^{-8}–10^{-5} M) agonists and antagonists and acetylcholine are ion transport through membranes, ATP synthesis, and motile reactions (membrane swelling and shrinking, leaf unrolling, and root pressure as processes dealing with contractile systems). The growth reactions of seedlings usually need high concentration ($\geq 5 \times 10^{-4}$ M) of acetylcholine and its agonists and antagonists, which makes it difficult to analyze cholinoreception in plants. Thus, the study of cholinoreception requires sensitive reactions to low concentrations of anticholinergic drugs and suitable models for the acetylcholine-cholinoreceptor interactions. Based on the conception of contacts for neurotransmitter functioning, pollen-pistil interaction system at fertilization was used as a model system of cell-cell interactions (Roshchina and Melnikova, 1998a, b). But intracellular interactions of acetylcholine and cholinoreceptor were also studied on the model system of isolated chloroplast. Below we will specially consider these models.

Pollen-pistil system. As described in Chapter 2, autofluorescence of pistil stigma is a sensitive fast reaction to acetylcholine and own pollen. Moreover some species of pollen contain acetylcholine (Chapter 1) and cholinesterase (Chapter 3, see below). On the basis of these facts experiments with antagonists of acetylcholine d-tubocurarine and atropine were carried out on the pollen and pistil of *Narcissus pseudonarcissus* (Roshchina, 1999a) and *Hippeastrum hybridum* (Roshchina and Melnikova, 1998a, b). The treatment of *Narcissus* flower with 10^{-7}–10^{-6} M atropine or atropine + acetylcholine decreased the fruit weight by 30–50% of control (Roshchina, 1999a).

The flower of *H. hybridum* is more suitable for separate treatment of both longest pistil (11–15 cm length) and pollen (Roshchina and Melnikova, 1996; 1998a, b), according to the following diagram:

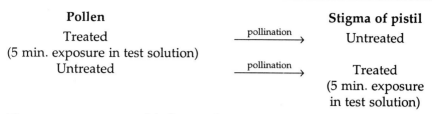

The responses to acetylcholine and its antagonists were estimated on autoflourescence of pollen and pistil stigma as well as the pollen germination *in vivo* and *in vitro*. If pollen grains were treated there was no marked effect of their autofluorescence and germination *in vitro* (on nutritional medium, 10% sucrose) or *in vivo* (on pistil stigma). A different picture was observed on pistil stigma (Fig. 29).

Atropine and d-tubocurarine prevented the stimulation of autofluorescence of the pistil stigma induced by acetylcholine. Flower behavior after the treatment and pollination were observed. The fertilization, estimated as flower closure after the pollination, proceeded more slowly 5–6 days after treatment with antagonists atropine and d-tubocurarine (Fig. 29) than in control (1–2 days). Similar effects were observed in variants atropine + acetylcholine and d-tubocurarine + acetylcholine. Later, although fruits formed, seeds developed poorly and did not mature. Thus, during a short exposure (5 min.) atropine and d-tubocurarine blocked both primary response to acetylcholine (pistil autofluorescence, i.e., pollen-pistil recognition) and the more long-term response, realization of the genetic program of seed maturation. If pollen is a source of acetylcholine, pistil stigma serves as receiver of this chemical signal, perhaps via functional analog of animal cholinoreceptor. Cholinoreception may be one of the mechanisms of cell-cell communication in phytocenosis (Roshchina, 1999a).

Chloroplast-cytoplasm system. Intracellular cytoplasm-organelle cholinoreception was also studied on isolated organelles: chloroplasts (Roshchina, 1987, 1989a, b; 1990a, b, c; 1991). During the course of evolution, the chloroplast, once a ancient separate organism, has become part of the cell, but it has stored internal structures (thylakoids) and an external envelope, analogous to the plasmatic membrane. Acetylcholine is found in both chloroplasts and cytoplasm (Roshchina, 1991a, b). So, the chloroplast envelope can interact with cytoplasmic acetylcholine, whereas thylakoids interact inside, with stromal chloroplast acetylcholine. On the basis of these arguments, systematic studies of cholinoreception were done on isolated chloroplasts, both intact with envelope and without envelope (thylakoids) (Table 16). The main features of cholinoreception are that low concentrations induced an effect, and that the concentration curves for acetylcholine were saturated.

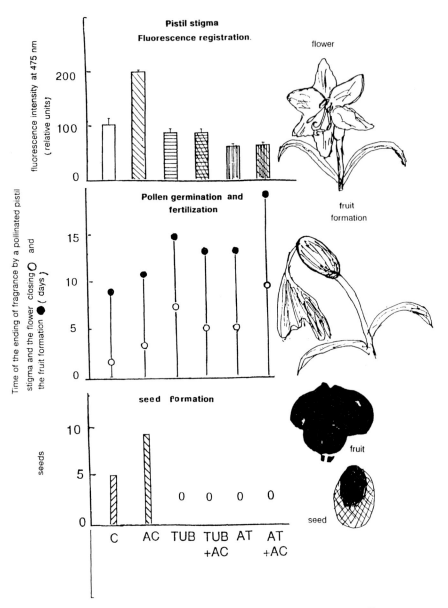

Fig. 29. The effects of acetylcholine antagonists on pistil stigma autofluorescence and fertilization of *Hippeastrum hybridum* (after Roshchina and Melnikova, 1998a, b). C, control; TUB, d-tubocurarine 10^{-5} M; TUB + AC, d-tubocurarine 10^{-5} M + acetylcholine 10^{-4} M; AT, atropine 10^{-5} M; AT + AC atropine 10^{-5} M + acetylcholine 10^{-4} M.

Low concentrations of agonists and antagonists were tested on the reactions most sensitive to acetylcholine: Na^+-efflux from intact chloroplasts or photophosphorylation in thylakoids (Roshchina, 1989a, b; 1990a; 1991a, b). As seen in Fig. 30, Na^+-efflux, stimulated by acetylcholine, was weakly inhibited by atropine, and completely inhibited by d-tubocurarine. α-Bungarotoxin acted like d-tubocurarine (Roshchina, 1991b), which showed in the occurrence of nicotinic type of cholinoreception, rather than muscarinic type in chloroplast envelope.

Table 16. Characteristics of cholinoreceptory features of chloroplast membranes (Roshchina, 1991a)

Reaction	Minimal agonist concen-, tration induced effect, M				Antagonist, blocked effect, induced by Ach	Saturation of the concentration curves for Ach
	ACh	M	AR	CCh		
Na^+ efflux from intact chloroplasts	(++) 10^{-9}	(++) 10^{-9}	(++) 10^{-9}	(+) 10^{-9}	At, Tub, Bung	yes
Photophosphorylation ($H_2O \rightarrow NADP^+$) in thylakoids	(++) 10^{-9}	(–)	(++) 10^{-10}	(+) 10^{-9}	At, Bung, Qub	yes

Ach, acetylcholine; M, muscarine; AR, arecoline; CCh, carbamoylcholine; At, atropine; Bung, α-bungarotoxin; Qub, quinuclidinyl benzylate; Tub, d-tubocurarine; (++); (+), strong and weak stimulation; (–), inhibition.

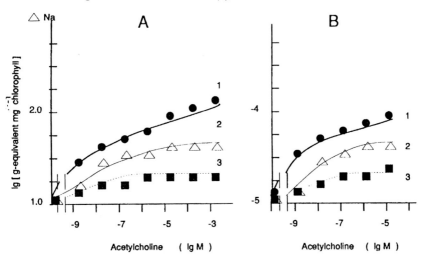

Fig. 30. The effects of acetylcholine antagonists on the Na^+ efflux from intact pea chloroplasts (Roshchina, 1987, 1989a, 1990c). (A) 1, without antagonist; 2, atropine 10^{-6} M; 3, d-tubocurarine 10^{-6} M. (B) 1, without antagonist; 2 and 3, relatively α-bungarotoxin 10^{-12} and 10^{-11} M.

Thylakoids are sensitive to acetylcholine agonists muscarine (Roshchina and Mukhin, 1987a) and arecoline (Roshchina, 1990c). Like acetylcholine, arecoline in low concentrations stimulated non-cyclic photophosphorylation and in high concentration inhibited it (Fig. 31). The curve is bell-shaped. The binding constants (K_B) for acetylcholine were obtained from the graphic representations (Roshchina, 1989a, 1990c). K_B for acetylcholine was 5×10^{-9} M and for arecoline 4×10^{-9} M.

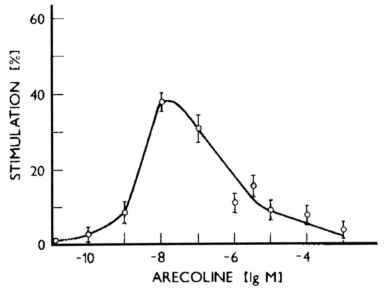

Fig. 31. The Concentration curve of the arecoline action as agonist of acetylcholine on non-cyclic photophosphorylation in pea chloroplasts (Roshchina, 1990c)

Thylakoids demonstrated different sensitivity of photochemical reactions as physiological response to acetylcholine antagonists (Figs. 32, 33). Electron transport from H_2O to $NADP^+$ was not sensitive to them, whereas coupled photophosphorylation was blocked by α-bungarotoxin or quinuclidinyl benzylate.

The affinity of antagonists to proposed receptor has been estimated as a constant of binding (K_B) and dissociation (K_D) according to the formula of Mikhelson and Zeimal (1973):

$$K_B = K_D = \frac{[S][A]}{[S_A]-[S]}$$

where (S) and (S_A) are concentrations of a mediator without and with antagonist in the medium that induce maximal or equal effect on the membrane model reaction, and (A) is the concentration of the antagonist

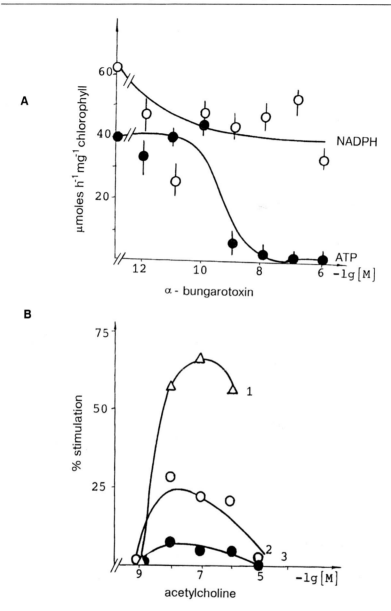

Fig. 32. Effects of α-bungarotoxin on the photochemical activity of thylakoids from chloroplasts of *Pisum sativum* (after Roshchina, 1991b). A: without acetylcholine in the medium. $NADP^+$-photoreduction ($NADP^+$) and coupled photophosphorylation (ATP). B: The acetylcholine concentration curves on coupled photophosphorylation without (1) and with preliminary treatment by α-bungarotoxin 10^{-12} M (2) and 10^{-10} M (3).

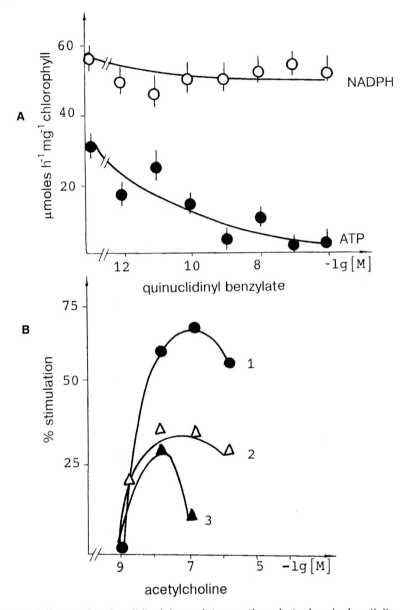

Fig. 33. Effects of quinuclidinyl benzylate on the photochemical activity of thylakoids from chloroplasts of *Pisum sativum* (after Roshchina, 1991b). A: without acetylcholine in the medium. NADP$^+$-photoreduction (NADP$^+$) and coupled photophosphorylation (ATP). B: The acetylcholine concentration curves on coupled photophosphorylation without (1) and with preliminary treatment by quinuclidinyl benzylate 10^{-12} M (2) and 10^{-10} M(3).

in the medium. K_B is formally equal to (A) the concentration of antagonist necessary for a two-fold increase in the concentration of mediator or its agonist in order to have the same response of the system as is observed without the antagonist $K_B = (A)$. The dose-effect curves for acetylcholine and its antagonists have also been plotted (Roshchina, 1990c, 1991a, b). According to the concept to antagonism (Mikhelson and Zeimal, 1973) the response-dose curves in the presence of an antagonist should be parallel to those obtained without the antagonist. This was found in our experiments (Fig. 34). Higher concentrations of the mediator displace the antagonist (A) from the complex with the receptor (RA). Another explanation of the phenomenon could be an interaction with molecules of the receptor (R) that are free (non-binding with the antagonist). Because of the interaction, the concentration of free receptors (R) is decreased, and the equilibrium in the reaction of formation of the receptor-antagonist complex is lost. As a result of the dissociation of (RA), the

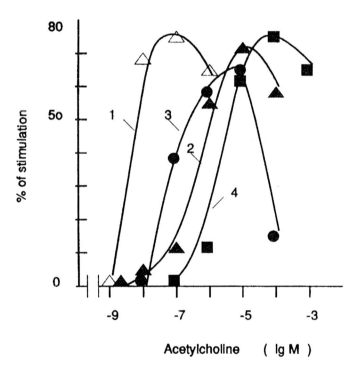

Fig. 34. Dose-effect concentration curves of acetylcholine action on non-cyclic photophosphorylation in isolated pea chloroplasts without (1) and with (2 to 4) its antagonist atropine (after Roshchina, 1990c). The antagonist concentrations were 10^{-8} M (2), 10^{-7} M (3), 10^{-6} M (4). Control rate of ATP synthesis 7 mmol $s^{-1}kg^{-1}$ of chlorophyll.

amount of free receptors increases. According to Roshchina (1990c, 1991b), K_B for atropine = 10^{-10} M. Low values of K_B demonstrated high affinity of the substance to the chloroplast cholinoreceptor, possibly of muscarinic type. It is possible that in thylakoids the cholinoreceptor is of a mixed type. There is some indication of this from the experiments on the binding ligand L-(^{3}H)-quinuclidinyl benzylate (muscarinic type of binding) with intact chloroplasts, thylakoids, and etioplasts (Roshchina, 1991b), according to the method of Sillard *et al.* (1985), in the Laboratory of Biochemistry of Tartu State University. There was non-specific binding of the ligand.

Thus elements of cholinoreception are found in plant cells. Further experiments in this line may give new evidence.

3.1.2. Choline Acetyltransferase

The activity of enzyme choline acetyltransferase or cholineacetylase (EC 2.3.1.6), participating in the synthesis of acetylcholine from choline and acetic acid, was first found by Barlow and Dixon (1973) in stinging hairs of common nettle *Urtica dioica*. These authors isolated the protein from acetonic extracts of leaves and partly purified it on polyacrylamide gel Bio-Gel P4. The rate of the acetylcholine formation by isolated enzyme was 1.7–6.7 nmoles min^{-1} g^{-1} of fresh mass of leaves or 0.76 – 1.12 nmoles min^{-1} mg^{-1} of protein. Then analogous enzyme was found in extracts of the embryonal buds of pea *Pisum sativum*, cabbage *Brassica oleracea* var. *botrytis*, and kidney bean *Phaseolus vulgaris* (Biro, 1978), and in seeds of onion *Allium altaicum* (Hadacova *et al.*, 1981). In some papers the purification methods are described. Most convincing are studies of Smallman and Maneckjee (1981) in which this enzyme was purified from five plants belonging to five different families (Table 17) on DEAE-Sephadex, Sefacryl S-200, and activated Sepharose 4B CoA SH. Affinity chromatography on Sepharose 4B-CoA SH permitted the authors to purify choline acetyltransferase 1500-fold. The determination of activity of this preparation by the inclusion of (^{3}H) acetyl-CoA into acetylcholine demonstrated that the enzyme from common nettle *Urtica dioica* had the highest capacity for labeling of cholinic ester. The rate of this process was similar to that of acetylcholine synthesis by housefly *Musca domestica*. Enzymes isolated from pea *Pisum sativum*, spinach *Spinacia oleracea* and sunflower *Helianthus annuus* synthesized acetylcholine at 10 to 100 times lower rates than protein from cyanobacteria *Oscillatoria agardhii*. Besides this work, there was an attempt to purify choline acetyltransferase from kidney bean (Ph.D. theses of Hoffman, 1982 and Hock, 1983, cited in Hartmann and Gupta, 1989). Unlike analogous enzyme from brain of rats (4000 nmoles of acetylcholine min^{-1} mg^{-1} protein) or pigs (135,000 nmoles min^{-1} mg^{-1} protein), all known plant preparations of the enzyme demonstrated 100 to 1000 times lower rates of acetylcholine synthesis.

Table 17. Choline acetyltransferase in living organisms
(Smallman and Maneckjee, 1981)

Organism	Acetylcholine nmoles/min/ml medium	Protein mg/ml	Specific activity nmoles/min/mg protein
Utrica dioica	1.067	1.0	1067.0
Pisum sativum	0.108	5.7	18.94
Spinacia oleracea	0.016	3.5	4.53
Helianthus annuus	0.026	4.1	6.26
Oscillatoria agardhii	0.016	101.0	0.16
Musca domestica	1.210	1.2	1008.00

Cholineacetylase isolated from seedlings of *Phaseolus vulgaris* has molecular mass about 80 kDa (Hartmann and Gupta, 1989), which is similar to that of analogous enzymes from animals: 66–68 kDa. K_m for the plant enzyme was about 0.2 mM, whereas for that of animals the values were 0.41–1.9 mM and 0.01–0.047 mM as estimated by determination of the choline decrease and the acetylcholine increase, respectively (Tucek, 1983). As in animal cells, choline acetyltransferase in plant cell is localized mainly in cytoplasm (Hartmann and Gupta, 1989) and easily associated with cell membranes in media with low ion strength (Tucek, 1983). Bacterial and mammalian choline acetyltransferase is inhibited by steryl pyridine analogues (White and Cavallito, 1970).

3.1.3. Cholinesterase

Enzyme-decomposed acetylcholine is called cholinesterase. Cholinesterase was first found by Loewi (1937) in experiments on hearts of amphibians. Oury and Bacq (1938) later demonstrated the presence of acetylcholine-like compound in fruit body of field mushroom *Lactarius blennius* and the capacity of the fungal tissue to hydrolyze this substance. In 1953 Goldstein and Goldstein (1953) discovered bacterial cholinesterase. So it became clear that the enzyme is present not only in animal organisms having a nervous system.

3.1.3.1 Main Characteristics of Animal Cholinesterases

Fifty years after the discovery of cholinesterase, methods were developed for its detection in the animal cell. The methods are based on detection of the products of acetylcholine hydrolysis, acetic acid and choline (Whittaker, 1963), or colored product of the reaction of thiocholine (if acetylthiocholine is used as substrate) with 5,5-dithiobis-2-nitrobenzoic acid (Ellman *et al.*, 1961).

The criterion of cholinesterase activity is the capacity of tissue to hydrolyze acetylcholine and other cholinic esters at high rates, as well as to be depressed by specific inhibitors neostigmine (proserine) and physostigmine (eserine) (Augustinsson, 1963; Munoz-Delgado and Vidal, 1987; Maelicke, 1991). As a serine-containing protein, cholinesterase is also inhibited by organophosphorus compounds, including insecticides.

Two types of cholinesterases are distinguished in animals. The enzyme specific to acetylcholine (hydrolyzing the substrate with higher rates than other cholinic esters) is called "true cholinesterase" or acetylcholinesterase (EC 3.1.1.7). The enzyme hydrolyzing not only acetylcholine with high rates, but also other cholinic esters such as butyrylcholine or propionylcholine is called pseudocholinesterase or simply cholinesterase or butyrylcholinesterase (EC 3.1.1.8). The activity of acetylcholinesterase is inhibited by high concentrations ($> 10^{-3}$ M) of substrate, whereas the activity of pseudocholinesterase is not. The curve of the substrate hydrolysis by acetylcholinesterase from nervous tissues, erythrocytes from mammals, electric organs of fishes, or analogous enzymes from plant tissue (see below) has a bell shape with maximum at a certain optimal concentration of substrate (Brestkin et al., 1973). Besides, antibodies against acetylcholinesterase are not binding with pseudocholinesterase (Toutant et al., 1985). Thus, cholinesterases are distinguished by substrate specificity, sensitivity to inhibitors, and the concentration curves of the substrate hydrolysis. These are taxon-specific and tissue-specific enzymes. Isolated enzymes from different tissues of the same animal differ in their catalytic features (Brestkin et al., 1973). The decrease in the rate of hydrolysis upon increase of the substrate amount is caused by the interaction of two or more molecules of acetylcholine with one catalytically active subunit of acetylcholinesterase, therefore they prevent each other from orienting exactly in the active center of the enzyme. The excess substrate molecules may also interact with allosteric centers of acetylcholinesterase (Brestkin et al., 1973).

There is a homology in primary structure between acetylcholinesterase and butyrylcholinesterase, thus confirming their common evolutionary origin (Soreq et al., 1992). Acetylcholinesterase and butyrylcholinesterase have a similar primary structure but differ in substrate specificity and hydrophobicity.

3.1.3.2. Occurrence of Cholinesterase Activity in Plants

Systematic studies of plant cholinesterases were started in the 1960s. Cholinesterase activity was first found in lower plants: in extracts of Characeae algae *Nitella* (Dettbarn, 1962), mycelium of fungi *Physarium polycephalum* (Nakajima and Hatano, 1962), and yeast *Saccharomyces cerevisiae* (Jacobsohn and Azevedo, 1962). Its activity was also demonstrated in lichen *Parmelia caperata* (Raineri and Modenesi, 1986) and in fungi *Aspergillus niger* (Roshchina and Alexandrova, 1991). A series of papers by Jaffe and coworkers (Riov and Jaffe, 1973a, b; Fluck and Jaffe, 1974b, c, d) on plant extracts showed that this phenomenon is peculiar to many plant species. Later Miura and coworkers (1982) also demonstrated cholinesterase activity in leaf disks of many terrestrial plants, although in the review of Hartmann and Gupta (1989) acetylcholine hydrolysis was absent in 65 plant species. The occurrence of cholinesterase in plants has also been discussed in some reviews (Fluck and Jaffe, 1975; Roshchina and Mukhin, 1986; Hartmann and Gupta, 1989; Tretyn and Kendrick, 1991) and in a monograph (Roshchina, 1991a) devoted to acetylcholine, as well as in special reviews on cholinesterases in plants (Maheshwari *et al.*, 1982; Gupta and Maheshwari, 1980; Roshchina and Semenova, 1990). The occurrence of cholinesterase was analyzed again in 118 terrestrial species (Gupta and Gupta, 1997) and 10 marine algae (Gupta *et al.*, 1998). Cholinesterase could be detected in 67 higher plant species and in all algae tested. The accumulated information on the presence of cholinesterase activity in plants is represented in Table 18.

As seen in Table 18, cholinesterase activity was determined with biochemical methods (Ellman *et al.*, 1961; Gorun *et al.*, 1978; Miura *et al.*, 1982; Roshchina and Alexandrova, 1991; Roshchina *et al.*, 1994) and histochemical methods (Karnovsky and Roots, 1964), on the basis of inhibition of the acetylcholine (acetylthiocholine) or ^{14}C-acetyl-β-methylcholine hydrolysis by both carbonate inhibitors (neostigmine and physostigmine) and organophosphate inhibitors (diisopropyl fluorophosphate).

From Table 18, it can be seen that cholinesterase occurs at all levels, from fungi and algae to gymnosperms and angiosperms. The enzyme is also found in bacteria (Fitch, 1963 a, b). Cholinesterase is widespread in the plant kingdom: 2 species of 2 families from Pteridophytes, 10 species of a families from marine algae, 10 species of 7 families from gymnosperms, and 172 species of 78 families from angiosperms (Table 18). As for more primitive families, most active cholinesterase is contained in Rhodomeliaceae and Codiaceae among algae (Gupta *et al.*, 1998), in Pteridaceae among Pteridophytes, and in Gingkoaceae, Cycadaceae, and Zamiaceae among gymnosperms (Gupta and Gupta, 1997). In angiosperms most species demonstrating cholinesterase activity belong to Fabaceae or Leguminosae (31), Solanaceae (12), Euphorbiaceae (11), and Convolvulaceae (5). Highest activity is observed in young leaves and embryonal buds, which reached from 7 nmoles min^{-1} g^{-1} of fresh mass in *Pisum sativum* (Vackova *et al.*, 1984) up to 134 nmoles min^{-1} g^{-1} of fresh mass in *Datura innoxia* and *Physalis minima* (Gupta and Gupta, 1997), as well as in pollen extracts, from 24 nmoles min^{-1} g^{-1} of fresh mass up to 90,000 nmoles min^{-1} g^{-1} of fresh mass, for example, in *Pinus sylvestris* (Roshchina *et al.*, 1994; Roshchina and Semenova, 1995; Roshchina, 1999a). The former case might be related to the period of pollen fertility, when microspores are ready to fertilize egg cell of pistil.

Table 18. Occurrence of cholinesterase (ChE) activity in plants

Family, species	Organ, where ChE activity found	Reference
ALGAE		
Caulerpaceae		
Caulerpa racemosa (Forsk.) van Bosse	Supernatant	Gupta *et al.*, 1998
Caulerpa scalpelliformis (R.Br.) van Bosse	Pellet, supernatant	Gupta *et al.*, 1998
Chaetangiaceae		
Galaxaura oblongata Lamour.	Pellet	Gupta *et al.*, 1998
Characeae		
Nitella sp.	Whole plant	Dettbarn, 1962
Cladoforaceae		
Spongomorpha indica Thivy and Visalakshmi	Pellet	Gupta *et al.*, 1998
Codiaceae		
Udotea indica A. and E.S. Gepp.	Pellet	Gupta *et al.*, 1998
Gracilariaceae		
Gracilaria corticata J. Ag.	Pellet, supernatant	Gupta *et al.*, 1998
Grateloupiaceae		
Halymenia venista Boergs.	Pellet	Gupta *et al.*, 1998
Helminthocladiaceae		
Helminthocladia calvadosii (Lamour.) Setchell f.	Pellet, supernatant	Gupta *et al.*, 1998
Punctariaceae		
Iyengaria stellata Boergs.	Pellet	Gupta *et al.*, 1998
Rhodomelaceae		
Laurencia pedicularioides Boergs.	Pellet, supernatant	Gupta *et al.*, 1998
PTERIDOPHYTES		
Adiantaceae		
Adiantum capillus-veneris L.	No activity in frond, stalk, rhizoid	Gupta and Gupta, 1997
Equisetaceae		
Equisetum ramosissimum Desf.	Rhizoid	Gupta and Gupta, 1997
Oleandraceae		
Nephrolepis biserrata (Sw.) Schott	Frond, stalk	Gupta and Gupta, 1997
Pteridaceae		
Pteris multifida Poir.	Frond, rhizoid	Gupta and Gupta, 1997
Salviniaceae		
Salvinia natans (L.) All.	No activity in frond, stalks, rhizoids	Gupta and Gupta, 1997

(Contd.)

Table 18. (*Contd.*)

Family, species	Organ, where ChE activity found	References
GYMNOSPERMS		
Cupressaceae		
Biota orientalis (D. Don) Endl.	Leaves	Gupta and Gupta, 1997
Cycadaceae		
Cycas revoluta Thunb.	Leaves	Gupta and Gupta, 1997
Ephedraceae	No activity in leaves,	Gupta and Gupta, 1997
Ephedra foliata Boiss.	branches, roots	
Gingkoaceae		
*Gingko biloba L.	Leaf disks	Miura *et al.*, 1982
Gingko biloba L.	Leaves	Gupta and Gupta, 1997
Pinaceae		
*Pinus abies L.	Leaf (needle) disks	Miura *et al.*, 1982
Pinus sylvestris L.	Pollen	Roshchina *et al.*, 1994; Roshchina, 1999a
Taxaceae		
*Taxus baccata L.	Leaf disks	Miura *et al.*, 1982
Zamiaceae		
Dioon edule Lindl.	Leaves	Gupta and Gupta, 1997
Dioon spinulosum Dyer.	Leaves	Gupta and Gupta, 1997
Zamia furfuracea L.f.	Leaves	Gupta and Gupta, 1997
ANGIOSPERMS		
Acanthaceae		
Justicia gendarussa Burm.	Leaves, branches	Gupta and Gupta, 1997
Peristrophe paniculata (Forsk.) Brummitt	No activity in leaves, branches and roots	Gupta and Gupta, 1997
Aceraceae		
*Acer negundo L.	Leaf disks	Miura *et al.*, 1982
Agavaceae	No activity in	
Dracaena deremensis Engl.	leaves, branches, roots	Gupta and Gupta, 1997
Aizoaceae		
Mesembryanthemum crystallinum L.	No activity in leaves, branches, roots	Gupta and Gupta, 1997
Trianthema portulacastrum L.	Leaves, branches, roots	Gupta and Gupta, 1997
Altingiaceae		
* Liquidambar marginata Lam.	Leaf disks	Miura *et al.*, 1982
Alliaceae		Hadacova *et al.*,
Allium altaicum Pall.	Seeds	1981, 1983
Amaryllidaceae		Semenova and
Clivia sp.	Pollen, anthers	Roshchina, 1993
Hippeastrum hybridum	Pollen, anthers	Roshchina *et al.*, 1994; Roshchina and Semenova, 1995

(*Contd.*)

Table 18. (*Contd.*)

Family, species	Organ, where ChE activity found	Reference
Narcissus pseudonarcissus L.	Pollen	Roshchina, unpublished data
Anacardiaceae		
Mangifera indicia L.	No activity in leaves, branches, roots	Gupta and Gupta, 1997
Rhus copallina L.	Leaf disks	Miura *et al.*, 1982
Annonaceae		
Polyaltia longifolia Thw.	No activity in leaves, branches, roots	Gupta and Gupta, 1997
Apiaceae		
Daucus carota L.	No activity in leaves, branches, roots	Gupta and Gupta, 1997
Apocinaceae		
Carissa carandus L.	No activity in leaves, branches, roots	Gupta and Gupta, 1997
Catharanthus roseus (L.) G. Don	Leaves	Gupta and Gupta, 1997
Tabernaermontana divaricata (L.) R. Br.	No activity in leaves, branches, roots	Gupta and Gupta, 1997
Thevetia peruviana (Pers.) Merr.	No activity in leaves, branches, roots	Gupta and Gupta, 1997
Aquifoliaceae		
Ilex crenata Thunb.	Leaf disks	Miura *et al.*, 1982
Ilex opaca Ait.	Leaf disks	Miura *et al.*, 1982
Araceae		
Calocasia esculenta (L.) Schott	No activity in leaves, branches, roots	Gupta and Gupta, 1997
˙Peltandra virginica Rafin	Leaf disks	Miura *et al.*, 1982
Arecaceae		
Caryota urens L.	No activity in leaves, branches, roots	Gupta and Gupta, 1997
˙Chamaedorea elegans Mart.	Leaf disks	Miura *et al.*, 1982
Livistona chinensis R. Br.	Leaves	Gupta and Gupta, 1997
Asclepiadaceae		
˙Asclepias syriaca L.	Leaf disks	Miura *et al.*, 1982
Calotropis procera (Ait.) R. Br.	No activity in leaves, branches, roots	Gupta and Gupta, 1997
Asparagaceae		
Asparagus racemosus Willd.	Leaves of seedlings and old plant	Gupta and Gupta, 1997
Ruscus aculeatus L.	Leaves	Gupta and Gupta, 1997
Aspidiaceae		
˙Onoclea sensibilis L.	No activity in leaves, branches, roots	Miura *et al.*, 1982
Asteraceae		
Ageratum conyzoides L.	No activity in leaves branches, roots	Gupta and Gupta, 1997
Bidens biternata (Lour.) Merr & Sheriff	No activity in leaves, branches, roots	Gupta and Gupta, 1997
Chrysanthemum cinerariaefolium Vis.	Leaves	Gupta and Gupta, 1997
Eclipta prostrata (L.) L.	No activity in leaves, branches, roots	Gupta and Gupta, 1997

(*Contd.*)

Table 18. (Contd.)

Family, species	Organ, where ChE activity found	Reference
Galinsoga quadriradiata Ruiz & Pav.	No activity in leaves, branches, roots	Gupta and Gupta, 1997
Gnaphalium indicum L.	No activity in leaves, branches, roots	Gupta and Gupta, 1997
Helianthus annuus L.	No activity in leaves, branches, roots	Gupta and Gupta, 1997
Matricaria chamomilla L.	Pollen	Roshchina et al., 1994
Parthenium hysterophorus L.	No activity in leaves, branches, roots	Gupta and Gupta, 1997
Sonchus oleraceus L.	No activity in leaves, branches, roots	Gupta and Gupta, 1997
Tagetes erecta L.	No activity in leaves, branches, roots	Gupta and Gupta, 1997
Taraxacum officinale Wigg	No activity in latex	Roshchina, unpublised data
Tridax procumbens L.	No activity in leaves, branches, roots	Gupta and Gupta, 1997
Vernonia conyzoides DC.	No activity in leaves, branches, roots	Gupta and Gupta, 1997
Vernonia elaeagnifolia DC.	No activity in leaves, branches, roots	Gupta and Gupta, 1997
Bombacaceae		
Bombax ceiba L.	Leaves of seedlings	Gupta and Gupta, 1997
Balsaminaceae		
Impatiens balsamina L.	Leaf disks	Miura et al., 1982
Betulaceae		
Betula pendula Roxb.	Leaf disks	Miura et al., 1982
Betula verrucosa Ehrh	No activity in pollen	Roshchina et al., 1994
Brassicaceae (Cruciferae)		
Brassica oleracea L.	Leaf disks	Miura et al., 1982
Coronopus didymus (L.) Sm.	No activity	Gupta and Gupta, 1997
Raphanus sativus L.	Leaves, roots	Momonoki and Momonoki, 1992; Fluck and Jaffe, 1974b
Sisymbrium irio L.	No activity in leaves, branches, roots	Gupta and Gupta, 1997
Burceraceae		
Commiphora wightii (Arnott.) Bhandari	Leaves	Gupta and Gupta, 1997
Cactaceae		
**Echinocereus pentalophus* Rump.	Pollen	Roshchina et al., 2001
**Epiphyllum hybridum*	Pollen, pistil	Roshchina and Semenova, 1995
**Gymnocalicium castellanosii* Brekby	Pollen	Roshchina et al., 2001

(Contd.)

Table 18. (*Contd.*)

Family, species	Organ, where ChE activity found	Reference
**Gymnocalicium zegarrae* Card.	Pollen	Roshchina *et al.*, 2001
Caesalpinaceae		
Caesalpinia pulcherrima (L.) Sw.	Leaves	Gupta and Gupta, 1997
Cassia occidentalis L.	Leaves, branches, roots	Gupta and Gupta, 1997
Cassia tora L.	Roots	Fluck and Jaffe, 1974b
Campanulaceae		
Campanula hybrida	Pistil	Roshchina and Semenova, 1995
Cannabiaceae		
Cannabis sativa L.	Branches, roots	Gupta and Gupta, 1997
Cannaceae	No activity in leaves,	Gupta and Gupta, 1997
Canna indica L.	branches, roots	
Caprifoliaceae		
Lonicera japonica Thunb.	Leaf disks	Miura *et al.*, 1982
Sambucus nigra L.	Branches	Gupta and Gupta, 1997
Viburnum dilatatum Thunb.	Leaf dishes	Miura *et al.*, 1982
Viburnum dentatum L.	Leaf disks	Miura *et al.*, 1982
Caricaceae		
Carica papaya L.	Leaves	Gupta and Gupta, 1997
Caryophyllaceae		
Stellaria media (L.) Vill.	Leaves	Gupta and Gupta, 1997
Ceratophyllaceae	No activity in leaves,	Gupta and Gupta, 1997
Ceratophyllum demersum L.	branches, roots	
Chenopodiaceae		
Spinacia oleracea L.	Leaf disks	Miura *et al.*, 1982
Combretaceae		
Quisqualis indicia L.	Leaves	Gupta and Gupta, 1997
Commelinaceae		
Commelina communis L.	Leaf disks	Miura *et al.*, 1982
Tradescantia virginiana L.	Leaves	Gupta and Gupta, 1997
Convolvulaceae		
Calystegia sepium (L.) R. Br.	Leaf disks	Miura *et al.*, 1982
Calystegia sepium R. Br	Leaves	Roshchina, 1991b
Convolvulus arvensis L.	Pistil, anthers	Roshchina and Semenova, 1995
Convolvulus arvensis L.	Leaves	Roshchina, 1990a
Evolvulus nurninulurius L.	No activity in leaves, branches, roots	Gupta and Gupta, 1997
Ipomaea abutiloides (Carnea)	Flowers, seeds	Villalobos *et al.*, 1974
***Ipomaea nil* (L.) Roth	Pistil, stigma	Bednarska and Tretyn, 1989
Ipomaea nil (L.) Roth	Branches	Gupta and Gupta, 1997

(*Contd.*)

Table 18. (*Contd.*)

Family, species	Organ, where ChE activity found	Reference
Cornaceae		
Cornus florida L.	Leaf disks	Miura et al., 1982
Crassulaceae		
Kalanchoe pinnata (Lamk.) Pers.	Leaves	Gupta and Gupta, 1997
Cucurbitaceae		
Coccinia cordifolia (L.) Cogn.	No activity in leaves, branches, roots	Gupta and Gupta, 1997
Cucurbita pepo L.	Leaf disks	Miura et al., 1982
Cyperaceae		
Cyperus strigosus L.	Leaf disks	Miura et al., 1982
Dennstaedtiaceae		
Dennstaedtia punctilobula T. Moore	No activity in leaf disks	Miura et al., 1982
Ebenaceae		
Diospyros virginiana L.	Leaf disks	Miura et al., 1982
Droseraceae		
**Drosera capensis* L.	Traps, slime hairs	Roshchina and Semenova, 1995
Euphorbiaceae		
Acalipha indica L.	Branches, roots	Gupta and Gupta, 1997
Codiaeum variegatum Blume	No activity in leaf disks	Miura et al., 1982
Codiaeum variegatum Blume	Leaves	Gupta and Gupta, 1997
Euphorbia hirta L.	Leaves, roots	Gupta and Gupta, 1997
Euphorbia milii Ch. des Moul.	Leaves	Gupta and Gupta, 1997
Euphorbia neriifolia L.	Branches	Gupta and Gupta, 1997
Euphorbia pulcherrima Willd. ex Klotz.	Leaves, roots	Gupta and Gupta, 1997
Euphorbia viminalis L.	No activity in latex	Roshchina, unpublished data
Jatropha integerrima Jacq.	Leaves, branches	Gupta and Gupta, 1997
Phyllanthus fraternus Webster	Leave, branches, roots	Gupta and Gupta, 1997
Putranjiva roxburghii Wall.	Leaves	Gupta and Gupta, 1997
Ricinus communis L.	Leaves	Gupta and Gupta, 1997
Synadenium grantii (L.) Hook	Latex	Govindappa et al., 1987
Fabaceae (Leguminosae)		
Aechynomene indica L.	Leaves	Gupta and Gupta, 1997
Caronilla varia L.	Roots	Fluck and Jaffe, 1974b
Crotalaria juncea L. Merr	Leaves	Gupta and Gupta, 1997
Cicer arietinum L.	Roots	Hartmann and Gupta, 1989
Gleditsia triacanthos L.	Leaves	Fluck and Jaffe, 1974b
Glycine max L. Merr	Roots	Fluck and Jaffe, 1974b
Glycine max L. Merr	Leaf disks	Miura et al., 1982
Lathyrus latifolia L.	Flowers	Fluck and Jaffe, 1974b
Lathyrus odoratus L.	Roots	Fluck and Jaffe, 1974b

(*Contd.*)

Table 18. (*Contd.*)

Family, species	Organ, where ChE activity found	Reference
Lathyrus sativus L.	Roots	Maheshwari *et al.*, 1982
Lens culinaris L. Medic.	Roots	Fluck and Jaffe, 1974b
Leucaena leucocephala (Lam̄k.) Wit.	Leaves	Gupta and Gupta, 1997
Medicago sativa L. CV (Vorthwest)	Roots	Fluck and Jaffe, 1974b
Melilotus albus Desr.	Root nodules	Maheshwari *et al.*, 1982
Phaseolus aureus Roxb	Root nodules	Fluck and Jaffe, 1974b
Phaseolus aureus Roxb	Roots	Maheshwari *et al.*, 1982
Phaseolus aureus Roxb	Leaf disks	Miura *et al.*, 1982
Phaseolus vulgaris L.	Cotyledons, leaves,	Lees and Thompson, 1975; Roshchina, 1991b;
Phaseolus vulgaris L.	roots	Fluck and Jaffe, 1974b
Phaseolus vulgaris L.	Leaf disks	Miura *et al.*, 1982
Pisum sativum L.	Roots,	Fluck and Jaffe, 1974b;
Pisum sativum L.	leaves	Roshchina and Mukhin 1984, 1985a
Pisum sativum L.	Stems	Vackova *et al.*, 1984
Pisum sativum L.	Leaf disks	Miura *et al.*, 1982
Psophocarpus tetragonolobus (L.) DC	Leaves, branches	Gupta and Gupta, 1997
Robinia pseudoacacia L.	Leaves	Roshchina, 1990a 1991a
Robinia pseudoacacia L.	Leaf disks	Miura *et al.*, 1982
Sesbania sesban (L.) Merr	Leaves, branches, roots	Gupta and Gupta, 1997
Trifolium pratense L.	Roots	Fluck and Jaffe, 1974b
Trifolium repens L.	Roots	Fluck and Jaffe, 1974b
Trigonella foenum-graecum L.	Roots	Fluck and Jaffe, 1974b
Vicia faba L.	Roots	Fluck and Jaffe, 1974b
Vicia faba L.	Pollen	Bednarska, 1992
Vicia faba L.	Guard cells (stomata)	Madhavan *et al.*, 1995
Vigna sinensis L. Savi	Roots	Fluck and Jaffe, 1974b
Graminae		
Avena sativa L.	Coleoptiles of etiolated seedlings	Kesy *et al.*, 1991
Zea mays L.	Leaves	Fluck and Jaffe, 1974b
Hidrocharitaceae		
Hydrylla verticulata (L.f) Royle	Branches	Gupta and Gupta, 1997
Hippocastanaceae		
Aesculus hippocastanum L.	Pollen	Roshchina *et al.*, 1994
Hydrocotylaceae		
Hydrocotyle verticillata Turcz.	Leaf disks	Miura *et al.*, 1982
Iridaceae		Semenova and
Gladiolus sp.	Anthers, pollen	Roshchina, 1993

(*Contd.*)

Table 18. (*Contd.*)

Family, species	Organ, where ChE activity found	Reference
Gladiolus sp.	No activity in pistils	Roshchina and Semenova, 1995
Funkia sp. (Hosta)	Anthers, pollen	Roshchina and Semenova, 1995
Iris viriginica L.	Leaf disks	Miura *et al.*, 1982
Lamiaceae		
Coleus blumei Benth.	Leaves	Gupta and Gupta, 1997
Glechoma hederacea L.	Leaf disks	Miura *et al.*, 1982
Ocimum sanctum L.	No activity in leaves, branches, roots	Gupta and Gupta, 1997
Plectranthus australis R. Br.	Leaf disks	Miura *et al.*, 1982
Lauraceae		
Sassafras albidium Nees.	Leaf disks	Miura *et al.*, 1982
Lentibulariaceae		
Utricularia sp.	Trap	Roshchina (unpublished data)
Liliaceae		
Funkia (Hosta) sp.	Anthers, pistils	Roshchina and Semenova, 1995
Hemerocallis fulva L.	Pollen	Roshchina *et al.*, 1994
Tulipa sp.	Pollen	Roshchina (unpublished data)
Magnoliaceae		
Liriodendron tulipifera L.	Leaf disks	Miura *et al.*, 1982
Magnolia grandiflora L.	No activity in leaves, branches, roots	Gupta and Gupta, 1997
Malvaceae		
Gossypium herbaceum L.	Leaves, branches	Gupta and Gupta, 1997
Hibiscus palustris L.	Leaf disks	Miura *et al.*, 1982
Hibiscus rosa-sinensis L.	No activity in leaves, branches, roots	Gupta and Gupta, 1997
Meliaceae		
Melia azedarach L.	No activity in leaves, branches, roots	Gupta and Gupta, 1997
Mimosaceae		
Albizzia julibrissin (Willd. Durazz).	Leaves, roots	Fluck and Jaffe, 1974b
Albizzia julibrissin (Willd. Durazz).	Leaf disks	Miura *et al.*, 1982
Macroptilium atropurpureum (DC) Urban	Leaves	Momonoki and Momonoki, 1992
Mimosa pudica L.	Roots	Fluck and Jaffe, 1974b
Moraceae		
Ficus benghalensis L.	No activity in leaves, branches, roots	Gupta and Gupta, 1997
Ficus benjamina L.	Leaf disks	Miura *et al.*, 1982
Ficus elastica Roxb.	Leaves, branches	Gupta and Gupta, 1997
Ficus krishnae C.DC.	Leaves	Gupta and Gupta, 1997
Ficus racemosa L.	No activity in leaves, branches, roots	Gupta and Gupta, 1997

(*Contd.*)

Table 18. (*Contd.*)

Family, species	Organ, where ChE activity found	Reference
Ficus religiosa L.	No activity in leaves, branches, roots	Gupta and Gupta, 1997
Morus alba L.	Leaf disks	Miura *et al.*, 1982
Musaceae	No activity in leaves, branches, roots	Gupta and Gupta, 1997
Musa paradisiaca L.		
Myrtaceae		
Callistemon lanceolatus DC.	Leaves	Gupta and Gupta, 1997
Psidium guajava L.	Leaves	Gupta and Gupta, 1997
Nyctaginaceae	Leaves	Gupta and Gupta, 1997
Boerhaavia diffusa L.	Leaves	Gupta and Gupta, 1997
Bougainvillea glabra Choisy	Branches	Gupta and Gupta, 1997
Oleaceae		
Forsynthia sp.	Leaf disks	Miura *et al.*, 1982
Fraxinus americana L.	No activity in leaf disks	Miura *et al.*, 1982
Osmundaceae		
Osmunda regalis L.	Leaf disks	Miura *et al.*, 1982
Oxalidaceae		
Oxalis corniculata L.	Leaves	Gupta and Gupta, 1997
Papaveraceae		
Argemone mexicana L.	Leaves	Gupta and Gupta, 1997
Chelidonium majus L.	No activity in latex	Roshchina, unpublished data
Papaver orientale L.	Pollen	Roshchina *et al.*, 1994; Roshchina, 1999a
Papaver somniferum L.	Leaves	Gupta and Gupta, 1997
Peperomiaceae		
Peperomia clusiaefolia Hook.	Leaf disks	Miura *et al.*, 1982
Piperaceae		
Piper betle L.	Leaves	Gupta and Gupta, 1997
Plantaginaceae		
Plantago major L.	Weak activity in anthers	Roshchina *et al.*, 1994
Plantago rugelli Decne.	Leaf disks	Miura *et al.*, 1982
Platanaceae		
Platanus occidentalis L.	Leaf disks	Miura *et al.*, 1982
Poaceae		
Cynodon dactylon (L.) Pers.	Branches	Gupta and Gupta, 1997
Paspalum ciliatifolium Michx	Leaf disks	Miura *et al.*, 1982
Polygonaceae		
Antigonon leptopus Hook & Arn.	No activity in leaves, branches, roots	Gupta and Gupta, 1997
Polygonum hydropiperoides Michx.	Leaf disks	Miura *et al.*, 1982
Pontederiaceae		
Eichhornia crassipes (Mart.) Solms.	Branches, roots	Gupta and Gupta, 1997

(*Contd.*)

Table 18. (*Contd.*)

Family, species	Organ, where ChE activity found	Reference
Pontederia cordata L.	Leaf disks	Miura *et al.*, 1982
Portulacaceae	No activity in leaves, branches, roots	Gupta and Gupta, 1997
Portulaca quadrifida L.		
Proteaceae	No activity in leaves, branches, roots	Gupta and Gupta, 1997
Grevillea robusta A. Cunn.		
Punicaceae		
Punica granatum L.	Leaves	Gupta and Gupta, 1997
Rosaceae		
Malus domestica Borh.	No activity in pollen	Roshchina *et al.*, 1994
Prunus serotina Ehrh.	Leaf disks	Miura *et al.*, 1982
Rosa sp.	Leaf disks	Miura *et al.*, 1982
Rutaceae		
Citrus aurantifolia (Christm) Swingle.	Leaves, branches	Gupta and Gupta, 1997
Vitis vinifera L.	Leaves	Gupta and Gupta, 1997
Saxifragaceae	Pollen	Roshchina and Semenova, 1995
Philadelphus grandiflorus Willd.		
Saxifraga stolonifera Meerb.	Leaf disks	Miura *et al.*, 1982
Scrophulariaceae		
Mazus pumilus (Burm. f) Steen.	Leaves	Gupta and Gupta, 1997
Verbascum chinense (L.) Santapau	No activity in leaves, branches, roots	Gupta and Gupta, 1997
Salicaceae		
Populus balsamifera L.	No activity in pollen	Roshchina *et al.*, 1994; Roshchina and Semenova, 1995
Populus grandidentata Michx.	Leaf disks	Miura *et al.*, 1982
Salix caprea L.	Pollen	Roshchina and Semenova, 1995
Solanaceae		
Datura innoxia Mill.	Leaves, branches, roots	Gupta and Gupta, 1997
Lycopersicon esculentum L. (Mill.)	Leaves, roots	Fluck and Jaffe, 1974b
Nicotiana glauca Graham.	Guard cells (stomata)	Madhavan *et al.*, 1995
Nicotiana plumbaginifolia Viv.	No activity in leaves, branches, roots	Gupta and Gupta, 1997
Nicotiana rustica L.	Leaves	Gupta and Gupta, 1997
Petunia hybrida Vilm	Leaf disks	Miura *et al.*, 1982
Petunia hybrida Vilm	Pollen	Roshchina *et al.*, 1994; Roshchina and Semenova, 1995; Roshchina, 1999a

(*Contd.*)

Table 18. (*Contd.*)

Family, species	Organ, where ChE activity found	Reference
Petunia hybrida Vilm	Pistil	Kovaleva and Roshchina, 1997; Roshchina, 1999a
Physalis minima L.	Leaves, branches, roots	Gupta and Gupta, 1997
Solanum melongena L.	Leaves, roots	Fluck and Jaffe, 1974b
Solanum melongena L.	Leaf disks (weak activity)	Miura *et al.*, 1982
Solanum nigrum L.	Leaves	Gupta and Gupta, 1997
Solanum tuberosum L.	Leaves	Fluck and Jaffe, 1974b
Withania somnifera (L.) Dunal.	Leaves, branches	Gupta and Gupta, 1997
Tiliaceae		
Corchorus aestuans L.	Leaves	Gupta and Gupta, 1997
Tilia cordata Mill	No activity in pollen	Roshchina *et al.*, 1994
Tropaeolaceae		
Tropaeolum majus L.	Leaves, roots	Gupta and Gupta, 1997
Urticaceae		
Boehmeria cylindrica Sw.	Leaf disks	Miura *et al.*, 1982
Urtica dioica L.	Leaves	Roshchina and Mukhin, 1985a; Roshchina, 1988a, b; 1990b;
Vaccinaceae		
Vaccinium corymbosum L.	Leaf disks	Miura *et al.*, 1982
Verbenaceae		
Lantana camara L.	No activity in leaves, branches, roots	Gupta and Gupta, 1997
Nyctanthes arbor-tristis L.	No activity in leaves, branches, roots	Gupta and Gupta, 1997
Violaceae	Leaf disks	Miura *et al.*, 1982
Viola papilionacea Pursh.		
Vitidaceae		
Parthenocissus quinquefolia Planch.	Leaf disks	Miura *et al.*, 1982
FUNGI		
Ascomycetes		Roshchina and Alexandrova, 1991
Aspergillus niger	Mycelium	

* The acetylcholine hydrolysis estimated according to the method of Miura *et al.* (1982), inhibited by diisopropyl phosphofluoridate. In another variant the hydrolysis determined according to methods of Ellman *et al.* (1961) or Gorun *et al.* (1978) with Ellman reagent or method of Roshchina *et al.* (1994) with red analog of Ellman reagent (see Appendix 2) is inhibited by neostigmine and/or physostigmine.

** Histochemical reaction on cholinesterase, according to Karnovski-Roots method (1964) or method of Roshchina *et al.* (1994), Roshchina (1999a).

Remark: Different authors relate genus *Populus* to Populaceae or Salicaceae and genus *Philadelphus*-to Saxifragaceae or Hydrangeaceae.

According to Fluck and Jaffe (1976), highest cholinesterase activity takes place mainly among representatives of the family Fabaceae, especially in roots. Miura and coworkers (1982), analyzing the activity on leaf disks, have shown that the highest rate of acetylcholine hydrolysis was characteristic for family Brassicaceae (Cruciferae), and the lowest rate for gymnosperms plants, especially families Gingkoaceae and Taxaceae. Cholinesterase was inhibited by organophosphorus compound diisopropyl phosphofluoridate. The level of inhibition of the enzymes in plants belonging to families Chenopodiaceae, Platanaceae, and Solanaceae was not high, whereas in Caprifoliaceae, Iridaceae, and Vaccinaceae it was 100%. Species belonging to Solanaceae demonstrated highest activity in crude extracts, which is similar to acetylcholinesterase in the human brain (Gupta and Gupta, 1997).

Cholinesterase activity was demonstrated to differ from other plant esterases, such as citric acetylesterase (Jansen *et al.*, 1948; 1974; Riov and Jaffe, 1973b), esterase from wheat seeds (Jansen *et al.*, 1948), and pectin methylesterase (Fluck and Jaffe, 1974d), and it is not a product of bacterial contamination (Fluck and Jaffe, 1974d). Unlike sinapine esterase, cholinesterase has some special features (Tzagolov, 1963a, b), namely hydrolysis of sinapine weakly inhibited by physostigmine.

The absence of acetylcholine hydrolysis was often observed in the family Asteraceae; this could be associated with the terpenoids contained in those plants, which may be potential inhibitors of cholinesterase. A similar observation is made in other families. For instance, acetylcholine is found in *Helianthus annuus*, although its hydrolysis was not detected. The cholinesterase activity in Asteraceae may be blocked insecticidic terpenoids, like pyrethrine in *Chrysanthemum cinerariaefolium*. Gupta and Gupta (1997) also discussed a correlation between the presence of anticholinesterase drugs and cholinesterase activity in a certain family. The family Euphorbiaceae tested positive for acetylcholine hydrolysis, but at a very low level. On the contrary, the family Solanaceae, a rich source of cholinesterase, also contains α-chaconine, solanine, nicotine, scopolamine, atropine, and hyoscyamine, known as anticholinergic drugs. Moreover, allicin, an active antimollusk agent of garlic *Allium sativum* (Alliaceae), is known as a cholinesterase inhibitor (Singh and Singh, 1996), although the genus *Allium* possesses high activity of the enzyme (Hadacova *et al.*, 1981, 1983). Perhaps different compartmentation of anticholinergic drugs and cholinesterase takes place in these families.

Comparison of the data in Tables 1 and 18 shows that acetylcholine has been detected in 67 species, whereas cholinesterase was found in three times as many species. Thus, cholinesterase detection could serve as additional indicator of acetylcholine presence.

All plant parts—leaves, roots, stems, seeds, flowers—have a capacity to hydrolyze cholinic esters (Table 18). Moreover, in pea there is an oscillation of cholinesterase activity with periodicity of the maximum at 3-6-9 days (Vackova et al., 1984). At the end of this cycle the rise in enzymic activity in stems could reach 100%, whereas in roots, on the contrary, a drop in the rate of the process has been marked after the fifth day of seed germination, and it is 3.5 times lower than in stems.

The lack of cholinesterase activity might also be due to methodological difficulties. The Ellman method (Ellman et al., 1961)), based on the formation of yellow (A = 415 nm) product with thiocholine, is used most often (see Appendix 2, Methods). This assay, however, is not useful in case of high concentration of thiols in extracts of certain plants, because they might interfere with Ellman reagent 5, 5-dithio-bis(2-nitrobenzoate). Another factor preventing enzyme determination appears to be yellow or green-yellow pigments in plant preparations (Roshchina, 1988a, b; 1991a). In these cases other assays are perferable, for instance the radioisotope method (labeled acetyl-β-methylcholine) by which Miura and coworkers (1982) have found cholinesterase activity in leaf disks of 64 plant species from 50 families (Table 18). For yellow and green plant extracts assay with red analog of Ellman reagent 2,2-dithio-bis-(n-phenyleneazo)-bis (1-oxy-8-chlor-3,6-) disulfonic acid sodium salt, which forms a blue color product with thiocholine (A_{620}), may be recommended (Roshchina and Alexandrova, 1991; Roshchina et al., 1994) (see Appendix 2, Methods).

Unsuccessful determination of cholinesterase is also due to the fact that various plant organs might contain different amounts of the enzyme and/or contain inhibitors of cholinesterases that are liberated from disrupted compartments when the plant extract is prepared. Lack of cholinesterase activity could be explained as a screening of the enzyme expression at a certain physiological stage (Kasturi, 1979).

The capacity to hydrolyze cholinic esters arises at the earliest stages of plant development (Lees and Thompson, 1975; Kasturi, 1978). According to Lees and Thompson (1975), the enzymic hydrolysis of acetylcholine by cotyledons of *Phaseolus vulgaris* increased 63-fold between second and third days of seed germination and was high for up to 6 days. Kasturi (1978) marked that the seeds of *Pisum sativum* start to show acetylcholinesterase activity just after germination. In the first 24 hours this activity is rather high, and then it decreases. Between 48 and 72 hours, after beginning of growth, a lag phase was observed. After that, synthesis of the enzyme *de novo* starts, and its activity increases simultaneously, exceeding the original level three-fold. The primary roots of the plant already contain a marked quantity of newly synthesized enzyme, approximately 4–10% of total protein. Cholinesterase activity was observed even in embryos and cells of aleuronic layer in seeds of wheat, oat, and gourd (Tretyn et al., 1986). It was observed in the stage of root and stem differentiation in epidermis, phloem, cambium, and apical meristems of these plants. The enzymic activity in chickpea *Cicer arietinum* decreases in stems as far as the plant grows, and more than 90% of this activity is concentrated in roots (Hartmann and Gupta, 1989).

Cholinesterase activity was found in vegetative specialized cells as well as in stomata by histochemical method (Gorska-Brylass and Smolinski, 1992) and biochemical assay (Madhavan et al., 1995). Moreover, the same activity was observed in carnivorous plants: in slime hairs of *Drosera capensis* (Roshchina and Semenova, 1995) and in trap of *Utricularia* sp. (Roshchina, unpublished data).

The participation of plant cholinesterase in the recognition of pollen belonging to the same species by the pistil stigma in flowers of *Pharbitis nil* was hypothesized in a paper by Bednarska and Tretyn (1989). Using electron microscopic histochemistry the authors showed that the enzymic activity is localized in the pellicle, the thin film covering the outer surface of the stigma papilla, and it has also been found in pollen grains and tubes of this plant. The enzyme is supposed to be excreted by the stigma papillae. Cholinesterase activity of pollen is found by histochemical method in *Vicia faba* (Bednarska, 1992).

Using histochemical and biochemical methods, the presence of cholinesterase in pollen has also been demonstrated in extracts of 15 species of plants and pollen-loads collected by bees (Roshchina et al., 1994; Roshchina and Semenova, 1995). Most active was the enzyme of pollen belonging to *Pinus sylvestris*, *Hemerocallis fulva*, *Aesculus hippocastanum*, *Epiphyllum* sp., and *Papaver orientale*. The rate of acetylcholine hydrolysis depended also on the time of pollen storage. Freshly collected microspores showed 3-fold higher enzymic activity than microspores stored from 1 week to 1 month (Kovaleva and Roshchina, 1997). Moreover, pollen and pistil of self-incompatible clones of *Petunia hybrida* demonstrated lower activity than self-compatible clone (Kovaleva and Roshchina, 1997). Thus, the cholinesterase occurrence in generative organs could be connected with mechanisms of plant breeding.

3.1.3.3. Kinetic Parameters of Cholinesterases Found in Plant Extracts

Kinetic characteristics of cholinesterase not isolated from plant extracts are usually studied when purification is impossible because of the small amount of crude material available, for example, the generative organs of flower.

Table 19 shows that the cholinesterase activity concentrates in anthers with pollen grains (male gametophytes) and pistil (female gametophyte),

Table 19. Rates of AThCh and BuThCh (both 1 mM) hydrolysis (μmol h^{-1} g^{-1} of fresh weight) by water extracts of intact parts of flower. Mean for three to five experiments ± SE at a 5% significance level. Ellman reagent was for ChE assay (Roshchina and Semenova, 1995).

Plant species	Part of flower	AThCh without inhibitor	AThCh + neostigmine 0.01 mM	AThCh + physostigmine 0.01 mM	BuThCh without inhibitor
Amaryllidaceae	petals	0	0	0	
Hippeastrum	sepals	0	0	0	
hybridum	anthers	15.12 ± 0.32	13.32	7.56	
Convolvulaceae	petals	0	0	0	0
Convolvulus	sepals	0	0	0	0
arvensis	anthers	468 ± 18.72	0	0	127.8 ± 14.4
	pistils	900 ± 36	0	262.8	1152 ± 25.2
Hydrangeaceae	petals	0	0	0	0
Philadelphus	sepals	0	0	0	0
grandiflorus	anthers	25.56 ± 0.07	0	0	8.28 ± 0.14
	pistils	0	0	0	0
Iridaceae	petals	0	0	0	0
Gladiolus sp.	sepals	0	0	0	0
	anthers	58.32 ± 0.07	8.64	20.9	43.2 ± 0.04
	pistils	0	0	0	0
Liliaceae	petals	0	0	0	0
Hosta sp.	sepals	0	0	0	0
	anthers	2.77 ± 0.04	0	0	0
	pistils	1.66 ± 0.07	1.66	0.94	0
Plantaginaceae	petals	0	0	0	
Plantago major	sepals	0	0	0	
	anthers	0.32 ± 0.03	0	0	
	pistils	0	0	0	

unlike sepals and petals. In extracts of anthers of some species of Liliaceae cholinesterase activity has been demonstrated too in *Lilium tenuifolium, Gladiolus hybridum, Hippeastrum hybridum,* and *Clivia* sp. (Semenova and Roshchina, 1993; Roshchina and Semenova, 1995). Some kinetic characteristics for anther cholinesterases are represented in Table 20. The highest rate of hydrolysis among cholinic esters tested was for acetyl-β-methylthiocholine, while K_M (Michaelis-Menten constant) of the reaction was lowest for acetylthiocholine. The rate of the acetylthiocholine hydrolysis was higher than those of butyrylthiocholine (Table 19). Figure 35 demonstrates inhibition of the cholinesterase activity by high concentration of the substrate for anther of *Convolvulus arvensis*. The bell-shaped concentration curves are characteristic for animal acetylcholinesterase (see 3.1.1).

Table 20. Kinetic characteristics of cholinesterase of anthers from Hippeastrum and gladiolus (Semenova and Roshchina, 1993; Roshchina and Semenova, 1995).

Plant	Substrate	K_M, µM	V_{max}, µmol g^{-1} of fresh mass h^{-1}
Hippeastrum hybridium	Acetylthiocholine	1.17	11.27
	Acetyl-β-methylthiocholine	4.86	15.10
	Propionylthiocholine	4.77	8.90
	Butyrylthiocholine	3.33	13.48
	Benzoylthiocholine	No hydrolysis	
Gladiolus sp.	Acetylthiocholine	3.73	8.33

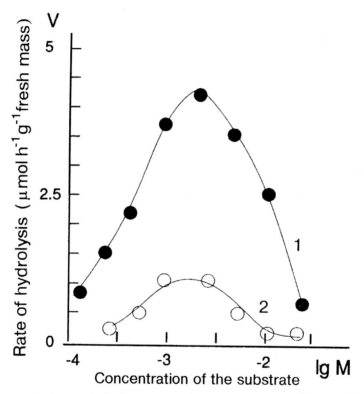

Fig. 35. Hydrolysis of cholinic esters by water extracts from anthers of *Convolvulus arvensis* (Roshchina, unpublished data). 1, acetylthiocholine; 2, butyrylthiocholine.

Tables 21 and 22 demonstrated the rates and K_M of hydrolysis of cholinic esters by extracts from intact anthers and pollen. Inhibition of the processes by physostigmine or neostigmine was higher in pollen

Table 21. Hydrolysis rates (μmol h^{-1} g^{-1} of fresh wt × 10^{-2}) of AthCh and BuThCh (both 1 mM) by water extracts of pollen grains. Mean values are presented for three to five experiments with error of 1–3%. Ellman reagent or its red analog (*) was used for ChE assay (Roshchina et al., 1994; Roshchina and Semenova, 1995).

Plant species	Hydrolysis of AthCh			Hydrolysis of BuThCh	AthCh/ BuThCh	Substrate inhibition
	without inhibitor	+ physostigmine 0.01 mM	+ neostigmine 0.01 mM			
Alliaceae						
Allium cepa	16.56*	0.43*	0*	16.96*	AThCh = BuThCh	No
Amaryllidaceae						
Hippeastrum hybridum	57.60	72.00	28.80	45.00	AThCh > BuThCh	No
Asteraceae						
Matricaria chamomilla	10.08* 16.20	3.49* 4.61	2.92* 6.23	17.28	AThCh < BuThCh	No
Betulaceae						
Betula verrucosa	10.37	10.30	10.08			
Cactaceae						
Epiphyllum sp.	403.2	0	0			
Geraniaceae						
Geranium pratense	4.46	4.32	4.61			
Hydrangeaceae						
Philadelphus grandiflorus	3.82* 2.84	0.015* 0	0.005* 0	1.12*	AThCh > BuThCh	No
Hippocastanaceae						
Aesculus hippocastanum	244.8	115.2	57.60	230.4	AThCh = BuThCh	Yes
Liliaceae						
Hemerocallis fulva	11.34	0	0	7.92	AThCh > BuThCh	No
Tulipa sp.	432.0	86.40	86.40			
Papaveraceae						
Papaver orientale	17.95	3.84	8.10			
Pinaceae						
Pinus sylvestris	540.0	221.8	209.9			
Plantaginaceae						
Plantago major	35.28	32.04	31.68			No
Rosaceae						
Malus domestica	50.40	45.00	61.20			No
Salicaceae						
Populus balsamifera	5.33	5.36	5.36			
Salix caprea	0.018*	0*	0.001*			
Solanaceae						
Petunia hybrida	7.56* 9.50	5.62* 5.40	6.48* 6.12	6.48 6.01	AThCh ≥ BuThCh	No
Tiliaceae						
Tilia cordata	0	0	0			

Table 22. The kinetic parameters of cholinesterase from water extracts of intact anthers and pollen. Mean values for three to fine experiments with the error of 1 3% are presented. Ellman reagent or its red analogue (*) was used for ChE assay (Roshchina and Semenova, 1995).

Plant species		Substrate	K_m, mM	V_{max}, μmol h^{-1} g^{-1} fresh wt
Aesculus	pollen	AThCh	0.25	25.2
hippocastanum		BuThCh	0.31	22.7
Allium cepa	pollen	AThCh	0.60*	842.4*
		BuThCh	0.50*	334.8
Convolvulus	anthers	AThCh	0.50	18.7
arvensis		BuThCh	0.70	3.6
Petunia hybrida	pollen	AThCh	0.60	842.8
		BuThCh	0.50	334.8
Philadelphus	pollen	AThCh	0.40	5.8
grandiflorus		BuThCh	0.70	1.8

extracts from *Allium cepa, Epiphyllum* sp., *Philadelphus grandiflorus*, and *Hemerocallis fulva*. There was no cholinesterase activity in the pollen grains of *Betula verrucosa, Geranium pratense, Malus domestica,* and *Tilia cordata*. Figure 36 shows different forms of the concentration curves of pollen from various species. Cholinesterase activity inhibited by high amounts of substrate is observed only for pollen of *Aesculus hippocastanum*, while pollen extracts of *Philadelphus grandiflorus, Allium cepa*, and *Petunia hybrida* have a saturation in the concentration curves for both acetylthiocholine and butyrylthiocholine, without any inhibition. Although inhibition by high concentrations of a substrate is peculiar to animal acetylcholinesterase, whereas lack of the inhibition is peculiar to pseudocholinesterase, it is difficult to classify to enzymes in pollen grains. The main difference lies in the rate of hydrolysis of cholinic esters. Acetylthiocholine is hydrolyzed with higher rates than butyrylthiocholine in the species with the concentration saturation, while bell-shaped curves of *Aesculus* are equal in the rates of hydrolysis of both substrates. It is more accurate to consider pollen cholinesterases as a mixed type of cholinesterase, peculiar to the enzymes of lower animals (Roshchina and Semenova, 1990).

Acetylcholine, formylcholine, and propionylcholine are also hydrolyzed by cholinesterase of honey (Goldschmidt and Burkert, 1955). The hydrolysis of acetylcholine in honey solution is largely retarded in contrast to that of a pure aqueous acetylcholine solution of equal concentration. The retarding effects of honey at a given pH depend on organic acid of honey, in particular on malic acid, which has the strongest retarding effect. The rates of acetylcholine hydrolysis were highest for the sequence butyrylcholine > propionylcholine > acetylcholine > formylcholine.

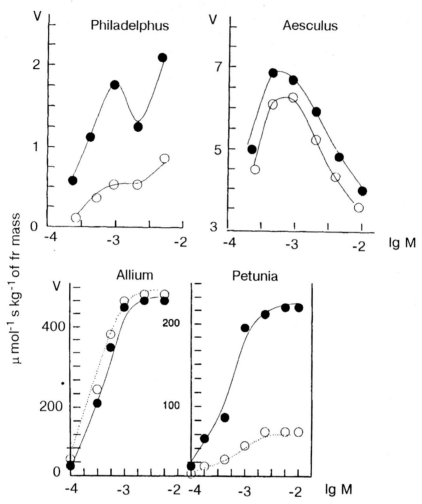

Fig. 36. Concentration curves of cholinesterase activity in excretions from pollen grains (after Roshchina and Semenova, 1995; Roshchina, 1999a). ●—●, acetylthiocholine, o—o, butyrylthiocholine

3.1.3.4. Localization of Cholinesterase in Plant Cells

There is information about the localization of cholinesterase within animal and plant cells. In synapse of animals, acetylcholinesterase is found mainly in the postsynaptic membrane. Acetylcholinesterase activity has also been found in endoplasmic reticulum and Golgi apparatus, nuclei, microsomes, and mitochondria of brain cells (de Robertis, 1967). As was first demonstrated on plants by electron microscopic histochemistry, cholinesterase is localized in plasmalemma, cell wall, and partly the cytoplasm of root cells of *Phaseolus aureus* (Fluck and Jaffe, 1974c). In lichen *Parmelia caperata*, acetylcholinesterase is concentrated in cell walls and/or in plasmatic membranes of both symbiotic partners, fungi and algae (Raineri and Modenesi, 1986). Information about localization of the enzyme in various cellular compartments depending on plant organ and tissue is compiled in Table 23.

Table 23. Localization of cholinesterase in plant cell

Compartment	Organ, tissue	Plant	ChE_c/ChE_w	Reference
Cell wall	Hypocotyls, roots, callus	*Phaseolus vulgaris*	58–94	Hartmann and Gupta, 1989
	Callus	*Glycine max*	(+)	Kieffer, 1986 (cited in Hartmann and Gupta, 1989)
	Root	*Phaseolus aureus*	95	Fluck and Jaffe, 1974c
	Root	*Pisum sativum*	(+)	Maheshwari et al., 1982
	Cotyledon	*Phaseolus vulgaris*	(+)	Lees and Thompson, 1975
	Thallus	*Parmelia caperata*	(+)	Raineri and Modenesi, 1986
	Etiolated seedlings	*Avena sativa*	50	Kim et al., 1990
	Stomata	*Marchantia polymorpha*	(+)	Gorska-Brylass and Smolinski, 1992
	Pistil stigma	*Pharbitis nil*	(+)	Bednarska and Tretyn, 1989
Exine	Pollen	*Vicia faba*	(+)	Bednarska, 1992
	Pollen	*Hippeastrum hybridum*	(+)	Roshchina et al., 1994; Roshchina and Melnikova, 1999
	Pollen	*Epiphyllum hybridum*	(+)	Roshchina et al., 2001
Plasmalemma	Root	*Phaseolus aureus*	(+)	Fluck and Jaffe, 1974c
	Cotyledon	*Phaseolus vulgaris*	(+)	Lees and Thompson, 1975
	Stomata	*Marchantia polymorpha*	(+)	Gorska-Brylass and Smolinski, 1992
	Pollen tube	*Vicia faba*	(+)	Bednarska, 1992
	Pistil stigma	*Pharbitis nil*	(+)	Bednarska and Tretyn, 1989
	Thallus	*Parmelia caperata*	(+)	Raineri and Modenesi, 1986

(Contd.)

Table 23. (Contd.)

Compartment	Organ, tissue	Plant	ChE$_c$/ChE$_w$	Reference
Nucleus	Root	Pisum sativum	(+)	Maheshwari et al., 1982
Chloroplast	Callus of leaf	Phaseolus vulgaris	7	Hock, 1983 (cited in Hartmann and Gupta, 1989)
	Leaf	Pisum sativum	90	Roshchina and Mukhin, 1984
	Leaf	Urtica dioica	90	Roshchina and Mukhin, 1984
	Leaf	Convolvulus arvensis	90	Roshchina, 1990a
	Leaf	Phaseolus aureus	90	Roshchina, 1990a
	Leaf	Robinia pseudoacacia	90	Roshchina, 1990a
	Leaf	Zea mays	50	Roshchina, 1990a
Cytoplasm	Root	Phaseolus aureus	(+)	Fluck and Jaffe, 1974c
	Cotyledon	Phaseolus vulgaris	(+)	Lees and Thompson, 1975
	Etiolated seedlings	Avena sativa	42	Kim et al., 1990

ChE$_c$/ChE$_w$ ratio of the cholinesterase activity in compartment and in homogenate of tissue. (+): There are no quantitative data.

The enzyme was mainly found in the contact sites: outside the plasmalemma, in the cell between cell wall and plasmalemma, and in the plasmalemma itself. Fluck and Jaffe (1974c) demonstrated especially electron-dense material (histochemical staining) in joints of cell-cell contacts in root tissue, while Gorska-Brylass and Smolinski (1992) demonstrated it in cell wall and plasmalemma of guard cells. Exine, the outer cover of pollen, also contains cholinesterase (Bednarska, 1992; Roshchina et al., 1994). In *Vicia faba* pollen, as shown by Bednarska (1992), the activity is localized in hydrated pollen grains, mainly in exine, plasmatic membrane, pollen surface, and aperture, whereas non-specific activity is localized mainly in intine, cytoplasm, vesicles, and spherosomes. Germinating pollen grain contains cholinesterase in aperture, sporoderm, and pollen surface, and non-specific esterase in exine. Moreover, pollen tube demonstrates cholinesterase in vesicles and cell wall of the tip, and non-specific esterase in pectine and cellulose cell wall of the tip. In *Pharbitis nil* pistil, cholinesterase is located in pellicule and plasmalemma (Bednarska and Tretyn, 1989), which suggests that the enzyme is related to pollen-pistil contacts. This is supported by the histochemical staining of cholinesterase in pollen exine of *Hippeastrum hybridum* and microelectrophoresis of pollen excretion (Fig. 37).

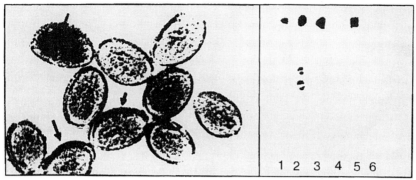

Fig. 37. Cholinesterase activity of *Hippeastrum hybridum* pollen (after Roshchina et al., 1994; Roshchina, 1999b). Left, histochemical staining with red analog of Ellman reagent. Blue color is seen on the surface and in exine (shown with arrows). Right, microelectrophoresis (7% PAGE) of proteins. 1, 3, 4, marker animal protein, acetylcholinesterase from *Electrophorus electricus* (2.5 µg of protein/*gel hole*). 2, 5, 6, of 10 min, water extract from the pollen of *Hippeastrum hybridum* (0.4 µl of extract from 10 mg of microspores/*gel hole*. 1 and 2, gels colored with Coumassi R250. 3–6, gels colored with red analog of Ellman reagent (3, 5 without and 4, 6 with preliminary treatment by the cholinesterase inhibitor neostigmine 10^{-5} M repeating during 30 min. Blue color was absent if gel was treated with the inhibitor).

Carnivorous plants excrete slime to attract insects. The histochemical staining on cholinesterase is observed on the surface of secreting trap of *Utricularia* sp. (Fig. 38). The possible involvement of the enzyme in recognition of insect at contact is not excluded. Figure 38 also shows that staining on cholinesterase is seen in the joints between cells of the trap.

Histochemical staining on acetylcholinesterase activity in motile elements of plants was also analyzed on slides of *Macroptilium atropurpureum* made from petioles, stem, root, and primary and secondary pulvini (Momonoki and Momonoki, 1993a). The enzyme was only present in the primary and secondary pulvini, mainly in the endodermal cells around the vascular system. According to the results, the primary pulvinus is associated with acetylcholine function. The motile system of stele-cortex in *Zea mays* was studied with $(1-^{14}C)$ acetylcholine transported into mesocotyl cortex (Momonoki, 1992). This hydrolysis was inhibited by neostigmine at the interface between stele and cortex, mainly at the junction of cell walls.

Moving guard cells of stomata also showed acetylcholinesterase activity in protoplasts (Madhavan et al., 1995). One of the mechanisms may be a liberation of H^+ and consequent acidification of the internal medium, leading to regulation of K^+/Na^+ efflux and uptake.

124 *Neurotransmitters in Plant Life*

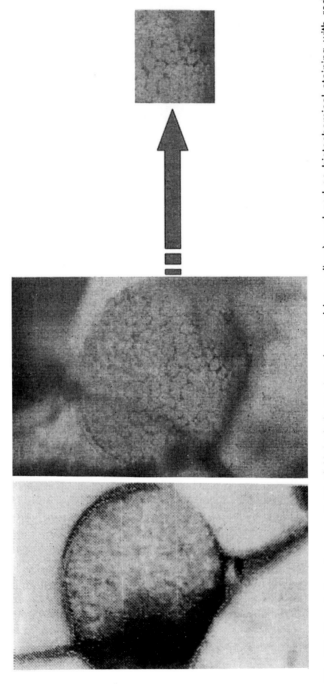

Fig. 38. The cholinesterase activity of *Utricularia* sp. trap (covered by a slime) analyzed as histochemical staining with red analog of Ellman reagent (Roshchina, unpublished data). Left, photo the common view of the trap surface, 1 cm bar = 0.8 mm; right, one part of the trap, 1 cm bar = 0.05 mm. Blue color is seen on the surface, especially concentrated on the entrance (left) and in the joints of cell walls (right) of stacking cells.

The cholinesterase activity is located within the cell also (Table 23), in nucleus, cytoplasm, and chloroplast. The contacts between organelles may involve the enzyme.

When the hydrolysis of acetylcholine by the fractions of leaves and chloroplasts of pea and common nettle was analyzed (Roshchina and Mukhin, 1984; 1987a), it was shown that highest hydrolyzing activity is concentrated in chloroplasts and is inhibited by specific inhibitors of animal cholinesterases neostigmine and physostigmine (Table 24). Particular block of the hydrolysis of acetylcholine in leaf homogenates by these inhibitors can be connected with the presence of not only cholinesterase, but non-specific esterases as well.

Table 24. The rate of hydrolysis of acetylthiocholine by various leaf fractions (from Roshchina and Mukhin, 1984)

Plant	Fraction	Hydrolysis of acetylthiocholine, kat kg^{-1} of protein		
		without inhibitor	+ neostigmine	+ physostigmine
Pisum sativum	Leaf homogenate	0.043	0.025	0–0.020
Pisum sativum	Intact chloroplasts	0.060	0.0	0.0
Urtica dioica	Leaf homogenate	0.090	0.007	0.015
Urtica dioica	Intact chloroplasts	0.090	0.0	0.0

SE = 1–2%. Neostigmine $2.5 \cdot 10^{-5}$ M, physostigmine 10^{-5} M.

In order to demonstrate where the enzyme is localized in plastids the chloroplast fractions in the sucrose density gradient were separated (Roshchina, 1989a). Cholinesterase activity was found in fractions of outer membranes and thylakoids (Table 25). Moreover, the activity of the enzyme in thylakoids was approximately seven-fold higher than in chloroplast envelope. The concentration curves of the rate of hydrolysis of cholinic esters in a dependence on substrate show that chloroplast cholinesterase hydrolyzes acetylcholine with a higher rate than butyrylcholine, and that excess of substrate depresses the cholinesterase

Table 25. Cholinesterase activity in fraction of pea chloroplasts (kat kg^{-1} of protein) (Roshchina, 1989a)

Fraction	Hydrolysis of acetylthiocholine		
	without inhibitor	+ neostigmine ($25 \cdot 10^{-6}$ M)	+ physostigmine (10^{-5} M)
Intact chloroplasts	0.06	0.0	0.0
Thylakoids	0.09	0.0	0.0
Outer membranes of chloroplasts	0.013	0.008	0.007

activity (Roshchina and Mukhin, 1984). Therefore, according to the classification followed for animal cholinesterases, the enzyme can be considered true cholinesterase or acetylcholinesterase (EC 3.1.1.7). It is supposed (Vackova et al., 1984) that in plants there are cholinesterases of both types and one of them may prevail.

3.1.3.5. Isolation and Purification of Cholinesterases

Plant cholinesterases have been isolated and purified and their characteristics have been analysed. Riov and Jaffe (1973b) first isolated acetylcholine-hydrolyzing protein from mush bean *Phaseolus aureus* by gel chromatography on Sephadex G-200. Unlike non-specific esterases, higher rates were peculiar for the enzyme ($K_m = 0.84 \times 10^{-4}$ M). Non-specific esterases have higher $K_m > 1$ M (Mansfield et al., 1978). The enzymatic activity was inhibited by neostigmine and physostigmine. Molecular mass of the enzyme is higher than 200 kDa (subunit 80 kDa), which distinguishes it from non-specific esterases. Gel chromatography for purification of plants cholinesterases was usually done on various Sephadexes, Sepharoses 6B and CL-6B (Ernst and Hartmann, 1980; Roshchina, 1986, 1988a, b), Toyopearl HW-55 (Roshchina, 1990a), affinity chromatography on MAS-Sepharose 2B (1-methyl-9(N-β-(ε-aminohexanoyl)-β-aminopropylaminol acridinium bromide ligand-binding Sepharose 2B) (Mansfield et al., 1978), or Sepharose 4B with immobilized ligand n-aminophenyltrimethyl ammonium iodide, known as a competitive inhibitor of cholinesterases (Solyakov et al., 1989).

The elution profiles of cholinesterases purified by different methods are shown in Figs. 39 and 40. Analysis of known data (Table 26) demonstrates that the highest rate of acetylcholine hydrolysis was characteristic for cholinesterase from roots of kidney bean *Phaseolus vulgaris*, near 3.7 kat kg^{-1} of protein, which is 2000-fold lower than is observed for analogous protein from electric ray. This enzyme was purified by affinity chromatography on Sepharose 2B with acridinium-based ligand (Mansfield et al., 1978).

By the use of biospecific affinity chromatography, Solyakov and coworkers (1989) made highly purified preparations of weakly bound (free) cholinesterase (specific activity 650 nmol acetylcholine per mg of protein per min.) and partly purified preparations of membrane-bound form of the same protein (specific activity 350 nmol of acetylcholine per mg of protein per min) from pea roots. It is significant that the rate of acetylcholine hydrolysis by pea root cholinesterase purified by affinity chormatography (Solaykov et al., 1989) was lower than was marked for the same enzyme from pea chloroplasts after purification by usual gel chromatography on Sepharose CL-6B or 6B, as well as on Toyopearl HW 55

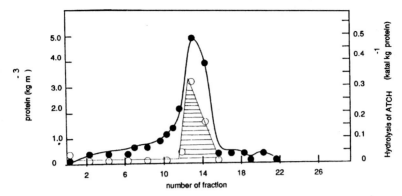

Fig. 39. The elution profile of the cholinesterase purification on Toyopearl HW-55 (after Roshchina, 1990a). ●-●, protein, o-o, rate of acetylthiocholine hydrolysis. The shaded fractions have the activity inhibited by neostigmine 2.5×10^{-5} M.

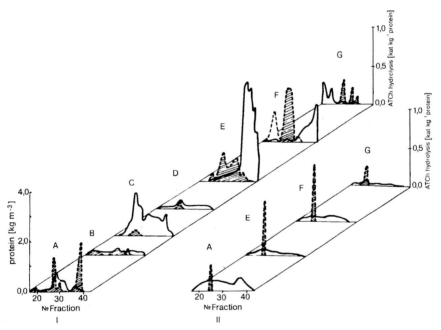

Fig. 40. The elution profile of the chloroplasts' cholinesterases from Sepharose CL-6B (Roshchina, 1988b). I, weakly bound (free) fractions, II, membrane-bound fractions. The shaded fractions have activity inhibited by neostigmine 10^{-5} to 2.5×10^{-5} M. A, pea *Pisum sativum*, B, etiolated pea chloroplasts; C, kidney bean *Phaseolus aureus;* D, maize *Zea mays*, E, common nettle *Urtica diocia*, F, false acacia *Robinia pseudoacacia;* G, cornbind, *Convolvulus arvensis.*

Table 26. Characteristics of cholinesterases isolated from plants

Family and species	Organ	Hydrolysis of cholinic esters	V_{max} kat $\times$ kg^{-1} of protein	Km $\times 10^{-4}$ (M)	Inhibition by substrate	$K^s_i \times 10^{-3}$ (μ)	Optimum pH	Optimum T, °C	Reference
Alliaceae *Allium altaicum*	Seeds	ATCh> BTCh	—	—	—	—	8.0	—	Hadacova et al., 1981
Chenopodiaceae *Spinacia oleracea* L.	Leaves	ATCh > BTCh	0.11	—	yes	—	—	—	Roshchina and Semenova, 1990
Convolvulaceae *Calystegia sepium* R.Br.	Leaves	ATCh >BTCh	0.23	5.6	yes	1.5	—	—	Roshchina, 1991a, c
Convolvulus arvensis L.	Leaves	ATCh >BTCh	0.09	4.0	yes	1.5	—	—	Roshchina, 1990a
Euphorbiaceae *Synadenium grantii*	Latex of laticifers	ATCh> PTCh> BTCh	1.7	0.36 to 0.47	yes	7–8	8.5	40	Govindappa et al., 1987
Fabaceae (Leguminosae) *Cicer arietinum* L.	Roots	ATCh> PTCh> BTCh	0.0058	1.5	no	8.0	8.0	30	Gupta and Maheshwari, 1980
Phaseolus aureus Roxb.	Leaves	ATCh> BTCh	0.07	1.2	yes	0.75	—	—	Roshchina, 1988b
	Roots	ATCh> PTCh> BTCh	0.0016	0.84	yes	5.3	8.5 to 8.7	—	Riov and Jaffe, 1973b
P. vulgaris	Roots	ATCh> PTCh> BTCh	3.7	0.56	yes	—	—	—	Mansfield et al., 1978
	Hypocotyls	ATCh> PTCh> BTCh	0.0021	4.6	yes	—	7.8 to 8.0	30 to 36	Ernst and Hartmann, 1980
Pisum sativum L.	Leaves	ATCh> BTCh	0.87	2.0 to 2.6	yes	1.0	7.8	30	Roshchina, 1986, 1988a, b
	Stems	ATCh> PTCh> BTCh	0.083	2.5	yes	—	8.5	37	Vackova et al., 1984
Pisum sativum L.	Roots	ATCh> PTCh> BTCh	0.00063	2.0	yes	—	9.0	30	Kasturi and Vasantharajan, 1976

(Contd.)

Table 26. (Contd.)

Family and species	Organ	Hydrolysis of cholinic esters	V_{max} kat $\times$ kg^{-1} of protein	Km $\times 10^{-4}$ (M)	Inhibition by substrate	$K^s_i \times 10^{-3}$ (μ)	Optimum pH	T, °C	Reference
Pisum sativum L.	Roots	ATCh > PTCh > BTCh	0.0058 to 0.011	0.74 to 1.2	yes	—	9.0	37 to 38	Solyakov et al., 1989
Robinia pseudoacacia L.	Leaves	ATCh > BTCh	0.8	4.0	yes	3.0	—	—	Roshchina, 1988b
Gramineae Avena sativa L.	Coleoptiles	ATCh ~ PTCh > BTCh	0.0021	2.8	no	—	7.2	36	Kesy et al., 1991
Zea mays L.	Leaves	ATCh > BTCh	0.14	7.0	yes	3.0	—	—	Roshchina, 1988b
	Roots	ATCh = PTCh > BTCh	—	—	no	—	8.0	—	Fluck and Jaffe, 1975
Solanaceae Solanum tuberosum L.	Roots	ATCh < PTCh > BTCh	—	—	yes	8.0	—	—	Fluck and Jaffe, 1975
Urticaceae Urtica dioica L.	Leaves and Chloroplasts	ATCh > BTCh	0.8	5.0 to 5.8	yes	1.0 to 1.8	—	—	Roshchina, 1988b
Ascomycetes Aspergillus niger	Mycelium	ATCh > BTCh	0.84	7.0	yes	—	—	—	Roshchina and Alexandrova, 1991

—: No data. ATCh, PTCh, and BTCh respectively are acetylthiocholine, propionylthiocholine, butyrylthiocholine.

(Roshchina, 1988a,b). The level of purification of isolated protein (20–700 times), determined by comparison between the rates of acetylcholine hydrolysis in original crude homogenate of tissue and purified enzyme, as well as the appearance of plural forms, depends on the procedure of purification and taxonomic features of the plant.

According to Vackova and coworkers (1984), there are weakly bound or free fractions (solubilized by buffer) and membrane-bound fractions (solubilized by buffer containing 4.5% of ammonium sulfate) of cholinesterases. The same authors found isoforms of cholinesterases in extracts from pea roots. By polyacrylamide gel electrophoresis (PAGE), isoforms of the enzymes were also separated from extracts of *Allium altaicum* seeds (Hadacova et al., 1981) and pea seedlings (Nicolova et al., 1986). Weakly bound (free) and membrane-bound forms of cholinesterase were isolated from intact chloroplasts of various plants by Roshchina (1988a, b). Plural forms were found in free fraction of acetylcholinesterase (Table 27 and Fig. 40) isolated from leaves of pea, common nettle, false acacia, and European glorybind. In membrane-bound fractions of the enzyme from the species studied plural forms were not observed (Table 27 and Fig. 40, II). In pea etioplasts there was only free acetylcholinesterase, which was also represented by isoforms. Weakly bound enzyme from mung bean and maize leaves had no plural forms, whereas membrane-bound forms of the same enzymes were absent.

The data dealing with characteristics of purified acetylcholine-hydrolyzing enzymes from 14 plant species belonging to 7 families are analyzed on the basis of Tables 26 and 27. Cholinesterases were isolated mainly from roots and leaves, but there is also information about enzymes purified from seeds, stems, and latex. This enzyme has also been isolated from intact chloroplasts of six species of higher plants: pea, mung bean, false acacia, maize, and european glorybind (Roshchina, 1988a, b) which also shows localization of cholinesterase in these organelles. The yield of protein for the objects mentioned was from 2 to 16 mg per kg of fresh leaf mass. Cholinesterase was successfully isolated not only from mature chloroplasts, but also from pea etioplasts (Roshchina, 1988a). In etiolated plastids the amount of cholinesterases was three times less than in normal green chloroplasts, whereas their activity was five-fold lesser. The development of photosynthetic process after the greening of etioplasts seems to be accompanied by a burst in enzyme synthesis.

Most of the isolated enzymes hydrolyze acetylcholine and acetylthiocholine at higher rates than other cholinic esters (Table 26). The Michaelis-Menten constants (K_m) characterizing the rate of the substrate hydrolysis and affinity of active center of plant cholinesterases to the substrate (Table 26) were of the same order as that of cholinesterases from nervous tissue, erythrocytes, and other tissues of vertebrates and usually equal to $0.67-5.0 \times 10^{-4}$ M (Brestkin et al., 1973). The excess of substrate inhibits protein activity, and the concentration curves of the rates of hydrolysis of cholinic esters are bell-shaped (Fig. 41). Therefore, plant cholinesterases are mainly true cholinesterases or acetylcholinesterases. Like the enzymes from animal tissues, they have optimum pH at 8.0–8.3 and optimum temperature at 37–40°C; however, in most plants these parameters are smaller, accordingly optimum pH 7.8–8.0 and optimum temperature near 30°C, which are similar to those of animal

Table 27. Molecular masses of the cholinesterases isolated from plants

Family and species	Organ or organelle	Form of protein	Molecular mass (kDa) gel filtration	Molecular mass (kDa) electrophoresis (Na-DDS)	Reference
Chenopodiaceae					Roshchina and Semenova, 1990
Spinacia oleracea	Leaves	F + B	—	63.1	
Convolvulaceae					Roshchina, 1991a
Calystegia sepium R.Br.	Leaves	F + B	—	60.3	
Convolvulus arvensis L.	Leaves	F	600	81.3	Roshchina, 1991a
Convolvulaceae *Convolvulus arvensis* L.	Chloroplasts	F_1 F_2 F_3	2512 1259 600	45.0 45.0 45.0	Roshchina, 1991a
Euphorbiaceae					
Synadenium grantii	Latex of laticifers	F + B	70	35	Govindappa et al., 1987
Fabaceae (Leguminosae)					Gupta and Maheshwari, 1980
Cicer arietinum L.	Roots	F + B	—	—	
Phaseolus aureus Roxb.	Leaves	F + B	1200	63.1	Roshchina, 1988b
	Chloroplasts	F	2600	63.1	Roshchina, 1988b
	Roots	F + B	200	80.0	Riov and Jaffe, 1973b
P. vulgaris	Roots	F + B	73	72 – 73 61.0	Mansfield et al., 1978
	Hypocotyls	F + B	65	65	Ernst and Hartmann, 1980
Pisum sativum L.	Leaves	F + B	600	63.1	Roshchina 1988a, b
	Chloroplasts	F_1 F_2 F_3	2340 1312 698	63.1 63.1 63.1	Roshchina, 1986
	Chloroplasts	B	902	63.1	Roshchina, 1986
	Roots	F + B	200	—	Kasturi and Vasantharajan, 1976
	Roots	F	330	—	Solyakov et al., 1989

(Contd.)

Table 27. (Contd.)

Family and species	Organ or organelle	Form of protein	Molecular mass (kDa) gel filtration	Molecular mass (kDa) electrophoresis (Na-DDS)	Reference
Robinia pseudoacacia L.	Chloroplasts	F B	2600 900	63.1 63.1	Roshchina, 1988b
Gramineae					
Zea mays L.	Leaves	F + B	900	63.1	Roshchina, 1988b
	Chloroplasts	F B	2512 0	63.1 0	
Urticaceae					
Urtica dioica L.	Leaves	F+B	> 200	69.0 60.0 45.0 23.4	Roshchina 1988a, b
	Chloroplasts	F_1 F_2 F_3 B	2600 1300 680 902	63.1 31.6 63.1 63.1	
Ascomycetes					
Aspergillus niger	Mycelium	F B	600	63.1 44.0	Roshchina and Alexandrova, 1991

F and B, relatively free and bound forms of proteins; 0, lacking of the form; Na-DDS, sodium dodecyl sulfate.

pseudocholinesterases. However, it is supposed that in seeds of *Allium altaicum* there are not only acetylcholinesterases, but also pseudocholinesterases (Hadacova et al., 1983). For certain plants it is rather difficult to differentiate the cholinesterase type. Enzymes specific to acetylcholine and isolated from roots of *Cicer arietinum* and *Phaseolus vulgaris* are not inhibited by high concentrations of a substrate (Gupta and Maheshwari, 1980; Ernst and Hartmann, 1980) like butyrylcholinesterase. Enzymes from roots of *Solanum tuberosum* hydrolyze butyrylcholine at higher rates than acetylcholine (Fluck and Jaffe, 1975), which does not allow us to attribute them to a certain type of cholinesterase. Like cholinesterases from some invertebrates possessing the same feature (Chadwick, 1963; Silver, 1974), they appear to be in an intermediate position between cholinesterases of both types. The existence of similar "intermediate" enzymes is possibly a reflection of the phylogenetic evolution of cholinesterases (Gupta and Maheshwari, 1980).

The comparison of molecular masses (M_r) determined by gel filtration method (Table 27) showing the proteins isolated from plants are aggregates of high molecular weight with very large molecular masses from 330 to 2600 kDa. The arising of isoforms and the formation of aggregates of high molecular weight is peculiar to cholinesterases that are solubilized

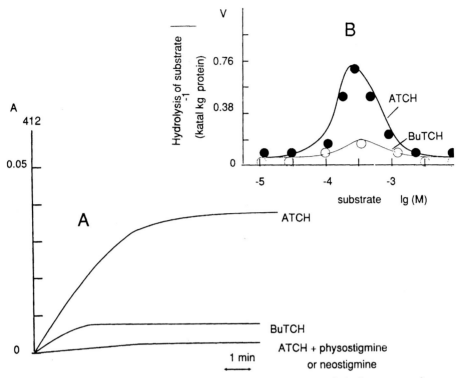

Fig. 41. The hydrolysis of cholinic esters by preparations of cholinesterase from pea leaves (Roshchina, 1988b)

in buffers with low ion strength. The increase in ion strength of the elution buffer leads to the break-up of the aggregates into smaller ones and disappearance of plural forms of the enzymes. The proteins isolated from plants are very similar to plural forms of analogous proteins from animals such as globular acetylcholinesterases from various objects with molecular masses 331–341 kDa (G_4 tetramers) and asymmetric acetylcholinesterases with M_r 1150–1062 kDa and 976–747 kDa (A_{12} and A_8 aggregates), which are connected with collagen (Brzin et al., 1983). Thus, animal acetylcholinesterases are also aggregates high molecular weight (Massoulie, 1986, 1992; Massoulie and Bon, 1982; Massoulie et al., 1992, 1993). Among them are aggregates extracted by detergents, and they are usually linked with collagen. Monomers of plant cholinesterases appear to be forms with molecular masses 60–80 kDa (Table 27), which are seen after 7.5% PAGE in the presence of 1% sodium dodecyl sulfate. Examples of electrophoregrams of the cholinesterases from chloroplasts of various plant species are shown in Figs. 42 and 43. Identical subunits are observed in all plural forms of the enzyme, which demonstrates that they belong to one and the same protein. Molecular masses of separated

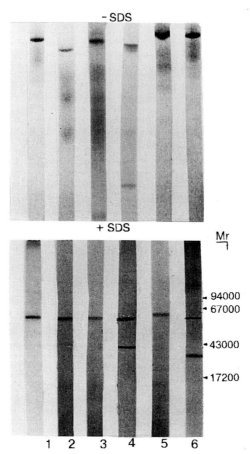

Fig. 42. Electrophoresis (7.5% PAGE) of cholinesterases isolated from chloroplasts and purified on Sepharose CL-6B (Roshchina, 1991a, c). 1, kidney bean *Phaseolus aureus*; 2, false acacia *Robinia pseudoacacia*; 3, pea *Pisum sativum*; 4, cornbind *Convolvulus arvensis*; 5, maize *Zea mays*; 6, common nettle *Urtica dioica*. –SDS and +SDS, without and with 1% sodium dodecyl sulfate.

multiple subunits of animal cholinesterases vary from 60 to 80 kDa. Based on the data of the electrophoretic motility for animal cholinesterases, coefficient of their sedimentation and immunological reactivity, it has been shown that each of both types of cholinesterases—acetylcholinesterase and butyrylcholinesterase—is represented by plural forms (Toutant et al., 1985; Massoulie, 1986; Massoulie et al., 1992). Cholinesterases from chloroplasts of the plants studied, except european glorybind, have subunits with molecular masses 31–63 kDa, which are similar to those of

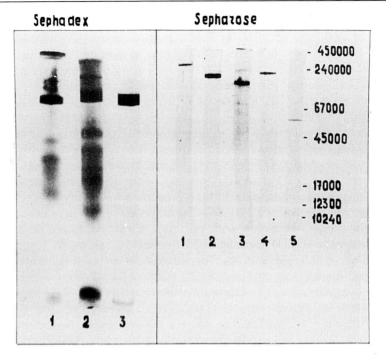

Fig. 43. Electrophoresis (7.5% PAGE without SDS) of cholinesterases isolated from pea leaves (Roshchina, 1988a). Left, crude extract of pea leaves (1) and purified on Sephadex G-25 (2) and then on Sephadex G-200 (3). Right, cholinesterases purified on Sepharose CL-6B. 1, 2, 3, free forms of cholinesterases; 4, bound form of cholinesterases; 5, with SDS.

cholinesterases isolated from kidney bean, 65 kDa (Ernst and Hartmann, 1980), and cholinesterase from rabbit microsomes, 65–70 kDa (Munoz-Delgado and Vidal, 1987). At 7.5% PAGE with the presence of sodium dodecyl sulfate (SDS), high-molecular-weight forms of plant cholinesterases from leaves sometimes are derived on subunits with much lower (multiple 15 kDa) molecular masses from 63.1 to 31.62 kDa (Roshchina, 1988b). Molecular masses of the cholinesterase subunits 35–70 kDa are characteristic for the protein of latex from laticifers of plant *Synadenium grantii* (Govindappa et al., 1987). The proteins with molecular mass 63.1 kDa appear to belong to monomers. Based on the study of animal cholinesterases, Small and Simpson (1988) supposed that smaller peptides (< 60 kDa) are products of the autolysis of the enzyme itself, if it is purified to homogenous preparation.

Acetylcholinesterase can act in processing of neuropeptides, as trypsin-like enzyme, and has carboxypeptidase activity (Small, 1988, 1989a, b). Usually animal acetylcholinesterases have molecular mass of subunits near 80 kDa, which form about six aggregates with molecular masses 150, 320, 500, and 650 kDa (150 kDa is most active) (Wenthold et al., 1974). Half-life of this acetylcholinesterase in mature rat brain is 2.84 days. The rate of hydrolysis is 4500 mmoles acetylthiocholine min^{-1} mg^{-1} protein. Unlike most plant and animal cholinesterases, molecular masses of other plant esterases such as from fruits of *Cucurbita pepo* are about 18–36 kDa (Nourse et al., 1989).

Amino acid analysis of free and membrane-bound forms of enzyme isolated from pea chloroplasts (Roshchina, 1986, 1988a) has been done by Dr. O.E. Trubetskaya (Table 28). High content of residues of asparagine, threonine, serine, glutamine, proline, and glycine is peculiar to cholinesterase of chloroplasts. If this is compared with acetylcholinesterase isolated from electric ray *Torpedo californica* (Ott, 1985), there is a similarity in the number of amino acids per mole of protein, based on average molecular mass 60 kDa (asparagine 59, threonine 22, serine 45.5, glutamine 58.5, proline 38.5, glycine 54). The differences are observed mainly in minor residues.

The enzyme from pea chloroplasts interacted with other plastid proteins, such as cytochrome f, plastocyanin, ferredoxin, and ribulose bisphosphate carboxylase, in model experiments (Roshchina and Karpilova, 1993). Its activity is stimulated by low concentrations of

Table 28. Amino acid composition of cholinesterase purified 200 times from pea chloroplasts

Amino acid	Amount of residues/mole of protein	Mole/100 moles
Asparagine	57	10.0
Threonine	37	6.6
Serine	50	9.0
Glutamine	68	12.0
Proline	32	5.6
Glycine	55	9.8
Alanine	46	8.2
Valine	33	5.8
Methionine	6–7	1.1
Isoleucine	20–21	3.8
Leucine	44–45	7.9
Tyrosine	14	2.6
Phenylalanine	20–21	3.6
Histidine	9–10	1.7
Lysine	35–36	6.3
Arginine	30–31	5.3

*Analysis has done by Dr. O.E. Trubetskaya from the Shemyakin Institute of Bioorganic Chemistry of Russian Academy of Sciences.

ribulose bisphosphate carboxylase and inhibited by higher ones (Table 29), whereas similar effects of electron carriers were observed only for Cu^{2+}-containing protein, plastocyanin from chloroplasts of Pisum sativum. As for cytochrome $f(C_{553})$ from Chlorella cells and ferredoxin from Pisum sativum, as well as bovine serum albumin, the proteins had no marked effects.

Table 29. Effects of water-soluble proteins of chloroplasts on pea acetylcholinesterase (AChE) activity, estimated as the hydrolysis of acetylthiocholine (AThCh). SE < 2 % (after Roshchina and Karpilova, 1993).

Added protein	Added protein/AChE ratio, kg/kg	AChE activity as AThCh hydrolysis, mmol s^{-1} kg^{-1} (protein)
No	—	43
Ribulose bisphosphate carboxylase	0.24	40
	0.50	42
	1.00	52
Plastocyanin	0.24	43
	0.50	60
	1.00	142
Cytochrome $f(C_{553})$	0.24	46
	0.50	58
	1.00	43
Ferredoxin	0.24	44
	0.50	40
	1.00	49
Bovine serum albumin	0.24	38
	0.50	38
	1.00	48

Purified acetylcholinesterase from latex Synadenium grantii (Govindappa et al., 1987) is glycoprotein and includes 14 neutral sugars. Among them are (as percentage of total number of neutral sugars) arabinose 49.3–67.2, xylose 27–34.3, galactose 4.2–7.1, and glucose 1.7–9.3.

The Stokes radius of cholinesterases isolated from plants is 3.6–4.0 nm (Mansfield et al., 1978; Govindappa et al., 1987), and the energy of activation is 9.82×10^3 J $mole^{-1}$ (Kasturi and Vasantharajan, 1976). Cholinesterases are mainly acidic proteins with isoelectric point 5.0–5.4 (Mansfield et al., 1978; Govindappa et al., 1987).

The substrate specificity of plant cholinesterases is observed for acetylcholine and acetylthiocholine. The rate of hydrolysis of other cholinic and non-cholinic esters by the enzymes may be expressed in percentages of variant with acetylcholine based on some papers (Riov and Jaffe, 1973b;

Kasturi and Vasantharajan, 1976; Vackova et al., 1984; Roshchina, 1988a, b), as follows:

propionylthiocholine	41%
butyrylthiocholine	0–38%
indoxyl acetate	67%
indophenyl acetate	52.3–62%
a-naphthyl acetate	0–17%
adenosine triphosphate	0.13%

Cholinesterases are also known to hydrolyze sinapine choline (Kutacek et al., 1981).

When animal and plant cholinesterases are stored at 20 – 37°C for 6 hours or at 4°C for one week, the enzymes can be decomposed on smaller peptides, which appears to demonstrate their capacity for autoproteolysis (Small and Simpson, 1988; Roshchina, 1988b). The peptidase activity of cholinesterases in relation to foreign proteins is found as well (Small et al., 1986). The character of cholinesterase interaction with its substrates cholinic esters is clear for the most part (Massoulie and Bon, 1982) and is shown in the following diagram:

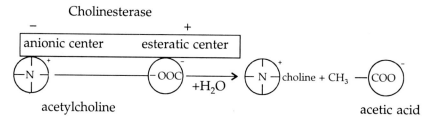

The substrate is usually linked with two parts of cholinesterase. Electrophilic –CO–O–group of cholinic esters connects via hydrogen bonds to the esteratic center of the enzyme, which itself plays a hydrolyzing function. At the same time, $-\overset{|}{\underset{|}{N}}{}^{+}-$ group of acetylcholine (cationic head) combines with the anionic (negatively charged) center of cholinesterase, which serves only for exact and true orientation of the substrate.

The concept described above helps to elucidate the high sensitivity of cholinesterases to quaternary nitrogen-containing and organophosphate compounds. Quaternary ammonium substances, in particular low concentrations (10^{-5}–10^{-6} M) of salts of carbaminic acid, neostigmine and physostigmine, depress the activity of the enzymes, hydrolyzing cholinic esters. Amino groups of the inhibitors are linked with anionic center of cholinesterase, but not with esteratic center. The inhibition has a competitive character and may be reversible. Organophosphorus compounds are binding with serine of the esteratic center of cholinesterase via $-O-\overset{\overset{O}{\|}}{\underset{\underset{O}{|}}{P}}-O-$ group (Silver, 1974). These compounds act as competitive inhibitors and in most cases their effects are irreversible.

3.1.3.6. Regulation of Cholinesterase Activity

Like animal cholinesterases, plant cholinesterases are sensitive to derivatives of carbaminic acid—neostigmine and physostigmine and organophosphorus compounds—diisopropyl phosphofluoridate, dichlorvos, phosphon, and quaternary ammonium compounds (see formulae in Fig. 44).

Table 30 shows concentrations that inhibit plant cholinesterases by 50% (I_{50}) and the rate constant of inhibition of the enzymes by these inhibitors (k_1 or k_2). Lowest value of I_{50} and highest constant of inhibition by neostigmine is observed for protein from latex of laticifers of *Synadenium grantii*, which shows a high affinity to this inhibitor. The values for other plant cholinesterases were accordingly 1–2 and 6 orders lower. I_{50} for the neostigmine inhibition of plant cholinesterases and of cholinesterases from blood erythrocytes (0.67×10^{-6} M) is similar and comparable, but the I_{50} and k_2 for other cholinesterses of mammals are approximately 10^{-7}–10^{-8} M and 17×10^4–10^5 moles^{-1} s^{-1}, respectively (Mikhelson and Zeimal, 1973). As a whole, the sensitivity of plant cholinesterases to neostigmine is lower than that of animal tissues. The affinity of plant cholinesterases to physostigmine was 10-100 times lower than to neostigmine, and sometimes less, whereas it was higher for acetylcholinesterases from erythrocytes ($I_{50} = 1.40 \times 10^{-6}$ M). Among plant cholinesterases highest sensitivity to carbaminic inhibitors is marked for cholinesterase from latex of *Synadenium grantii*. It is possible that this enzyme of laticifers, sensitive to poisons, plays a defensive role. There are also cases when plant cholinesterases can be inhibited by neostigmine, but not by physostigmine, in particular enzyme from cotyledons of *Phaseolus vulgaris* (Lees and Thompson, 1975).

In Table 31 I_{50} of the pollen cholinesterase inhibition are represented. Most sensitive to carbamate inhibitors are pollen of *Hemerocallis fulva*, *Tulipa sp.*, *Salix caprea*, and *Allium cepa*. No inhibition was found for acetylthiocholine hydrolysis in pollen extracts from *Plantago major*, *Malus domestica*, *Populus balsamifera*, and *Betula verrucosa*, which are, except *Malus*, wind-pollinated species. There are differences between pollen sensitivity to physostigmine and to neostigmine for certain species;

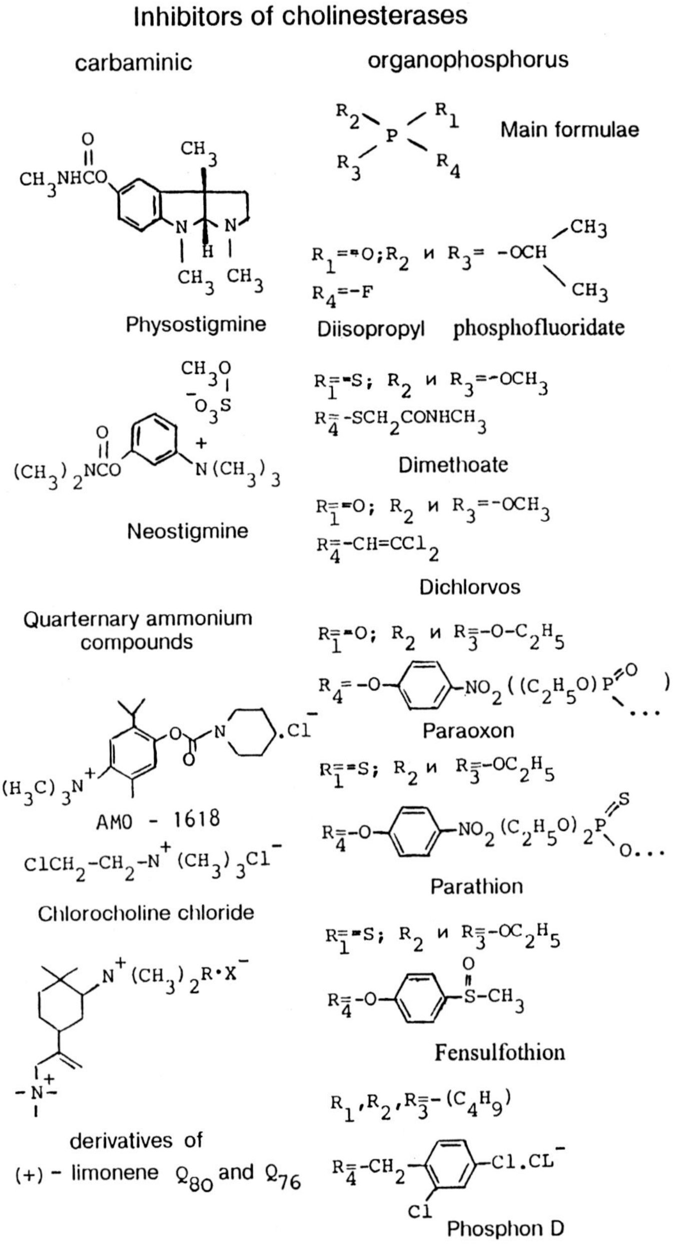

Inhibitors of cholinesterases

carbaminic

Physostigmine

Neostigmine

Quarternary ammonium compounds

AMO - 1618

$ClCH_2-CH_2-N^+(CH_3)_3 Cl^-$
Chlorocholine chloride

derivatives of
(+) - limonene Q_{80} and Q_{76}

organophosphorus

Main formulae

$R_1=O; R_2$ и $R_3= -OCH(CH_3)_2$
$R_4=-F$
Diisopropyl phosphofluoridate

$R_1=S; R_2$ и $R_3=-OCH_3$
$R_4=-SCH_2CONHCH_3$
Dimethoate

$R_1=O; R_2$ и $R_3=-OCH_3$
$R_4=-CH=CCl_2$
Dichlorvos

$R_1=O; R_2$ и $R_3=-O-C_2H_5$
$R_4=-O-C_6H_4-NO_2$ $((C_2H_5O)_2P(O)...)$
Paraoxon

$R_1=S; R_2$ и $R_3=-OC_2H_5$
$R_4=-O-C_6H_4-NO_2$ $(C_2H_5O)_2P(S)...$
Parathion

$R_1=S; R_2$ и $R_3=-OC_2H_5$
$R_4=-O-C_6H_4-S(O)-CH_3$
Fensulfothion

$R_1, R_2, R_3 = -(C_4H_9)$
$R_4=-CH_2-C_6H_3(Cl)-Cl \cdot CL^-$
Phosphon D

Fig. 44. Formulae of the cholinesterase inhibitors

Table 30. Inhibition of plant cholinesterase by derivatives of carbaminic acid

Plant from which cholinesterase was isolated	Organ	Neostigmine I_{50} or $K_i \times 10^{-6}$ (M)	Neostigmine K_1 or K_2 (moles^{-1} s^{-1})	Physostigmine I_{50} or $K_i \times 10^{-6}$ (M)	Physostigmine K_1 or K_2 (moles^{-1} s^{-1})	Reference
Avena sativa	Coleoptiles of etiolated seedlings	2.5	—	280	—	Kesy et al., 1991
Ciçer arietinum	Roots	3.0	0	6000	—	Gupta and Maheshwari, 1980
Phaseolus aureus	Roots	0.5	—	6000	—	Riov and Jaffe, 1973b
Phaseolus vulgaris	Roots	5.0	—	100	—	Ernst and Hartmann, 1980
Pisum sativum	Roots	0.6	—	900	—	Kasturi and Vasantharajan, 1976
Pisum sativum	Roots	2.0	—	250	—	Vackova et al., 1984
Pisum sativum	Roots	9.6*	1.7	3100*	2.5	Solyakov et al., 1989
Pisum sativum	Chloroplasts	1.12*	17.0	—	—	Roshchina, 1991a, b
Synadenium grantii	Latex of laticifers	0.012–0.014	600,000–880,000	1.35–1.50	7700–8500	Govindappa et al., 1987

I_{50}, concentration of the inhibitor decreasing the hydrolysis of the substrate by 50%; K^*_i, constant of inhibition; K_1 and K_2, constants of rates of inhibition in reaction of the hydrolysis of acetylthiocholine.

for instance, sensitivity is lesser to the first inhibitor in *Hippeastrum* and *Petunia* pollen. On comparison of I_{50} for carbamate inhibitors between cholinesterases from plant and animal tissue (Table 31), pollen cholinesterases seem to be most sensitive among the plant enzymes, except of latex of *Synadenium grantii*, especially to neostigmine. Sensitivity to physostigmine is lower in plant tissues and abdominal ganglion (Invertebrata) than sensitivity to neostigmine, unlike for Vertebrata and some Invertebrata, such as *Musca domestica*.

Organophosphorus compounds bind with any serine-containing proteins, including cholinesterases (Tables 32 and 33). For instance, (^{3}H) diisopropyl phosphofluoridate labeled not only subunits of cholinesterases with molecular masses 61, 23, and 17 kDa from mush bean *Phaseolus aureus* (Mansfield et al., 1978), but also serine protease with molecular

Table 31. I_{50} for pollen ChE inhibition by physostigmine and neostigmine. Mean values for three to five experiments are presented. Ellman reagent or its red analog (*) was used for ChE assay (Roshchina et al., 1994).

Plant species	I_{50}, μM	
	Physostigmine	Neostigmine
Alliaceae *Allium cepa*	1.05	1.25
Amaryllidaceae *Hippeastrum hybridum*	100.0	10.0
Asteraceae *Matricaria chamomilla*	1.6	5.0
	2.8*	2.2*
Betulaceae *Betula verrucosa*	No inhibition	No inhibition
Hydrangeaceae *Philadelphus grandiflorus*	1.6	1.6
Hippocastanaceae *Aesculus hippocastanum*	6.3	4.2
Liliaceae *Hemerocallis fulva*	1.03	1.02
Tulipa sp.	1.0	1.0
Papaveraceae *Papaver orientale*	2.0	2.0
Pinaceae *Pinus sylvestris*	6.3	4.0
Plantaginaceae *Plantago major*	No inhibition	No inhibition
Rosaceae *Malus domestica*	No inhibition	No inhibition
Salicaceae *Polulus balsamifera*	No inhibition	No inhibition
Salix caprea	1.0	1.3
Solanaceae *Petunia hybrida*	40.0	6.3

Other plant and animal tissues (from review of Roshchina and Semenova, 1990)

Plants	Roots	*Cicer arietinum*	6000	3.0
		Phaseolus aureus	420	0.6
		Phaseolus vulgaris	100	5.0
		Pisum sativum	25	2.0
	Chloroplasts	*Pisum sativum*		1.12
	Latex	*Synadenium grantii*	1.35–1.5	0.012–0.014

Animals, Vertebrata

Acetylcholinesterase, mammalian erythrocytes	0.03 – 1.60	0.016 – 0.5
Butyrylcholinesterase, mammalian serum	0.01 – 0.2	0.050 – 1.0
Animals, Invertebrata		
Brain tissue, *Musca domestica*	0.02	
Abdominal ganglion, *Aplysia californica*	1800	

mass 38 kDa from thylakoids of spinach *Spinacia oleracea* (Kawasaki and Takeuchi, 1989). Organophosphates demonstrate lower specificity to cholinesterases than carbaminic inhibitors. However, some organophosphorus compounds are often used for the study of active center of cholinesterases. Roshchina and Semenova (1990) have shown that the substances depress cholinesterase activity in leaves and chloroplasts isolated from leaves (Table 32). This indicates the presence

Table 32. Constants of the inhibition rate (K_2) of the cholinesterase from pea chloroplasts by organophosphorus compounds (Roshchina and Semenova, 1990).

Organophosphorus compound	Structural formula	K_2 (M^{-1} min^{-1}) $P < 0.95$
GD-42	C_2H_5O \ P / O ; CH_3 / P \ $S\text{-}CH_2CH_2S^+C_2H_5 \cdot CH_3SO_4$ \ CH	$(5.58 \pm 1.75) \times 10^4$ $n = 15$
GD-7	C_2H_5O \ P / O ; CH_3 / P \ $S\text{-}CH_2CH_2SC_2H_5$	$(1.32 \pm 0.21) \times 10^3$ $n = 14$
isoOMPA	$(iC_3H_7NH)_2 P(=O)\text{-}O\text{-}P(=O)(iC_3H_7NH)_2$	$(8.01 \pm 1.90) \times 10^1$ $n = 7$

GD-42, methylsulfomethylate 0-ethyl-S(β-ethylmercaptoethyl)methylthiophosphonate; GD-7, 0-ethyl-S(β-ethylmercaptoethyl)methylthiophosphonate, non-charged analog of GD-42; isoOMPA, tetraisopropylpyrophosphamide.

of serine in active center that is characteristic for cholinesterases. Specific inhibition of pseudocholinesterase inhibitor iso-OMPA acted in high concentrations $>10^{-3}$ M, whereas inhibitors GD-42 and GD-7 acted in 100-fold lower concentrations. Pea cholinesterase obviously does not belong to pseudocholinesterases. Mikhelson and Zeimal (1973) showed that the values of constants of rates of inhibition by GD-42 known for brain cholinesterases and pseudocholinesterases are respectively 2.5×10^8 and 3.7×10^6, whereas the same constants for GD-7 are respectively 6.0×10^4 and 6.8×10^3. The affinity of enzyme to an inhibitor is greater when this constant is higher. It is interesting that k_2 values for animal cholinesterases inhibited by positively charged GD-42 are 100–1000 times higher than in analogous enzyme of chloroplasts, while values of k_2 for neutral GD-7 are of comparable order. The sensitivity of the anionic part of active center of cholinesterase to organophosphorus compounds is usually estimated from the ratio (k_2 for GD-42)/(k_2 for GD-7), which is 42.3 for cholinesterase from pea chloroplasts, 9694–10,100 for acetylcholinesterase from bovine erythrocytes, 530 for butyrylcholinesterase of horse blood serum, and 13–14 for starfish (Roshchina and Semenova, 1990). Thus, plant cholinesterases have lower affinity of anionic center to organophosphorus compounds than analogous enzymes of mammals, but the affinity is similar to that known for invertebrate cholinesterases, in particular of starfishes.

The values of I_{50} and k_2 for the inhibition of cholinesterases purified from plants by organophosphates are shown in Table 33. As a whole, the enzymes are more sensitive to higher concentrations of organophosphorus compounds than to derivatives of carbaminic acid (I_{50} = 10^{-4}–10^{-3} M). Cholinesterase from latex of *Synadenium grantii* has greater sensitivity to organophosphates than analogous enzymes from other plant objects. Animal cholinesterase, in particular from erythrocytes, have k_i = 1.93×10^{-3} M and k_2 ~ 0.080 moles^{-1} s^{-1}, which were of the same order as the values for plant cholinesterases. It is interesting to note (Semenova and Roshchina, 1993) that cholinesterase of anther of *Hippeastrum* is sensitive only to organophosphate GD-42, but not to GD-7 or iso-OMPA.

All these carbamates and organophosphate compounds are also strong pesticides, mainly insecticides. The action of many pesticides is based on the depression of pest cholinesterases in plant cultivars. The possible inhibition of plant cholinesterases as well is usually not taken into

Table 33. Inhibition of plant cholinesterases by organophosphorus (OP) compounds and retardants

OP or retardant	$I_{50} \times 10^{-3}$ (M)	$K_2 \times 10^3$ (M^{-1} s^{-1})	Reference
	Roots of *Cicer arietium*		
Phosfon-D	1.7	—	Gupta and Maheshwari, 1980
	Roots of *Phaseolus aureus*		
Phosfon-D	0.033	—	Riov and Jaffe, 1973a
AMO-1618	0.21	—	
Q_{80}	0.23	—	
Q_{76}	3.5	—	
CCC	90.0	—	
	Roots of *Pisum sativum*		
Dimethoate	1.0	—	Kasturi and Vasantharajan, 1976
Parathion	0.04	—	
Fensulfothion	0.50	—	
	Latex of *Synadenium grantii*		
Dichlorvos	0.08–0.10	1.1–1.4	Govindappa et al., 1987
Paraoxon	0.12–0.14	0.85–0.97	
AMO-1618	0.12–0.14	0.85–0.97	

AMO-1618, retardant, 2-isopropyl-4-dimethylamino-5-methylphenyl-1-piperidine carboxylate methyl chloride. CCC, 2-chloroethyl-trimethylammonium chloride. Dimethoate, 0,0-dimethyl-S-(N-methylcarbamidomethyl)dithiophosphate. Dichlorvos, 0,0-dimethyl-0-(2,2-dichlorovinyl)phosphate.
Fensulfothion, [0,0-diethyl(p-methylsulfinylphenyl)phosphorothioate]. Paraoxon, 0,0-diethyl-0-4-nitrophenyl ester of phosphoric acid. Parathion, 0,0-(n-nitrophenyl) thiophosphate. Phosfon-D, tributyl-2,4-dichlorobenzylphos-phonium chloride. Q_{80}-1-p-mentanol, 2-dimethylamino-2,5-dimethylbenzyl chloride. Q_{76}-1-p-mentanol, 2-dimethylamino-4-bromobenzyl bromide.

account, although it could be one of the main causes of decline in plant growth. Some retardants, either derivatives of carbaminic acid or quaternary amines AMO-1618, chlorine choline chloride, or derivatives of limonene Q_{80} and Q_{76} (Table 33), act on plant cholinesterase by a similar mode (Riov and Jaffe, 1973a, b, c). Greatest plant sensitivity to AMO-1618 and Q_{80} was observed, whereas other retardants were effective in concentrations 10–100 times higher. Therefore, assessment of the effect of new pesticides or growth regulators on plant cholinesterase activity may be important for agriculture and environment monitoring, because it is possible to manufacture of highly specific and non-toxic (or weakly toxic) compounds (Zukovskii and Evstigneeva, 1983; Solyakov et al., 1989). One option is to use compounds synthesized by the plant and having anticholinesterase features, such as physostigmine (eserine) (Fluck and Jaffe, 1976), permethrin (Bandyopadhyay, 1982), capsaicin, cicutotoxin, (Roshchina, 1988a; Roshchina and Mukhin, 1989), and juglone (Sastry and Sadavongvivad, 1979).

Light and physiologically active compounds of plants can regulate the level of the inclusion of radioactively labeled amino acids to pea root cholinesterase (Kasturi, 1979). In the dark the rate of the enzyme synthesis is lower than in the light. Light quality is also an important factor. Under far red radiation the rate of cholinesterase synthesis is higher than under red light. However, catalytic activity of the enzyme does not depend on light conditions. Synthesis of cholinesterase is stimulated by indoleacetic acid, but in higher concentrations it is inhibited by acetylcholine, gibberellic acid, and kinetin.

The cholinesterase activity according to temperature has also been analyzed (Table 26). The optimums of temperature for most plant cholinesterases are 28–30°C, except cholinesterase from *Synadenium grantii*, which is 40°C; this could be related to the tropical origin of this species. Optimums of pH (Table 26) for most plant cholinesterases are 7.8 – 8.0, but for the enzyme from pea roots it is near 9.0, which is characteristic of neither animal nor plant proteins.

Ions of univalent and bivalent metals also influence cholinesterase. According to Vackova and coworkers (1984), the enzyme activity (as percentage of control) was stimulated (+) or inhibited (–) by those ions as follows:

Ions	Cholinesterase activity	Ions	Cholinesterase activity
Li^+	+3	Sr^{2+}	+8
Na^+	+55	Mn^{2+}	–28
K^+	+57	Co^{2+}	–10
Rb^+	+1	Cu^{2+}	–48
Be^{2+}	0	Cd^{2+}	–34
Mg^{2+}	+27	Al^{3+}	–13
Ca^{2+}	0		

These data show that Na^+, K^+, and Mg^{2+} stimulate cholinesterase activity in pea shoots by 30–60%, whereas heavy metals Cu^{2+} and Cd^{2+} in the same degree depress this process. Ca^{2+} and Be^{2+} have no action. However, high concentrations (1–10 mM) of Ca^{2+} and Mn^{2+} inhibit cholinesterase from kidney bean roots (Ernst and Hartmann, 1980), but Mg^{2+} does not. Cation choline$^+$ (> 1 mM) stimulates the enzyme activity of mung bean and kidney bean (Riov and Jaffe, 1973b; Ernst and Hartmann, 1980).

Unlike sinapine esterase, cholinesterase has some special features (Tzagolov, 1963a, b). The rate of sinapine hydrolysis is 0.14–0.20 mg sinapine/mg of protein per hour. pH optimum of sinapine esterase is 9–12, higher than for cholinesterase. Sinapine esterase hydrolyzes acetylcholine with a rate 320 μl CO_2/30 min, benzoyl choline-39, phenylacetate 342. pH optimum for acetylcholine is 8, while for sinapine it is 10. Hydrolysis of sinapine was inhibited by physostigmine by 20%, but the hydrolysis of acetylcholine was inhibited by only 13.3%. The most effective substrate is sinapine; the lack of one methoxy group acid decreases activity already to approximately 50% compared with sinapine. The polar caffeoylcholine was the least suitable substrate: K_m is almost two times as high and V_{max} approximately 30% of the values for sinapine (Strack et al., 1980). Hydrolysis of sinapine occurs with participation of sinapine esterase in cotyledons of *Raphanus sativus* (Strack et al., 1980). This enzyme differs from cholinesterase (Tzagoloff, 1963b; Nurmann and Strack, 1979).

The activity of plant cholinesterases potentially could be regulated by the plant's own alkaloids and other plant metabolites. For instance, animal acetylcholinesterase is inhibited by steroidal glycoalkaloids, their aglycons (Roddick, 1989), potato glycoalkaloids (Wierenga and Hollingworth, 1992), dehydroevodiamene constituent of *Evodia rutaecarpa* (Park et al., 1996), and phenylacetic acid (Nicolova et al., 1986).

3.1.3.7. Effects of Pesticides and Growth Regulators with Anticholinesterase Activity on Plant Reactions

Organophosphorus pesticides work by inactivating the cholinesterases of pests of cultivated plants. In a review by Roshchina and Semenova (1990) the inhibiting effects of a number of pesticides on cholinesterase activity of animals and plants were compared, and little difference was found in their mechanisms of action. In biological media on oxidation of P=S to P=O occurs, and substances that themselves are the cholinesterase inhibitors are formed. A true estimate of the strength of the anticholinesterase effect of organophosphorus compounds is not provided by the constant k_2, whereas I_{50} in this case is a rather indefinite characteristic, since for irreversible-type inhibitors the value of the inhibiting effect depends on the time of interaction of the enzyme with the inhibitor.

There is a noteworthy correlation between pronounced anticholinesterase activity of certain plant growth retardants and their ability to decrease the rate of plant growth (Riov and Jaffe, 1973b). The most powerful of the investigated retardants, phosphon D (tributyl-2,4-dichlorobenzyl-phosphonium chloride), belongs to the group of quaternary phosphonium compounds, which are reversible inhibitors of acetylcholinesterase and butyrylcholinesterase of vertebrates. The insecticide fensulfothion also depressed pea root growth (Kasturi and Vasantharajan, 1976). Widely used in agriculture, growth regulator chlorocholine chloride ((2-chloroethyl)trimethylammonium chloride) inhibited the elongation of plant stem cells, although it is not related to organophosphorus compounds (see review of Roshchina and Semenova, 1990). Organophosphorus pesticides acted on photosynthesis and chlorophyll *a* synthesis of *Chlamydomonas reinhardtii* (Wong and Chang, 1988) and photosynthetic electron transport in higher plants (Roshchina, 1991a). Some pesticides, such as methylparathion and phosphamidon, inhibit transpiration in tomato, which is related to stomata movement (Patil and Kulkarni, 1989). Unlike inhibitory effects mentioned, organophosphorus insecticide rigor stimulated pollen development and then the yield in winter wheat (Izotova et al., 1973).

3.1.3.8. Special Roles of Cholinesterases

The possible universal role of the acetylcholine hydrolysis for the vital biological systems with participation of acetylcholinesterase is discussed from the position of energy transformation in such processes as synaptic transfer, muscle contraction, and photosynthesis (Borodyuk, 1990). Forming acetic acid produced higher concentration of H^+ ion and so regulated many membranous processes. Many proteins of eukaryotic cell associated with the outer surface of the cell membrane with the help of covalently bound carboxyterminals of glycolipids (Cross, 1990). Acetylcholinesterase can be anchored in a similar way with plasmatic membranes (Low, 1987), Cholinesterase can interact with other cellular components. In section 3.1.3.5 it was shown that redox protein plastocyanin and ribulose-1,5-bisphosphate carboxylase (both containing Cu^{2+}) can regulate cholinesterase activity (Roshchina and Karpilova, 1993). But acetylcholinesterase, in turn, may inhibit the ribulose-1,5-bisphosphate carboxylase activity by 70% (Table 34). The level of inhibition was higher in the presence of acetylcholinesterase inhibitors neostigmine and physostigmine, as well as the catecholamine noradrenaline. The interaction of animal acetylcholinesterase with dopamine has been studied (Klegeris et al., 1995). Acetylcholinesterase selectively inhibited the rate of quinone production from dopamine as well as accumulation of hydrogen peroxide, while the rate of generation of superoxide anion radical was

Table 34. Action of acetylcholinesterase (AChE) on pea ribulose-1, 5 bisphosphate carboxylase (RuBPC) activity(Roshchina and Karpilova, 1993)

Compound	$^{14}CO_2$-fixation (% of control)		
	The RuBPC/AChE ratio (kg/kg)		
	5.8	0.54	0.17
AChE	95.3	30.6	18.8
Physostigmine + AChE	88.9	26.3	18.0
Neostigmine + AChE	79.6	17.0	19.0
Noradrenaline + AChE	88.9	15.76	—

The control rate $^{14}CO_2$-fixation (impluse s^{-1})
without addition, 40.7;
with physostigmine 10^{-3} M, 40.1;
with neostigmine 10^{-4} M, 33.0;
with noradrenaline 10^{-4} M, 36.2.

increased. A new product is formed after mixing of acetylcholinesterase and dopamine in neutral pH. In all cases, butyrylcholinesterase was ineffective. At the same time, incubation of acetylcholinesterase with dopamine resulted in a significant decrease in the catalytic activity of the enzyme.

Addition of dopamine with superoxide dismutase diminished $O_2^{\cdot-}$ concentrations or addition of dopamine with catalase and peroxidase, decomposing only H_2O_2 and H_2O_2 with other organic peroxides, relatively, showed a direct interaction of quinone and (or) semiquinone oxidation product with catalytic center of acetylcholinesterase. Thus, cholinesterases can regulate the redox state of cellular medium.

The high cholinesterase activity is correlated with motility and contractile reactions. The rate of the substrate hydrolysis in guard cells of stomata in *Vicia faba* and *Nicotiana glauca* is 5–10 times higher than in extracts from mesophyll or whole leaf (Madhavan et al., 1995). This cholinesterase is related to true cholinesterase or acetylcholinesterase. Acetylcholine is supposed to have a physiological role in stomatal movement because it can induce a relatively rapid (< 5 min) closing of stomata. Cholinesterase may control the acetylcholine concentration in guard cells (Madhavan et al., 1995).

High activity of cholinesterase in roots (Fluck and Jaffe, 1976) correlated with acetylcholine-induced stimulation of root pressure (Zholkevich et al., 1990), which also could be related to contractile elements of the tissue.

Acetylcholinesterase is observed at the interface between stele and cortex of the mesocotyl of *Zea mays* (Momonoki, 1992). It is supposed to release or diffuse acetylcholine to the junction between stele and cortex on the outside of the endodermis, like the post-synaptic cell, and there bind to cholinoreceptor inducing propagated action potential via plasmodesmal channels. Cholinesterase, hydrolyzing acetylcholine, liberates the cholinoreceptor.

The important role of cholinesterase in plant excretions has an analog in animals. For instance, acetylcholinesterase is excreted to the incubation media by oocytes (Fournier *et al.*, 1992) or by parasites, induces human infection (Misra *et al.*, 1993; McKeand *et al.*, 1992), and appears to serve as protective element. Cholinesterases may hydrolyze proteins as protease to low-molecular-weight peptides (Small *et al.*, 1986). Acetylcholinesterase is also found in plasmodium of the myxomycete, which shows possible participation of the enzyme both in chemorecognition and in the defensive system of the organism (Nakajima and Hatano, 1962). Cellular and secreted forms of animal acetylcholinesterase may differ in molecular form and activity (Rubin *et al.*, 1985).

The cholinesterase activity found in plant excretions and secretory cells is of special interest. The excretions of pollen just after the moistening of dry microspores contain cholinesterase (Roshchina *et al.*, 1994; Roshchina and Semenova, 1995). The slime hairs of *Drosera capensis* also include cholinesterase (Roshchina and Semenova, 1995). In both cases cholinesterase may participate in recognition:
1. in pistil-pollen relations at fertilization or
2. in insect-plant interaction, where acetylcholine of pollen, pistil, or insect is used as substrate.

As shown for pollen of self-incompatible clone of *Petunia hybrida*, lower activity of cholinesterase occurred than in self-compatible clone (Table 35). This is an indication of the participation of cholinesterase in the mechanism of pollen-pistil recognition. Low cholinesterase activity in self-incompatible clone correlated with low sensitivity of the seed yield to eserine and proserine (Kovaleva and Roshchina, 1997). In a sterile clone of *Petunia*, cholinesterase activity was absent, therefore fertility appears to be related to cholinesterase expression (Roshchina, 1999a). Analysis of the pistil stigma or pollen *Hippeastrum hybridum* with 10^{-6}–10^{-4} M of cholinesterase inhibitors neostigmine, physostigmine, and trimethylammonium showed sensitivity of the generative organs (autofluorescence and pollen germination) to the substances (Roshchina and Melnikova, 1998a, b). After treatment of pistil with neostigmine, sometimes larger fruits with more seeds were formed in comparison with control. These facts also show the importance of cholinesterase in fertilization. Thus, cholinesterase participates in breeding mechanisms. As for recognition in carnivorous plants, the slime, containing cholinesterase, may use acetylcholine as a chemosignal of insect at first, and then cholinesterase decomposes acetylcholine and simultaneously works as protease and non-specific esterase for insect tissue disruption.

Table 35. Maximal rate of acetylthiocholine hydrolysis (V_{max}) in water extracts from pollen and pistils of *Petunia hybrida* (after Kovaleva and Roshchina, 1997; Roshchina, 1999a)

Plant material,	V_{max}, µmol s^{-1} kg^{-1} of fresh mass		
	control	+physostigmine	+ neostigmine
Self-compatible clone			
Pollen	1400	93	0
Pistil	492	307	330
Self-incompatible clone			
Pollen	330	280	175
Pistil	50	22	0

Concentrations of physostigmine and neostigmine were 10^{-5} M; n = 5; SE = 2–3%.

Non-synaptic functions of cholinesterases in plants should be the subject of further studies, especially in relation to secretion and recognition.

3.2. OTHER REGULATORY SYSTEMS (ADRENERGIC, DOPAMINERGIC, SEROTONINERGIC, HISTAMINERGIC)

Like acetylcholine, catecholamines, serotonin, and histamine are included in a composition of complex systems of animal regulation. Biogenic amines are known to interact with membranes of animal cells via binding with specific receptors, which conformational changes activate intracellular enzymic systems (adenylate cyclase or guanylate cyclase), regulating the synthesis of secondary messengers and regulators cyclic AMP (cAMP) and cyclic GMP (cGMP), inositol triphosphate, or the accumulation of Ca^{2+}-ions. The interaction components belonging to the above-mentioned regulatory systems is represented in Fig. 45 and will be described below.

3.2.1. Reception of Catecholamines, Serotonin and Histamine

3.2.1.1. Main Features of Aminoreception

There are special animal receptors: (1) of dopamine D_1 and D_2, dopaminic receptors (Miller *et al.*, 1988); (2) of adrenaline and noradrenaline α and β, adrenoreceptors (Malborn *et al.*, 1987); (3) of serotonin M, D, T, serotoninic receptors (Kats and Lavretskaya, 1986; Baumgarten and Yöthert, 1997); and (4) of histamine H_1, H_2, and H_3, histaminic receptors (Vaisfeld and Kassil, 1981; Goot *et al.*, 1991; Ganellin *et al.*, 1995). Each group of receptors has its own agonists and antagonists. Agonists of dopamine (Fig. 46) are represented by analogs of phenolic nature, in which, mainly, NH_2-group is changed on Se, S-, or methyl-groups (Miller *et al.*, 1988). Some analogs similar to apomorphine can be also active. Antagonists of dopamine are derivatives of either chlorpromazine or

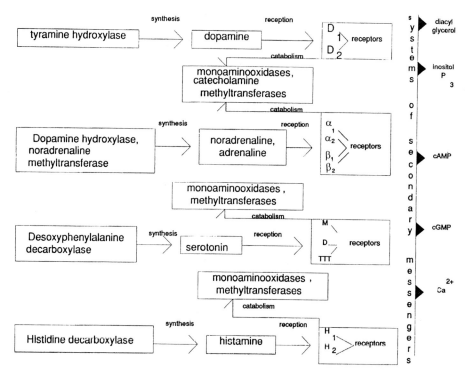

Fig. 45. Aminoergic system of regulation in animals

sulpiride (Fig. 47). Chlorpromazine and sulpiride themselves are not antagonists of dopamine. Agonists of noradrenaline and adrenaline are oxidized methylic derivatives of catecholamines, among which alkaloid ephedrine is known (Fig. 48). Antagonists are represented by complex phenolic compounds, in particular alkaloid yohimbine (Fig. 48). Agonists of serotonin (Fig. 49) are mainly tryptophan derivatives, including the product of serotonin methylation, bufotenine (Katz and Lavretskaya, 1986). Known antagonists are chlorpromazine, its derivatives, and fluoridated peridoles. Such antagonists of serotonin as the natural compound lysergic acid and its derivative LSD_{25} are hallucinogenic toxins. Known agonists of histamine (Figs. 50, 51) are its methylic metabolites. Diphenols are antagonists binding with H_1-histaminic receptors, while amides linked with NH_2-group of histamine represent antagonists sensitive to H_2-histaminic receptors (Fig. 50) (Vaisfeld and Kassil, 1981) and H_3-histamine receptor (Fig. 51) (Ganellin et al., 1995).

Some aminoreceptors are isolated from animal tissues. D_2-dopaminic receptor with molecular mass 92 kDa has been isolated from animal tissue and purified 21,500-fold (Lew et al., 1985; Ramwani and Mishra, 1986; Amlaiky and Caron, 1986; Elazar et al., 1988). β-Adrenoreceptors have also been purified from mammal cells (Benovic et al., 1984; Cubero and Malborn, 1984; Malborn et al., 1987). There are proteins with molecular masses equal to 62–67 kDa or 76 kDa and with specific activity 4–8 pmoles of receptor/mg of protein. All types of receptors for biogenic amines are localized mainly on the surface of animal plasmatic membranes.

3.2.1.2. Aminoreception in Plants

The presence of receptors for biogenic amines in plants is supposed from experimental work with agonists and antagonists, summarized in Table 36.

Table 36. Reactions of plants sensitive to biogenic amines, their agonists and antagonists

Reaction	Agonist	Antagonist	Reference
Changes in membrane potential	Noradrenaline, adrenaline, isadrin	Inderol	Oniani et al., 1974, 1977a, b
Rate of cytoplasm movement	Noradrenaline, adrenaline, isadrin	Inderol	Oniani et al., 1974, 1977a, b
Stimulation of flowering	Noradrenaline, adrenaline, L-isoproterenol	Propranolol	Khurana et al., 1987
Stimulation of photophosphorylation in chloroplasts	Dopamine, noradrenaline, adrenaline, 6,7-diOHATN (2-amino-6,7-dihydroxy-1,2,3,4-tetrahydronaphthalene)	Yohimbine	Roshchina, 1989a
Ca^{2+} and Mg^{2+} efflux from chloroplasts	Dopamine, noradrenaline, adrenaline, 6,7-diOHATN	Yohimbine	Roshchina, 1989a, b, 1990b, 1992a, c
Seed germination	Serotonin	Kur-14, inmecarb, trifluoroperazine	Roshchina, 1991a, 1992, 1994
Growth of seedlings	Histamine, 4-methylhistamine		Haartmann et al., 1966
Pollen germination in vitro	Serotonin		Roshchina et al., 1998a, b
Pollen germination in vitro	Dopamine, noradrenaline		Roshchina et al., 1998a, b
Pollen germination in vitro	Histamine		Roshchina and Melnikova, 1998a, b
Pollen germination in vivo (on pistil stigma)	Histamine	Tavegyl (clemastin)	Roshchina and Melnikova, 1998a, b

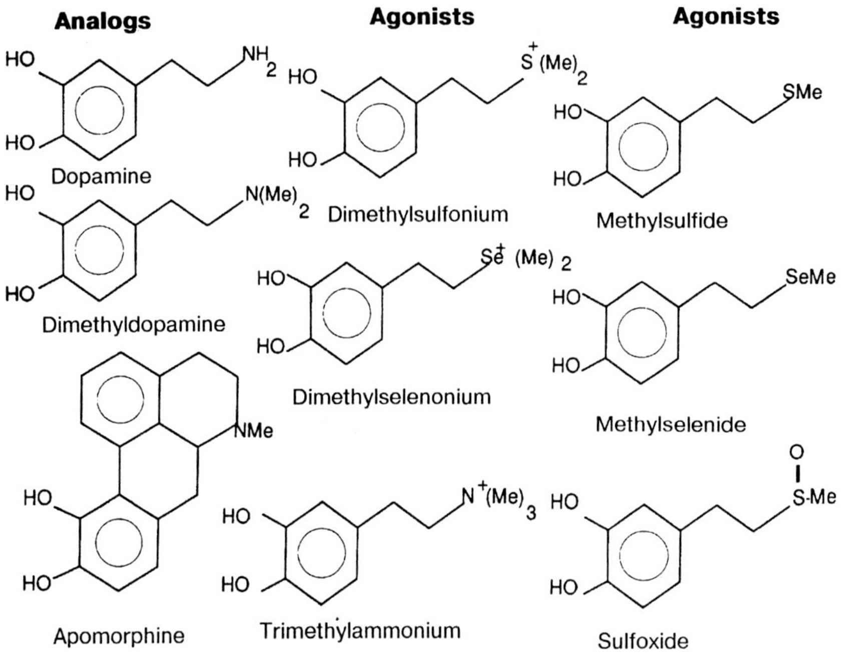

Fig. 46. Agonists of dopamine

Oniani and coworkers (1974, 1977a, b) first showed that the rates of cytoplasm movement and membrane potential of green algae *Nitella* cells change in the presence of catecholamines, isadrin (isoproterenol) and adrenoblocker inderal (propranolol). Although these data were not adequate and high concentrations (> 10^{-5} M) of adrenoblockers were used, it was a first step in the search for plant adrenoreceptors. Photomorphogenesis is sensitive to catecholamines too. Noradrenaline, adrenaline (10^{-6}–10^{-4} M), and their agonist L-isoproterenol stimulated the flowering of *Lemna paucicostata* under a photoperiod of 8 hours light and 16 hours dark (Khurana et al., 1987). Propranolol, antagonist of catecholamines, blocker of β-adrenoreceptor, partly depresses the flowering induced by catecholamines. The seed germination stimulated by noradrenaline or serotonin is also sensitive to their antagonists (Roshchina, 1992). If seeds of *Raphanus sativus* are first treated with 10^{-6} M noradrenaline antagonist yohimbine or 10^{-5} M antagonist of serotonin Kur-14 (5-methyl-2-α-dimethylaminoethyl-3-β-ethylindole), the stimulation is decreased by 50–70% (Fig. 52). The data can be explained as a competition of antagonist and neuromediator, if there is binding in the same part of plasmatic membranes. The pollen germination *in vitro*, stimulated by dopamine, serotonin, and histamine, was also treated with some of their agonists

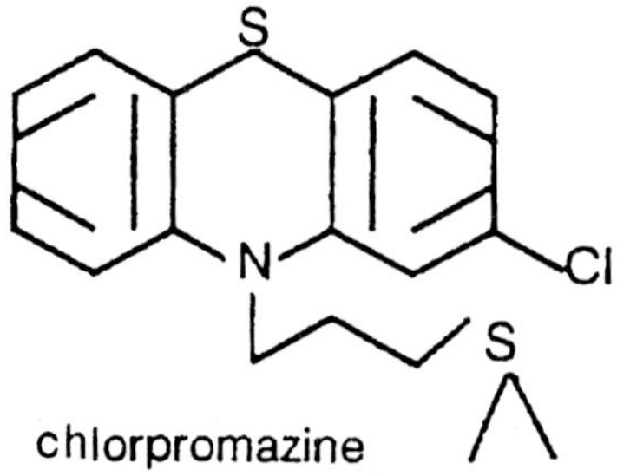

chlorpromazine dimethylsulfonium

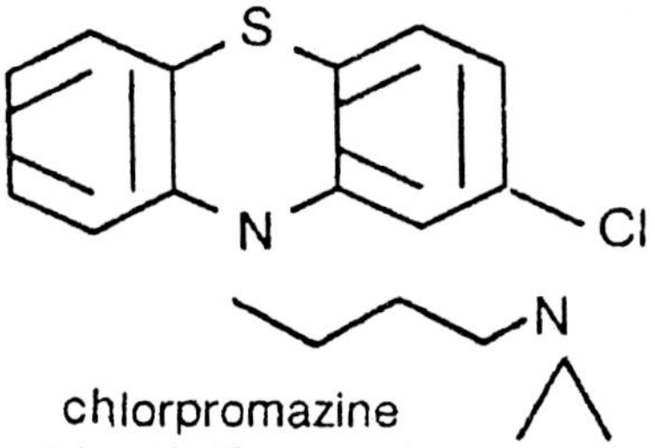

chlorpromazine trimethyllammonium

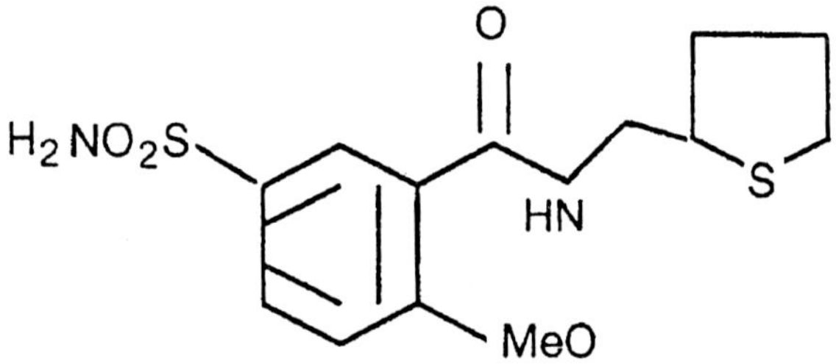

sulpiride sulfonium

Fig. 47. Antagonists of dopamine

and antagonists (Table 37). The effect of the dopamine agonist 6,7-diOHATN or 2-amino-6,7-dihydroxy-1,2,3,4-tetrahydronaphthalene hydrobromide was similar to the stimulatory effect of dopamine itself. However, antagonists demonstrated weak inhibitory effects. The pollen germination *in vivo* (on a pistil stigma of *Hippeastrum hybridum*) was not inhibited by the antagonists of noradrenaline (Table 37). The results may be explained as non-receptor binding with plasmatic membranes.

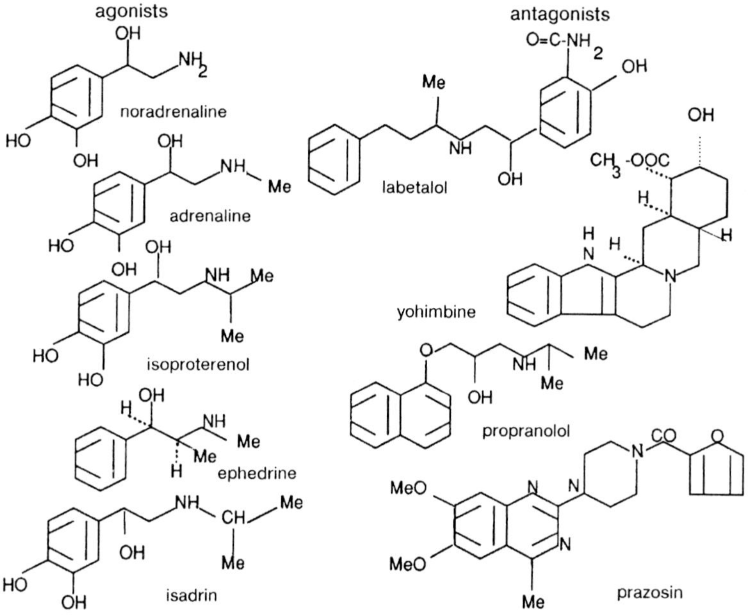

Fig. 48. Agonists and antagonists of noradrenaline

Table 37. The germination index of pollen *Hippeastrum hybridum in vivo* under the influence of the pistil treatment with biogenic amines and their antagonists. Index = ratio of germinated to total estimated pollen grains.

Variant	Pollen germination *in vitro* (each 50 pollen grains estimated)		Seeds	Weight of seeds, g	Weight of one seed, g
	Index	% of control			
Control	11.9 ± 1.3	100	56 ± 4 (100% ± 7.7)	1.8 ± 0.2 (100% ± 13)	0.03
Noradrenaline 1.2×10^{-5} M	9.5 ± 1.0	79.8*	45 ± 6	1.0 ± 0.2	0.023
6, 7-diOHATN 1.2×10^{-4} M	20.9 ± 1.4	175.6	64 ± 15	1.2 ± 0.4	0.02
Dopamine 10^{-4} M	13.1 ± 3.0	110.5*	51 ± 10	3.5 ± 0.8	0.07
Propranolol 10^{-4} M	9.02 ± 1.2	75.6*	31 ± 4	0.8 ± 0.5	0.026
Yohimbine 10^{-4} M	10.2 ± 1.0	85.7*	51 ± 15	2 ± 0.3	0.04
Histamine 10^{-5} M	12.4 ± 2.0	115	55 ± 15*	—	—
Tavegyl 2×10^{-5} M	12.9 ± 1.0	120	No	No	No

*Not verified. —no data

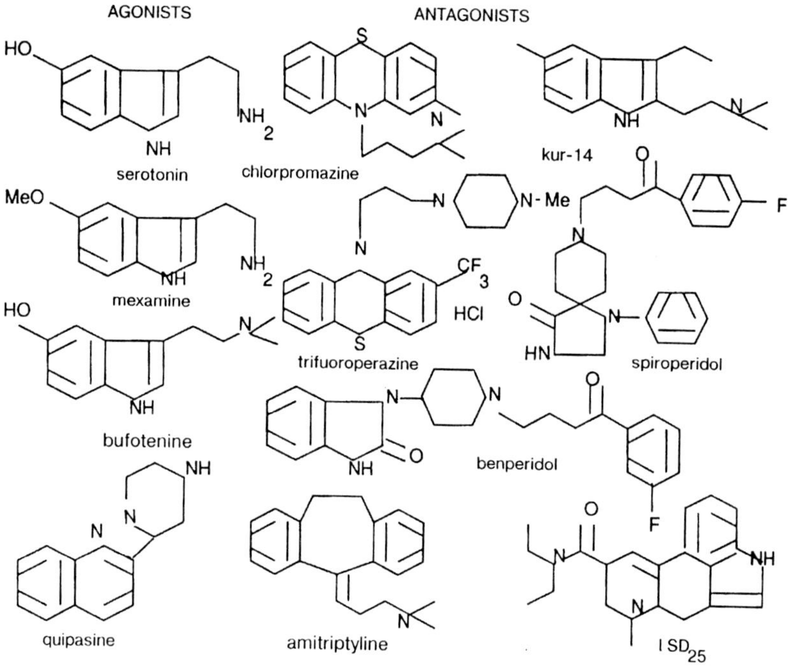

Fig. 49. Agonists and antagonists of serotonin

Low concentrations ($< 10^{-5}$ M) of adrenaline, noradrenaline, and dopamine stimulate photophosphorylation and Ca^{2+}/Mg^{2+} permeability in pea chloroplasts (Roshchina, 1989a, b). Based on the phenomenon, Roshchina (1990c, d) have carried out experiments with some agonists and antagonists of catecholamines where stimulation of photophosphorylation in isolated pea chloroplasts was considered as physiological response. Selective agonist of dopamine 6,7-diOHATN or 2-amino-6,7-dihydroxy-1,2,3,4-tetrahydronaphthalene hydrobromide (Horn and Rodgers, 1980) in concentrations 10^{-8}–10^{-7} M induced the stimulation of ATP synthesis (Fig. 53). In the same concentrations it imitates the effect observed for catecholamines. Constants of binding (K_b) for dopamine and its agonist are 10^{-9} M and 2×10^{-9} M, respectively. Antagonist of catecholamines and adrenoblocker yohimbine (10^{-10}–10^{-6} M) itself did not inhibit the synthesis of ATP. The inhibition was observed only when higher concentrations of yohimbine were used. The concentration curve of noradrenaline action on photophosphorylation in yohimbine-treated chloroplasts (Fig. 54) shows that the maximum of ATP stimulation by noradrenaline is shifted to a higher concentration of the mediator. Under treatment with 10^{-5} M yohimbine, the pool of possible

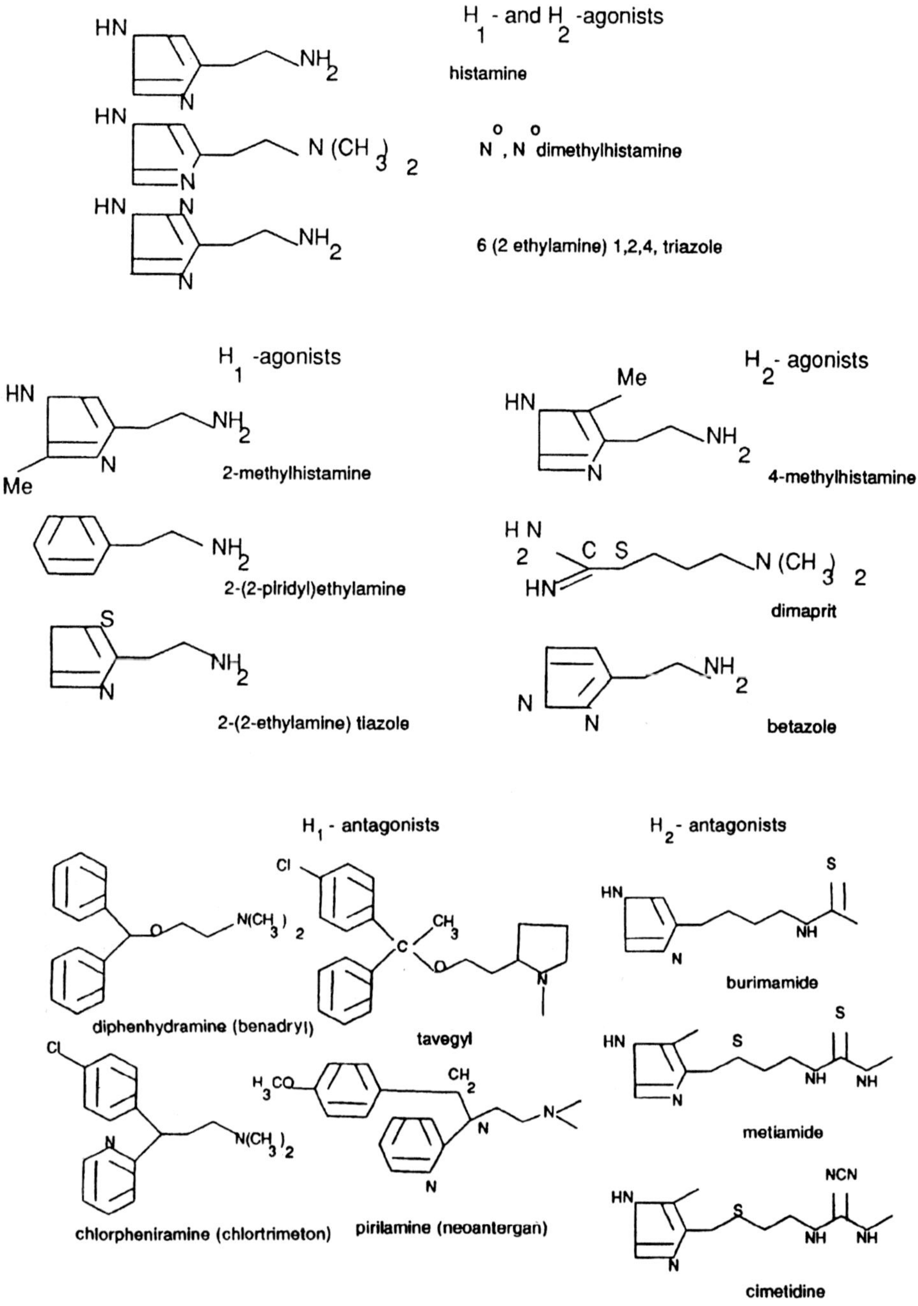

Fig. 50. Agonists and antagonists of histamine, binding with H_1 and H_2 receptors

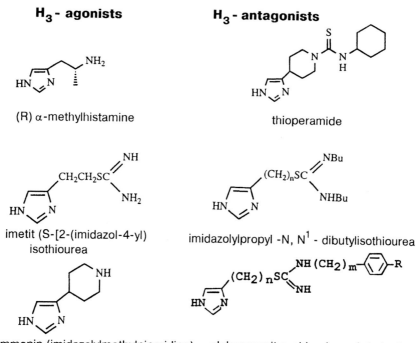

Fig. 51. Agonists and antagonists of histamine, binding with H_3 receptors

receptors is completely full. As a result, stimulation of photophosphorylation was not observed. The constants of binding of noradrenaline, its agonist adrenaline, and antagonist yohimbine with chloroplasts are 10^{-8}, 5×10^{-9} and 1.1×10^{-9} M, respectively. The specificity of effects observed for catecholamines and their agonists, their low active concentrations, the saturation in the sites of binding of these compounds, and blocking of physiological reactions by antagonists show in the presence of α-adrenoreceptors in chloroplasts. Perhaps α-adrenoreceptors of chloroplasts include or are linked with actin-like proteins as contractile elements, since blocker of actomyosin filaments cytochalasin B prevented the stimulation of photophosphorylation by noradrenaline, whereas blocker of microtubules colchicine had no similar effect (Roshchina, 1990c, d).

As for antagonists of serotonin, Roshchina (1992) has shown that they inhibit seed germination, largely in medium without serotonin and to a lesser extent in medium with serotonin. The germination of radish seeds is retarded by antagonists of serotonin 5-methyl-2-α-dimethylaminoethyl-3-β-ethylindole (Kur-14) (Fig. 52b) and inmecarb. Blockers of serotoninic receptor chlorpromazine and trifluoroperazine are simultaneously inhibitors of calmodulin. Stimulated by serotonin, the germination of radish

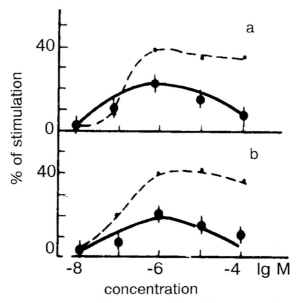

Fig. 52. Concentration curves of neurotransmitter effects on seed gemination of *Raphanus sativus* after preliminary treatment with their anagonists. (– –), control, without antagonist. a, noradrenaline after preliminary treatment with yohimbine (10^{-6} M); b, serotonin after preliminary treatment with 10^{-5} M methyl-2-α-dimethylaminoethyl-3-β-ethylindole (Kur-14).

seeds was also inhibited by trifluoroperazine (10^{-7}–10^{-5} M) by 50–60% of control. Trifluoroperazine and chlorpromazine (10^{-6} M) blocked Ca^{2+}-dependent and acetylcholine-induced swelling of protoplasts of wheat leaves (Tretyn et al., 1990a).

As for the presence of histaminic receptors, 4-methyl histidine (agonist of H_1-type of histaminic receptor and inhibitor of histidine carboxylase) can significantly decrease the growth of spinach seedlings to 30% of control (Haartmann et al., 1966). This may indicate especially that the H_1-histaminic receptor blocker tavegyl (clemastin), when added on the pistil stigma surface of *Hippeastrum hybridum* before pollination, inhibited the formation of seeds to maturity (Roshchina and Melnikova, 1998a, b). The same antagonist of histamine also acted on the autofluorescence of secretory hairs of *Lycopersicon esculentum* (induced by ultraviolet irradiation 360–380 nm) and the secretion efflux from the trichome (Fig. 55).

3.2.2. Enzymes Participating in Biosynthesis of Biogenic Amines

Biosynthesis of biogenic amines in animals and plants involves decarboxylation and hydroxylation of corresponding amino acids. The

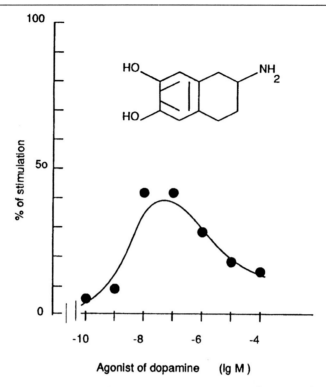

Fig. 53. Stimulation by the dopamine antagonist 6,7-diOHATN of non-cyclic photophosphorylation in isolated pea chloroplasts (Roshchina, 1990c)

diagram of biosynthesis of biogenic amines (Chapter 1, Figs. 3, 9, 10) shows the participation of key enzymes hydroxylases and decarboxylases.

Hydroxylases. Hydroxylases convert tyrosine to DOPA and tryptophan to 5-hydroxytryptophan. First enzyme tyrosine hydroxylase (EC 1.14.16.2) has been isolated from animal tissues. It is shown to hydroxylate tyrosine in the presence of tetrahydrobiopterin and oxygen (Kuhn and Lovenberg, 1983). Their hydroxylase is stimulated by ions of bivalent iron and has molecular mass from 32 to 220 kDa . Native protein has molecular masses 180–220 kDa. Tryptophan-5-hydroxylases (EC 1.14.16.4) hydroxylate tryptophan by similar mode. Tyrosine hydroxylase and tryptophan hydroxylase appear to be activated by phosphorylation with participation of cyclic nucleotides. Molecular mass of native tryptophan hydrohylase is 220 kDa for native form, whereas after PAGE with Na-DDS or treatment by trypsin monomeric subunits 55–60.9 kDa arise (Kuhn and Lovenberg, 1983). The process of hydroxylation can be expressed as follows:

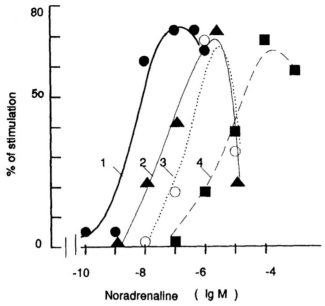

Fig. 54. Dose-effect concentration curves of noradrenaline action on non-cyclic photophosphorylation in isolated pea chloroplasts without (1) and with (2 to 4) its antagonist yohimbine (after Roshchina, 1990c). The antagonist concentrations were 10^{-8} M (2), 10^{-7} M (3), 10^{-6} M (4). Control rate of ATP synthesis 7 mmol s^{-1} kg^{-1} of chlorophyll.

tyrosine + tetrahydrobiopterin + O_2 → 5-dihydroxyphenylalanine
tryptophan   hydroxytryptophan
+ quinone dihydrobiopterin + H_2O

The intermediate stage of this reaction appears to be the formation of hydrogen peroxide.

Besides tyrosine and tryptophan hydroxylases, there is a third type of hydroxylase called dopamine-β-hydroxylase, converting phenylethylamines to noradrenaline and adrenaline:

**phenylethylamine + ascorbate + O_2 → noradrenaline + dehydroascorbate + H_2O
(dopamine)**

Native enzyme (EC 1.14.17.1) has molecular mass 290 kDa, whereas subunits of denaturated protein are about 75 kDa (Kuhn and Lovenberg, 1983). The enzyme contains copper in prosthetic group. Another enzyme hydroxylating tyramine to dopamine is found in fruits of banana *Musa sapientum*, partly purified and characterized (Deacon and Marsh, 1971). It is similar to dopamine-β-hydroxylase in its characteristics and in a whole scheme of the reaction of hydroxylation. The enzyme has pH optimum at 6.0. Ascorbate and oxygen are needed for the hydroxylation.

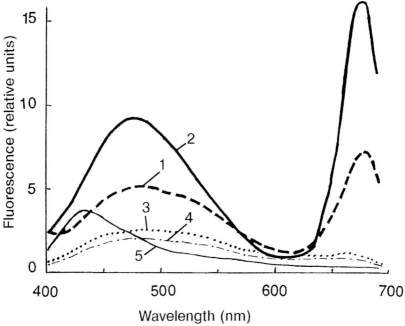

Fig. 55. Autofluorescence spectra of the secretory hair of *Lycopersicon esculentum* leaf without (1) and with (2) the addition of 10^{-5} M tavegyl, antagoinst of histamine. (3) empty hair; (4) 5×10^{-5} M tavegyl; (5) secretion evacuated into water by 5×10^{-5} M tavegyl.

Enzyme tryptophan-5-hydroxylase (EC 1.14.16.4), which converts tryptamine to serotonin, is found in all organs of the West African leguminous plant *Griffonia simplicifolia* (Fellows and Bell, 1970).

Decarboxylases. The group of decarboxylases derived from aromatic amino acid decarboxylases (Lovenberg *et al.*, 1962) includes dihydroxyphenylalanine decarboxylase converting dihydroxyphenylalanine to dopamine, decarboxylase (decarboxylase of aromatic amino acids [EC 4.1.1.26/27) catalyzing the formation of serotonin from 5-hydroxytryptophan] and histidine decarboxylase (EC 4.1.1.22) decarboxylating histidine to histamine (Wu, 1983).

Dihydroxyphenylalanine decarboxylase has been found in *Citisus scoparius* (Tocher and Tocher, 1972), L-tyrosine decarboxylase or carboxylyase (EC 4.1.1.26/27) in *Hordeum vulgare* (Gallon and Butt, 1971; Hosoi *et al.*, 1972; Hosoi, 1974), and histidine decarboxylase in *Spinacia* (Werle and Raub, 1948). L-tyrosine carboxylyase (EC 4.1.1.25) has been found in barley roots (Gallon and Butt, 1971; Hosoi, 1974). The enzyme was isolated and purified 80-fold (Gallon and Butt, 1971) and 25-fold (Hosoi, 1974). L-DOPA, m-tyrosine, and o-tyrosine, but not D-tyrosine, can serve as substrates for this catalyst. Maximal activity of tyrosine decarboxylase was observed for L-tyrosine as a substrate, 93.3–134 nmoles h^{-1} mg^{-1} of protien (K_m = 4.5 × 10^{-4} M), and for L-DOPA it was only 22.1 nmoles h^{-1} mg^{-1} of protein (unlike similar animal enzyme, where L-DOPA is a better substrate than others). Whereas optimum pH of mammalian and bacteria tyrosine decarboxylase was 8–9 optimum pH of plant tyrosine decarboxylase was 7.3. Apart from barley roots, highest activity of tyrosine decarboxylase has been found in roots of maize *Zea mays* and in flowers of *C. scoparius*. The enzymic activity was also found in roots of wheat *Triticum aestivum*. It was not found in shoots of barley as well as in *Sorghum vulgare, Panicum crusgalli, Phaseolus mungo,* and *Pisum sativum*. Tyrosine decarboxylase from barley includes pyridoxal phosphate in active center because this substance stimulated the enzymic activity. Isolated enzyme was inhibited by various substances, especially by iproniazid phosphate, hydroxylamine hydrochloride, $CuSO_4$, $HgCl_2$, DOPA, p-coumaric and caffeic acids, as well as SH-reagents. The enzyme differs from DOPA decarboxylase from *C. scoparius* in its greater sensitivity to p-chloromercurium benzoate. Tyrosine decarboxylase from barley roots (Gallon and Butt, 1971; Hosoi, 1974) is competitively inhibited by phenolic acids—coumaric (K_i = 0.07 M), ferulic (K_i = 0.06 M), caffeic, and cinnamic acids—as well as p-hydroxyphenylpyruvate, whereas inhibition by DOPA was a mixed type. The barley root tyrosine decarboxylase is also stimulated by S-adenosylmenthionine (0.1–1 mM) (Gallon and Butt, 1971).

Biosynthesis of serotonin from tryptophan (Fig. 9) includes the participation of two key enzymes: decarboxylase of aromatic acids (EC 4.1.1.26/27) converting 5-hydroxytryptophan to serotonin or tryptophan decarboxylase (EC 4.1.1.27) converting tryptophan to tryptamine (Reed, 1968). The 5-hydroxytryptophan decarboxylase activity of first enzyme is found in seeds of species belonging to genus *Juglans* (Grosse, 1982; Grosse et al., 1983). Another enzyme has also been discovered in fruits of walnut *Juglans regia* (Klapheck and Grosse, 1980; Grosse et al., 1983). It has been isolated from shoots of tomato *Lycopersicon esculentum* and partly purified (Gibson et al., 1972). The tryptophan decarboxylase activity of pea epicotyls is inhibited by o-phenols (1–40 mM) such as kaempferol, quercetin, and chlorogenic and caffeic acids as well as catechols (100 mM) and resorcinol (Reed, 1968). The synthesis of tryptophan decarboxylase (EC 4.1.1.27) *de novo* in cotyledons of *J. regia* L. was accelerated at the earlier stage of development by harvesting immature fruits (Lachmann and Grosse, 1990). Phytohormones were included in the regulation of this process. Gibberellic acid (10^{-8} M) suppressed it by 50%, whereas abscisic acid was much less effective in physiological concentrations. Unlike these hormones, indole-3-acetic acid (10^{-7} M) stimulated the synthesis of the enzyme.

The participation of histidine decarboxylase in the formation of histamine in plants was first demonstrated by Werle and Raub (1948) on spinach leaves. Haartmann and coworkers (1966) demonstrated that histamine was formed only after infusion of L-histidine to native leaf or slices of spinach seedlings, whereas in cell-free extracts there is no reaction. Activity of histidine decarboxylase is also observed in germinating seeds of cotton (Lloyd and Nicholis, 1965). Histidine decarboxylase is isolated from bacteria and mammals and has a molecular mass of 210 kDa (Vaisfeld and Kassil, 1981; Wu, 1983), including subunits 145 and 66 kDa.

Methyltransferases. Synthesis of adrenaline from noradrenaline in animals occurs with participation of enzyme phenylethanolamine-N-methyltransferase (EC 2.1.1.28). The reaction is as follows:

noradrenaline + S-adenosyl-L-methionine → adrenaline + S-adenosyl-L-homocysteine

The presence of similar enzyme was supposed in plants as well. In *Nerine bowdenii* Mann and coworkers (1963) found and partly purified the enzyme similar with catechol-o-methyltransferase of animal (Vidgren *et al.,* 1994). It can methylate dopamine and alkaloid norbelladine and, to a lesser degree, catechol.

3.2.3. Enzymes Participating in Catabolism of Biogenic Amines

In Chapter 1, catabolism of biogenic amines with participation of two key enzymes aminooxidases and methyltransferases was considered. Reactions of oxidative desamination occur with the help of monoamine oxidases (EC 1.4.3.4) and in some cases of diamine oxidases (EC 1.4.3.6). Both types of the enzymes are copper-containing proteins and catalyze predominantly decomposition of mono- or diamines (Finberg and Youdim, 1983). In plants, enzyme participating in oxidative desamination was first found in seedlings of various plants and identified as diamine oxidases (Werle and Zabel, 1948; Werle and Pechmann, 1949; Werle and Roewer, 1950). Extracts from *Trifolium pratense* and *Trifolium repens* were able to use noradrenaline and other monoamines as substrates. Later it was shown that not all diamine oxidases can utilize biogenic monoamines. They predominantly oxidize monoamines histamine and diamines putrescine, cadaverine, and spermidine (Kenten and Mann, 1952; Pec and Frebort 1997). Table 38 summarizes data about diamine oxidases catalyzing desamination of histamine and catecholamines in 21 species belonging to 7 families. Diamine oxidase of pea seedlings is induced in embryo (Werle *et al.,* 1961). This enzyme was first isolated and purified from seedlings of pea *Pisum sativum* (Mann, 1955, 1961; Werle *et al.,* 1961). The rates of the desaminating oxidation varied according to the substrate. If the activity with cadaverine is believed to be 1000, the substrate specificity is the following:

Table 38. Diamine oxidases in plants

Family, species	Organ	Substrate	Reference
Compositae			
Artemisia sp.	Leaves	Histamine	Werle and Zabel, 1948
Taraxacum officinale	Leaves	Histamine	Werle and Zabel, 1948
Fabaceae			
Glycine max	Roots (purified tyramine enzyme)	Histamine Tryptamine	Werle and Pechmann, 1949
Lupinus luteus	Seedlings (purified enzyme)	Tryptamine	Werle and Pechmann, 1949; Schutte et al., 1966
Medicago sativa	seedlings	Histamine, noradrenaline, tyramine, tryptamine	Werle and Roewer, 1950
Pisum sativum	Seedlings (purified enzyme)	Monoamines, histamine, diamines	Mann, 1955; Uspenskaya and Goryachenkova, 1958; Hill and Mann, 1964
Trifolium incarnatum	Seedlings (purified enzyme)	Monoamines, histamine, diamines	Suzuki, 1966
T. pratense	Seedlings, leaves, flowers, stems	Histamine	Werle and Zabel, 1948
T. repens	Seedlings (purified enzyme)	Histamine	Delhaize et al., 1986
T. subterraneum	Seedlings (purified enzyme)	Histamine	Delhaize et al., 1986
Vicia sp.		Histamine	Werle and Pechmann, 1949
Hydrangeaceae	Leaves	Histamine	Werle and Zabel, 1948
Hydrangea sp.			
Labiatae			
Lavandula spica	Leaves, stems	Histamine	Werle and Zabel, 1948
L. vera	Leaves	Histamine	
Rosmarinus officinalis	Leaves	Histamine	
Thymus vulgaris	Leaves	Histamine	Werle and Zabel, 1948
Oleaceae			
Jasminum sp.	Leaves	Histamine	Werle and Zabel, 1948
Rosaceae			
Fragaria sp.	Leaves	Histamine	Werle and Zabel, 1948
Rubus sp.	Leaves	Histamine	Werle and Zabel, 1948
Umbelliferae			
Aegopodium sp.	Leaves	Histamine	Werle and Zabel, 1948

spermidine 570 > tyramine 104 > spermine 85 > histamine 44 > dopamine 34 > tryptamine 27 > noradrenaline 20 > serotonin 8 > adrenaline 0

The activity of the enzyme was inhibited by cyanide, semicarbazide, diethyldithiocarbamate, hydroxylamines, 3-oxyquinoline, and o-phenantroline. A characteristic maximum of 500 nm is observed in absorbance spectra of isolated rose-colored protein, which is related to the presence of 0.08–0.09% copper (Hill and Mann, 1964). Copper activates the enzyme but is not included in its active center. When diamine oxidase interacts with a substrate, yellow complex arises in anaerobic conditions, perhaps because of the formation of copper chelates. Under oxygenation the rose color of the enzyme is restored. The active center in various diamine oxidases is represented as either flavine adenine dinucleotide or pyridoxal phosphate.

Diamine oxidases are also found in apoplast of Fabaceae seedlings (Angelini and Federico, 1986), which shows the significance of the role of the enzyme in intercellular metabolism (including ammonium detoxication and polyamine catabolism). The presence of the enzyme was studied by immunohistochemical and biochemical methods (Federico and Angelini, 1986; Angelini and Federico, 1986; Federico et al., 1988).

There is a functional correlation between diamine oxidase and peroxidase activities, and their dependence upon de-etiolation and wounding in chickpea stems (Angelini et al., 1990). Polyphenoloxidases and tyrosinases also participate in a cascade of oxidative reactions with aminooxidases in plants (Robb et al., 1966; Mayer and Harel, 1979; Mayer, 1987).

Oxidation of histamine in animals is due to the presence of histaminase, in plant extracts, which is similar to diamine oxidase and contains both the above-mentioned coenzymes in active center. Homogenous preparation of diamine oxidase from etiolated pea seedlings with molecular mass 185 kDa has two bands on electrophoregram in polyacrylamide gel. One of these bands has the capacity for oxidative desamination (Hassler et al., 1969).

The presence of monoamine oxidases in plants is not yet well studied. The main information about this enzyme is known from animal physiology. Monoamine oxidase or desaminating amino-oxygen-oxidase, which contains flavin (EC 1.4.3.4), desaminates and oxidizes amines to physiologically inactive products, in particular catecholamines to vanillic acid, serotonin to hydroxyindole acetic acid, and histamine to imidazole acetic acid (see Chapter 1). According to modern conception, copper, linked with protein via sulfhydryl groups, is included in monoamine oxidase

(Finberg and Youdim, 1983). Copper is supposed to be necessary for the activation of the reaction center, which includes two sulfhydryl groups, two histidine residues, and one covalently bound flavin adenine nucleotide. The molecular mass of subunits composing this enzyme isolated from animal cells is 100 kDA. However, there are subunits with molecular masses 89, 44, 75, and 52 kDa. This enzyme exists in plural forms. Depending on specificity to substrate, A and B forms are distinguished. The A type predominantly uses adrenaline, noradrenaline, and serotonin, whereas the B type uses other monoamines, such as tryptamine and methylhistamine. Both types of enzymes equally catalyze desamination and oxidation of tyramine and dopamine with high rates. The rate of reactions with participation of monoamine oxidase depends on pH. Inactivation of the enzyme leads to the accumulation of catecholamines and indole amines in brain of mammals, which induces various diseases (Finberg and Youdim, 1983). Chelating agents inactivate monoamine oxidase, binding iron, copper, and manganese, although these elements are supposed to be not included in active center. Deficit of copper in food does not influence enzyme activity, whereas deficit of iron decreases it. The basic localization of monoamine oxidases in animal cells is the mitochondria. Monoamine oxidase activity was first found in some plant species 40 years ago (Werle and Roewer, 1950). However, it should be noted that catecholamines, serotonin and histamine are not substrates for this enzyme. The precursor of serotonin tryptamine is oxidized and desaminated to indole acetic acid by monoamine oxidase of oat *Avena* coleoptiles (Thimann and Grochowska, 1968). Inhibitors of monoamine oxidases appear to act on metabolism of biogenic amines. Seed germination of *Raphanus sativus* was stimulated with serotonin (Roshchina, 1992), and this fact may be explained as the transformation of the neurotransmitter to indole acetic acid (plant phytohormone), catalyzing with monoamine oxidases. However, inhibitors of monooxidases transamine and ipraside had no depressive effects in the presence of serotonin in the seed growth medium (Roshchina, 1992).

Monoamine oxidase, acting on both tyramine and dopamine on phenolase activity by hydroxylating tyramine, tyrosine, and 4-hydroxyphenylacetaldehyde, is also present in the cells (Schutte, 1991). Monoamine oxidase oxidizes cinnamoylcholine (Williams *et al.*, 1992). DOPA-decarboxylase (in *Berberis* cell cultures) was potentially purified about 30-fold with an optimum pH of 7.5 and a molecular mass of 47 kDa. A phenoloxidase (cresolase) that can hydroxylate tyrosine (I) or tyramine (III) in meta position was purified 55-fold and has an optimum pH of 8.5 and molecular mass of 54 kDa. Transamination of tyrosine and DOPA was found to occur in a large number of cell cultures, all known to produce isoquinoline alkaloids. The enzymes of *Berberis stolonifera* have an optimum pH of 7.5 and were slightly less active on DOPA (84%) than tyrosine.

Tyrosinase (dihydroxyphenylalanine oxygen oxidoreductase, EC 1.14.18.1) is also found throughout plant and fungal species (Mayer and Harel, 1979; Mayer, 1987; Vamos-Vigyazo, 1981; Kumar and Flurkey, 1991).

Besides the above-mentioned enzymes, other oxidases are involved in decomposition of biogenic amines. For instance, catecholoxidase uses dopamine as substrate in grape fruits (Cash et al., 1976), mango fruits (Thomas and Janave, 1973), peach fruits (Luh and Phithakpol, 1972), and mushroom (Harrison et al., 1967). Adrenaline is used by catecholoxidase as substrate in broad bean leaves (Robb et al., 1966). Adrenaline and noradrenaline are also used as substrates in mushroom tyrosinase (Harrison et al., 1967). Mushroom tyrosinase (EC 1.10.3.1) is a copper-containing protein catalyst that exhibits both phenolhydrolase and polyphenoloxidase activities. Activity of the enzyme with various substrates decreases in the following order: catechol > L(-)-DOPA > dopamine > L (+)-adrenaline > D(–)-adrenaline > L (+)-noradrenaline > D(-)-noradrenaline (Harrison et al., 1967).

Amine oxidases which are capable to oxidize histamine as a substrate are found in *Glycine max*, *Pisum sativum* and *Vicia faba* (Medda et al., 1995).

Methyltransferases. Catechol-o-methyltransferase of *Nerine bowdenii* (Mann et al., 1963) can methylates catecholamines and their derivatives such as alkaloid norbelladine, and to a lesser degree catechol.

3.3. SYSTEMS OF SECONDARY MESSENGERS COUPLING WITH RECEPTORS

During so-called priming (primary response to the neurotransmitter or hormone interaction with the membrane receptor), conformational changes of receptors take place. Via the conformational changes the chemosignaling information transfers into the cell from the plasmic membrane surface. Then it follows two main pathways: (1) the triggering of the systems of secondary messengers such as cyclic 3',5'-adenosine monophosphate (cAMP), cyclic guanosine-3',5'-monophosphate (cGMP), inositol triphosphate, and Ca^{2+} ions (which can also function as secondary mediators or messengers) or (2) the changes in ion (K^+, Na^+, Ca^{2+}, Cl^-) permeability of membranes.

Synthesis of cAMP or cGMP from ATP or GTP occurs in cells with participation of cyclases such as adenylate cyclase or guanylate cyclase, respectively. Cyclic nucleotides can serve as intracellular mediators, if their concentration in cell (usually $\leq 10^{-6}$ M) is strictly controlled. Excessive accumulation of such substances is prevented by phosphodiesterases that hydrolyze cyclic nucleotides to noncyclic products, for instance cAMP to adenosine-5'-monophosphate. Synthesis and hydrolysis of cAMP are shown in Fig. 56.

The mechanism of receptor coupling with the synthesis of cyclic nucleotides has been mostly studied for the adenylate cyclase system. This process is believed to be as follows. The receptor perceiving signal

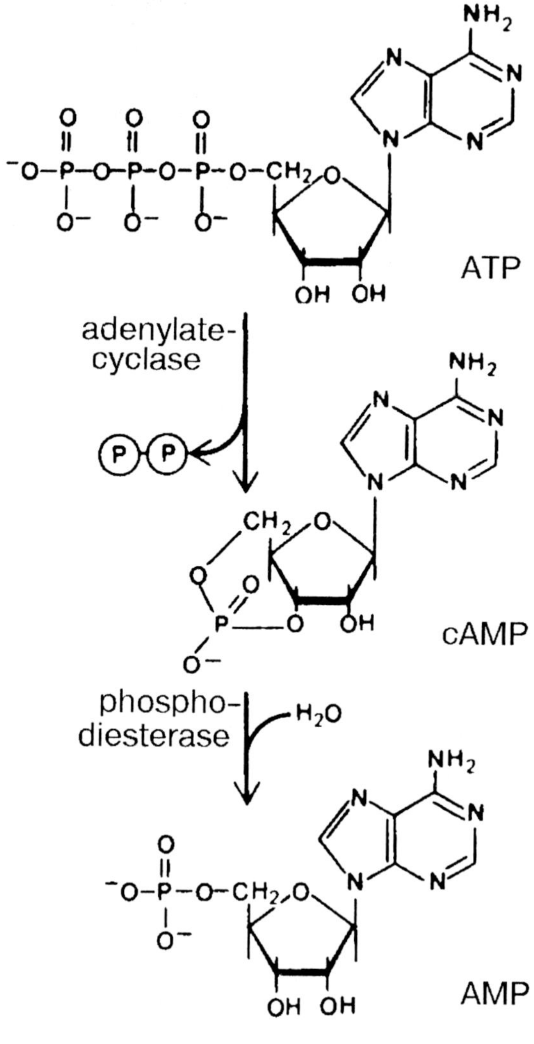

Fig. 56. Metabolism of cAMP

molecules (mediators) is located on the outer side of the plasmic membrane, whereas adenylate cyclase is located on the inner side. Similar localization permits the transmission of external signal into the cell, where intracellular processes are regulated with participation of secondary messengers. As has been demonstrated on animal tissues, adenylate cyclase is activated via various receptors by no less than six mediators and hormones, including adrenaline, noradrenaline, dopamine, serotonin, and histamine. Acetylcholine can also trigger the guanylate cyclase system. Adenylate cyclase is activated via a whole chain of biochemical reactions, beginning from conformational changes of receptors that arise as a result of interaction with mediator and become capable of contact with a regulatory protein, named G-protein or transducin, which binds GTP. G-proteins play a certain role in mechanisms of transmission of the signal from the surface to the interior of animal cells (Nederkoorn et al., 1997; Selbie and Hill, 1998; Haga and Berstein, 2000). Similar proteins were found in root plasmalemma of higher plants (Babakov and Abramycheva, 1989). Molecular masses of GTP-binding proteins in plants and animals approximately equal 90 kDa. The amount of proteins in plasmalemma of maize cells in near 0.4% of total membranous protein (Babakov and Abramycheva, 1989). In animal cells two G-proteins have been found: Gs and Gi. Under certain conditions Gs-protein interacts with catalytic component of adenylate cyclase and activates it. When the enzymes react with Gi-protein, the activity decreases. Thus, the interaction mediator with receptor induces the coupling in the adenylate cyclase system. As a result, the amount of cAMP, universal secondary messenger, is changed. Ion channels (mainly K^+ channels) are stimulated by G-protein in guard cells in fava bean (Fairley-Grenot and Assmann, 1991).

Secondary mediators not only participate in the transmission of external signal into cellular interior, but also strengthen the original signal. Each receptor molecule (Fig. 57), combining with signaling molecule, activates many molecules of adenylate cyclase that, in turn, catalyze the formation of a number of molecules of cAMP. In the end, the initial signals are strengthened about 10^7–10^8 times as it spreads all along the chain of signal transmission from receptor to observed cellular response. Thus, a few signaling molecules of mediators or hormones can change the functional or metabolic activity of the entire cell.

Cyclic AMP regulates intracellular reaction of any prokaryotic or eukaryotic cells. Its action is based on the activation of specific enzymes, cyclic AMP-dependent protein kinases, which phosphorylate various proteins, in particular ribosomal proteins, enzymes, the transport membranous proteins, and others. The phosphorylation of proteins is a mode of their activation. They become inactive again by dephosphorylation with the help of phosphoprotein phosphatase.

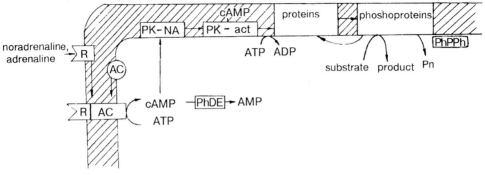

Fig. 57. The adenylate cyclase pathway of the intercellular regulation. R-receptor, AC-adenylate cyclase, ATP-adenosine triphosphate, ADP-adenosine diphosphate, AMP-adenosine monophosphate, PDE-phosphodiesterase, PK-protein kinase, PK-act-active and Pk-Na-non-active state of the enzyme, cAMP-cyclic adenosine monophosphate. PhPPh-phosphoprotein phosphatase, Pn-inorganic phosphate

Hormones and mediators can show their effect not only via cAMP synthesis, but via other intracellular transmitters as well. For instance, cyclic guanosine-3',5'-monophosphate is known as secondary messenger for some animal hormones. Its concentration in cell is 10-fold lesser than that of cAMP. Like cAMP, cGMP acts mainly via activation of certain protein kinases.

The secondary messengers whose formation can be induced by adrenaline and serotonin are diacylglycerol and inositol phosphates (Berridge and Irvine, 1984). They are formed at the hydrolysis of phosphatidylinositol-4,5-bisphosphate (Berridge, 1984; Munnik et al., 1998) (Fig. 58). Diacylglycerol activates protein kinase C, which phosphorylates proteins, whereas inositol phosphate stimulate Ca^{2+} efflux from cellular compartment in cytoplasm. Changes in calcium concentration regulate the activity of many enzymes, including kinases. Ions of Ca^{2+} themselves are known as secondary messengers. In a non-stimualted cell, their concentration is about 10^{-8}–10^{-7} M. When signaling molecule interacts with the receptor, concentration of Ca^{2+} increases 12-fold. As a result, some cellular biochemical processes are activated or inhibited. In most cases Ca^{2+} influences the intracellular processes via calcium-binding protein calmodulin, which is in all plant and animal cells (Dieter, 1984). Calmodulin is a polypeptide consisting of 148 amino acids. Connection with Ca^{2+} is accompanied by conformational changes of calmodulin and by this means activates calmodulin-dependent proteins, such as NAD-kinase, phosphodiesterase, and other enzymes.

Do similar events take place in plants? In order to answer this question it is necessary to consider components of regualtory systems of

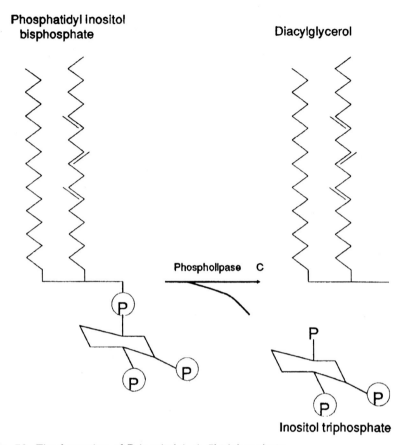

Fig. 58. The formation of D-inositol 1, 4, 5'-triphosphate.

secondary messengers. Some authors have analyzed the literature (Kessler and Levinstein, 1974; Doman and Fedenko, 1976; Brown and Newton, 1981) and confirmed evidence of the presence of cAMP in plant cells belonging to both dicotyledons and monocotyledons. In particular, this compound is found in *Vinca rosea* (Wood et al., 1972), *Zea mays* (Janistin, 1972), and some other plants, as well as in unicellular organisms such as yeast *Saccharomyces cerevisiae* (Levitzki, 1988). According to Kessler and Levinstein (1974), the amount of cAMP in plants varies from 0.06 to 1.27 nmoles g^{-1} of fresh mass, in particular in such species as *Pisum sativum* (etiolated seedlings 0.170–0.370), *Persea americana* (roots 0.520–1.300, leaves 0.826), *Lemna gibba* (shoots 0.540–1.040), *Nicotiana tabacum* (culture of tissue 0.06–0.122), *Citrus limettiodes* (leaves 0.230), and *Mangifera* (leaves 1.270). The presence of cAMP (light 4.6 nmoles g^{-1} of fresh mass, dark 0.6) and cGMP is observed in shoots and leaves of mung bean *Phaseolus aureus* (Brown et al., 1989). Chloroplasts from this species contain much more of these cyclic nucleotides: cAMP contains 38.3 nmoles g^{-1} of fresh mass under light and 12.2 in the dark, whereas cGMP contains near 20.2 nmoles g^{-1} (Brown et al., 1989).

Synthesis of cAMP in plants is closely connected with photophosphorylation and oxidative phosphorylation, because in these processes substrate ATP needed for the reaction is formed. As has been shown above, accumulation of cAMP in isolated chloroplasts and shoots of mung bean increases in light 3- to 7-fold in comparison with dark, which makes it evident that photophosphorylation plays a key role as the source of ATP required in the synthesis of this cyclic nucleotide (Brown et al., 1989). The experiments of Kirshner and coworkers (1975) demonstrated that cyclic AMP can function as secondary messenger in plants as well. Exogenous acetylcholine (10^{-11} M) and cAMP (10^{-9}–10^{-4} M) decreased the ATP amount in mung bean *Phaseolus aureus*. This perhaps occurs as a result of adenylate cyclase stimulation. On the contrary, blocker of cholinoreceptor atropine (10^{-4} M) sharply increased the ATP concentration, because the formation of cAMP does not occur. In this case atropine broke the coupling of cholinoreceptor with adenylate cyclase. The enzyme of synthesis of cAMP, adenylate cyclase (ATP pyrophosphate lyase, cyclizing EC 4.6.1.1), is found not only in animals, but also in plants. This membrane-binding enzyme is lipoproteid mainly with phospholipids. The protein isolated from roots of alfalfa *Medicago sativa* has molecular mass 84 kDa (Carricarte et al., 1988). It is activated by ions of Mg^{2+} and Ca^{2+}, as well as by calmodulin. Adenylate cyclases of animals have higher molecular mass (from 159 to 180–200 kDa) than those of plants. Perhaps these are dimers. Adenylate cyclases from both animals and plants are found in plasmic membranes (Brown and Newton, 1981) and in fractions of chloroplasts (Brown et al., 1989), nucleus, mitochondria, microsomes, and tonoplast (Brown and Newton, 1981, Ladror and Zielinski, 1989). Moreover, in chloroplast fractions, Brown and coworkers (1989) have found guanylate cyclase. It is possible that in cellular organelles there are receptors connected with adenylate cyclase.

In plant cells the phosphorylation of various proteins, mainly, enzymes, occurs. Among them there are controlling membranous processes such as opening of ion channels, in particular in plasmalemma (Shiina and Tazawa, 1986), activation of functioning of reaction centers of photosystems, and components of electron transport chain (Süss, 1981; Barber, 1986) as well as processing.

In plants there are also protein kinases (Ranjeva and Boudet, 1987; Viale et al., 1991). Protein kinase activity is found in many organelles: plasmalemma, tonoplast (Ladror and Zielinski, 1989), and chloroplasts (Brown and Newton, 1981; Barber, 1986). In plastids it is observed in envelope (Soll, 1988) and thylakoids (Staehelin and Arntzen, 1986). This is a peptide with molecular mass 63 kDa. Protein kinase participates in phosphorylation of reaction centers of photosystems (Staehelin and Arntzen, 1983), ATP synthetase (Barber, 1986) and cytochromes b and f (Süss, 1981). For some- time there was only indirect evidence that cAMP participates in protein phosphorylation in plants (Brown and Newton, 1981; Ranjeva and Boudet, 1987). However, Janistyn (1989) showed that cAMP is involved in endogenous phosphorylation of protein from dialyzed milky sap of coconut palm *Cocos nucifera* fruits. Excess of cAMP can be decomposed by phosphodiesterase (EC 3.1.4.17), which is found in pea (Lin and Varner, 1972), soybeans (Brewin and Northcote, 1973), artichoke, and other plants (Giannottasio et al., 1974). In animal tissues phosphodiesterase exists in free and bound forms, in particular in fractions of plasmatic, nuclear, and mitochondrial membranes, but mainly in cytosol. Dual localization of this enzyme in both cytoplasm and organelles (as well as in plasmalemma) has been found in higher plant cells (pea seedlings) and microorganisms (yeast) (Lin and Varner, 1972). Phosphodiesterase has been isolated from tissue culture of soybeans (Brewin and Northcote, 1973) and yeast *Saccharomyces cerevisiae*. Molecular mass of this enzyme is 350 KDa. Some of its characteristics differ from those of animal phosphodiesterases. First, enzyme from pea seedlings is not inhibited by methylxanthins, such as theophylline and caffeine (Lin and Varner, 1972); second, it is capable of hydrolyzing not only 3,5-cAMP, but also 2,3-cAMP, which also forms under the treatment of plant nucleases.

Brown and coworkers (1989) found cyclic AMP and GMP, adenylate cyclase and cAMP-phosphodisesterase activities, and guanylate cyclase and cGMP-phosphodiesterase activities in plant cells. They are localized in chloroplasts.

Besides cyclic nucleotides and related enzymes, components of inositol phosphate system of regulation (Fig. 58) in particular phosphatidylnositol phosphodiesterase or phosphodiesterase C, have been found in plants such as *Apium graviolens, Brassica oleracea, Allium cepa*, and *Narcissus* sp. (Irvine et al., 1980, 1992; Melin et al., 1987). The enzyme hydrolyzes phosphatidylinositol-4,5-diphosphate, forming diacylglycerol and inositol phosphates: IP_1, IP_2, IP_3 (Cote et al., 1989). The presence of messenger inositol triphosphate or calcium ions as well their effects were demonstrated in pollen of *Papaver rhoeas* (Franklin et al., 1994). There is also direct evidence of regulation in Ca^{2+} efflux in cytoplasm by micromolar concentrations of inositol-1,4,5-triphosphate (Drøbak and Ferguson, 1985).

Now the role of Ca^{2+} as secondary messenger in animals and plants is generally accepted (Dieter, 1984). Concentration of Ca^{2+} in cell is controlled by calcium-binding protein calmodulin, which is also thought to be a component of secondary systems of regulation (Cormier, 1983). Purified calmodulin of plants has different molecular masses in presence of Ca^{2+} and without it, 17–19 kDa and 14.5 kDa respectively (Marme and Dieter,

1983). The amino acid sequence of this enzyme was also demonstrated. Calmodulin is widespread in plants. Its quantity in cytoplasm is up to 10^{-3}–10^{-2} moles m^{-3} (Dieter, 1984; Jones and Halliwell, 1984). Calmodulin is present in chloroplasts as well (Roberts et al., 1983). Besides enzymes of adenylate cyclase system, Ca^{2+} and calmodulin control the activity of enzymes, participating in the photodestruction of water during photosynthesis (Barr et al., 1983), membrane depolarization, secretions of cell wall components, chloroplast movement in cytoplasmic stream and phototaxis of chloroplasts, leaf movement of *Mimosa*, and growth induced by hormones (Dieter, 1984; Roberts and Harmon, 1992).

In plant cells calcium as secondary messenger can be triggered by calmodulin on the level of organelles that bind with biogenic amines. For instance, noradrenaline, adrenaline, and serotonin stimulate Ca^{2+} efflux from intact chloroplasts (Roshchina, 1989a, b).

The attempt to show direct participation of cAMP system in intracellular signaling was demonstrated by use of agents (Fig. 59) acting on different

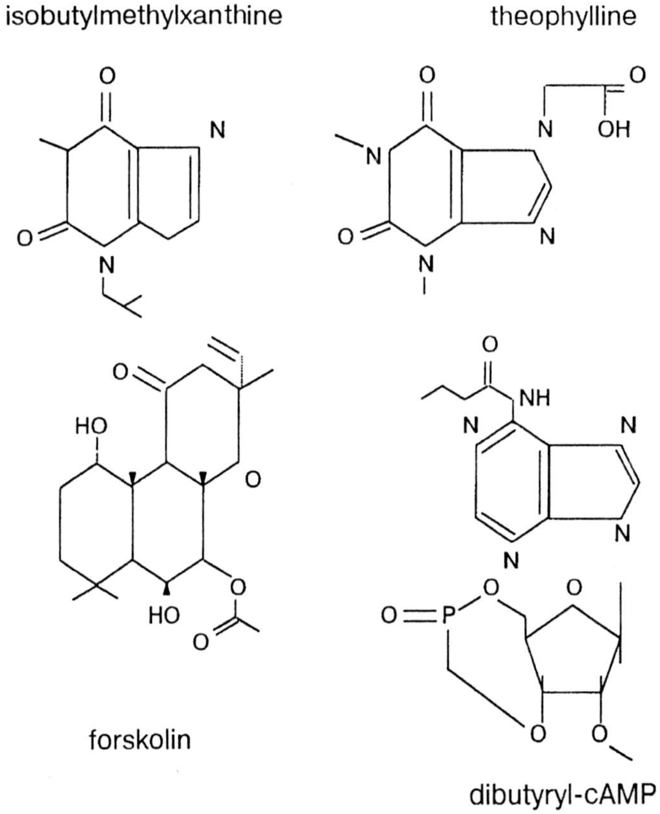

Fig. 59. Formula of activators and inhibitors of the cAMP system

sites of the system-activator of adenylate cyclase forscolin, activator of the intracellular cyclic AMP synthesis of dibutyryl cAMP, inhibitors of phosphodiesterase theophylline and isobutylmethylxanthine (Roshchina et al., 1998b, c). Figure 60 shows that forscolin and dibutyryl cAMP stimulated the germination of *Hippeastrum hybridum* pollen *in vitro*, whereas theophylline and isobutylmethylxanthine inhibited the process, and exogenous cAMP and cGMP practically had no effect. Therefore, endogenous cyclic nucleotides can participate in the transduction of the information from plasmalemma to nucleus. Stimulating effects of acetylcholine, histamine, and serotonin were not observed if the pollen was previously treated with theophylline and isobutylmethlyxanthine. Thus, these compounds can act on pollen germination via the cAMP system.

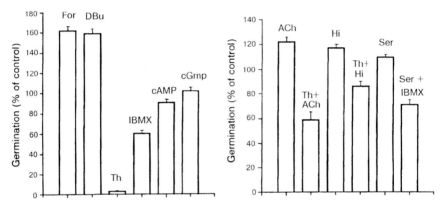

Fig. 60. Action of activators and inhibitors of the cAMP system on pollen germination of *Hippeastrum hybridum* (Roshchina et al., 1998b, c). For, forscolin 10^{-4} M; DBu, dibutyryl cAMP 10^{-5} M; Th, theophylline 10^{-5} M; IBMX, isobutylmethylxanthine 10^{-4} M; cAMP and cGMP, cyclic AMP and GMP 10^{-9} M; ACh, acetylcholine 10^{-5} M; Th + ACh, theophylline 10^{-5} M + acetylcholine 10^{-5} M; Hi, histamine 10^{-5} M; Th + Hi, theophylline 10^{-5} M + histamine 10^{-5} M; Ser, serotonin 10^{-5} M; Ser + IBMX, serotonin 10^{-5} M + isobutylmethylxanthine 10^{-4} M.

Active oxygen species superoxide radical, hydroxylradical, and peroxides, formed in plant cells containing biogenic amines (Fig. 8, Chapter 1), could also signal to nucleus via secondary messengers.

CONCLUSION

Various components of animal cholinergic and aminoergic systems of regulation are present in plants: neurotransmitters, enzymes of their synthesis (choline acetyltransferases, methyltransferases, hydroxylases, and decarboxylases), enzymes of their catabolism (cholinesterases, aminooxidases), and possible analogs of their receptors. Neurotransmitters are recepted by sensitive plant surfaces, mainly plasmalemma and membranes of organelles, via binding with special sites (sensors-receptors), and received information is transported into cell (from outside plant cell) or organelle (from cytoplasm) through transducins (G-proteins) and via systems of secondary messengers. Secondary messengers cyclic AMP and GMP, inositol triphosphate, and Ca^{2+} ions are also found in the plant cell and within organelles. The enzyme cholinesterase, catalyzing acetylcholine hydrolysis, participates in many plant functions, including fertilization, protein-protein interactions, and recognition. Enzymes of deamination of biogenic amines (aminooxidases) are also included in redox reactions of plant cells. Some results of experiments demonstrate participation of secondary messengers, such as inositol triphosphate, Ca^{2+} ions, and cAMP, in the transfer of information (received from neurotransmitter outside plant cell) within the cell.

Physiological Role of Neuromediators in Plants

Cellular recognition systems in plants have analogs in animal organisms, where they deal with information transfer between cells in synapses in the form of chemical signals such as transmitters or mediators. In plant excretions, nitrogen-containing compounds that function as neurotransmitters in neural and muscle tissues are also found. Among them are cholinic ester acetylcholine and biogenic amines dopamine, noradrenaline, serotonin, and histamine (see section 3.1), which participate in intercellular chemical signaling in animals.

Transmission of information in the form of chemical signals began in the early unicellular organisms. Their selective chemosensitivity enables chemoreception, which is based not only on food requirement, but also on participation in such relationships as attraction or repulsion, and partnership or competition with other species or persons of the same species. The evolution of chemoreception has evidently been directed toward development of information systems both in the biocenosis and within individual organisms:

Chemoreception of unicellulates

Biocenosis
Chemoreception of multicellular animals (olfaction)
Chemoreception of multicellular plants (allelopathic recognition)

Individual multicellular organism
Transmission of information by mediators inside organism (cholinoreception, adrenoreception, etc.)

The peak of evolution in animal organisms is the appearance of mediators-transmitters of information between cells of multicellular organisms such as acetylcholine and biogenic amines (see section 3.1). In the bacterial cell, recognition systems, for instance at chemotaxis, are stimulated by compounds of plant excretions or their analogs (Macnab, 1985). Among them are attractants (sugars, amino acids, amines, cyclic AMP, sesquiterpenes) and repellents (short-chain and unsaturated fatty acids, aliphatic C_2-C_4-alcohols, hydrophobic amino acids leucine, isoleucine, valine, tryptophan, indole, skatole, and aromatic acids).

4.1. MEDIATORY FUNCTION

Common principles. Excitability is the capacity to form impulses of excitation in response to changes in the environment. This response is peculiar to any living cell. The plasmalemma, which regulates the metabolic exchange between the cell and its surrounding medium, is the chief agent in sensing of external signals and spreading them. Simultaneously, plasmalemma plays the role of transmitter between cell-free surrounding medium and cellular organelles. Most cellular membranes, including mitochondria and chloroplast, are structurally similar to the plasmalemma. Common principles of chemical signalization are believed to exist between cells. Plasmalemma, when it receives external signal, transmits the impulse of excitation to membranes of separated organelles via chemical mediators, as occurs in intercellular synapses of animals, where acetylcholine, catecholamines, serotonin, histamine, amino acids, and other substances function as signaling agents. Interacting with cell membranes, chemical agents transmit information in this way, signaling about changes in the external medium. Signals are received by membranes of various organelles: nuclei, mitochondria, and, in plants, perhaps chloroplasts as well.

Rapid intracellular regulation of physiological processes using metabolites as chemical transmitters of the excitation impulse from plasmalemma to organelles first occurred in unicellular organisms, where plasmalemma itself received signals from the environment (Schiffman and Gallin, 1979).

Thus, intracellular and intercellular signalization as a means of transmitting information appears to have developed from unicellular to multicellular organisms. Chemical signalization between cells in multicellular animals has arisen as an evolutionary development of the adaptations made by prokaryotes in reaction to chemical changes in the environment (Koshland, 1981). This is evident from the fact that intracellular compounds such as acetylcholine, catecholamines, and serotonin are used in both unicellular and multicellular organisms (Strakhovskaya *et al.*, 1982; Levitzki, 1988). For instance, in mammals and yeast, cellular mediators binding with receptor appear to activate the enzymes adenylate cyclase and guanylate cyclase, which catalyze the synthesis of secondary intracellular regulators or secondary messengers, respectively cyclic AMP and GMP.

Location and transport of neurotransmitters within cell. One important problem is how acetylcholine and biogenic amines can move in plant cells and penetrate through membranes. Their water-soluble molecules are too hydrophilic to cross the lipid layers of membranes.

In animal cells, most hormones and local mediators are localized in special vesicles that differ from other similar structures in having a coat consisting of clathrin. Fibrillar clathrin with molecular mass 180 kDa, linked with polypeptide 35 kDa, forms a cover and the vesicles are thus seen as coated or fringed structures under the microscope. Coated vesicles are the transport organelles that via endo- and exocytosis can transfer substances between organelles and membrane recycling. Having crossed the membrane, the hormone, mediator, or other ligand binds with a special receptor, inducing the gemmation of vesicles from the membrane. Then within the cell the translocation and transformation of these vesicles take place. When the vesicle is transported to a membrane and fuses with it, the vesicle sheds the cover and liberates from clathrin, so that its contents can be released. ATPase with molecular mass 70 kDa takes part in this process. The biological significance of the formation of coated vesicles is still unclear. The clathrinic cover is supposed to prevent lysis of these structures and their content. Perhaps it promotes the coupling of vesicles with cytoskeleton following their transport into the cell.

Clathrin-coated vesicles are also found in plant cells (Coleman et al., 1988). They are observed in many cells of algae, higher plants, and fungi. In higher plants, coated vesicles are in meristematic tissues, actively growing and differentiating cells, root hairs, germinating pollen, differentiating vascular tissues, and cells of parasitic mycorrhiza. It is possible that these vesicles are found in any cell and have a universal function. Coleman and coworkers (1988) point out that under electron microscope coated vesicles in plant cells are seen near the plasmatic membrane, Golgi apparatus, and other organelles. The same authors proposed two pathways of transport for specific large molecules (ligands) and low-molecular-weight mediators in clathrinic vesicles. The first is between internal compartments and the second from internal compartments to plasmatic membrane and back. The transport pathways of coated vesicles are proposed for plant cells as well.

The translocation of clathrinic vesicles within cells is promoted by the organization of the cytoskeleton, including contractile systems of microtubules and microfilaments. Microtubules are observed not only in cytoplasm, but also inside organelles, in particular in chloroplasts (Vaughn and Wilson, 1981), which permits these vesicles to move within the plastid too. Since acetylcholine and catecholamines are found in chloroplasts (Roshchina, 1989a) (see Chapter 1), this may be an indication (parallel with other experimental evidences) of intracellular mediation.

Distinct mechanisms operate in different secretory systems of plants and animals. Neurotransmitters may be transported within the plant cell and outside it, as in animal cells. Among the mechanisms are: (1) free diffusion through the plasma membrane; (2) exocytosis, resulting from fusion of a secretory granule with the plasma membrane, and sorting of proteins in the secretory system, and (3) fleeting release from a granule through a transient pore without full fusion or the release through a specialized plasmalemma molecule such as the mediatophore. The last mechanism is proposed to occur in rapid synapses of animals in which the neurotransmitter is emitted as an abrupt chemical impulse of quantal composition (Dunant, 1994). The release of the secretion is momentarily signaled in the plasma membrane by large intramembrane particles (Dunant, 1994). Synaptic vesicles are also essential for regulation of this type of release. They fuse with the plasma membrane only late after activity and seem to be involved in calcium sequestration and extrusion. Cellular mechanisms of the excretion may be active or passive. The active mechanism is exocytosis. Now the concept of four fundamental mechanisms of membrane fusion in eukaryotic cells from yeast to mammalian neuron is considered. According to that concept, there are endosome-derived and Golgi-derived vesicular pathways to plasmatic membrane. Hydrophilic substances such as neurotransmitters and water-insoluble particles are full fusions via the secretory vesicles or secretory granules by exocytosis or the rupture of the plasmatic membrane. The secretion of the large granules is supposed to connect with the formation of a fusion pore between the interior of a granule and the extracellular space. A novel special group of proteins, neurotransporters through plasmic membrane, has also been found (Schloss et al., 1992; Kelly, 1993).

Neurosecretion uses mechanisms common to all eukaryotic membrane transport, and the process should be a model of the secretion as a whole (Bajjalieh and Scheller, 1995). Neurotransmitter release is the sum of many molecular processes. The release of neurotransmitter via regulated exocytosis (from 50 nm vesicles, forming in endosomal compartments, and then undergoing many rounds of fusion and recycling at presynaptic terminals) is the primary mode of communication in the nervous system. Synaptic vesicles fill with a transmitter after they are formed. Filling with the substance is mediated by specific transport molecules. Loaded vesicles are either sequestered in a reserve pool via interaction with cytoskeletal elements or cluster at the presynaptic terminal at specialized regions termed active zones. The fusion of these docked vesicles, with the plasma membrane occurs when intracellular calcium concentrations rise during an action potential. Calcium regulates one of the final events in synaptic vesicle fusion because the neurotransmitter is released in less than 1 ms following calcium elevation. Docking and fusion of the vesicles occur through a series of protein-protein interactions. First of all there is a targeting of the vesicle on the acceptor membrane. The transport vesicle contains a specific targeting protein that is capable of associating only with a receptor protein specific to the appropriate acceptor membrane. The first stage is docking, where small GTP-binding proteins are required for vesicle targeting and docking. Then acidic phospholipids are necessary for both constitutive and regulated secretion. The actin cytoskeleton also participates in the secretion. The calcium regulation of the process is supposed to be associated with the occurrence of a calcium sensor in the membrane.

Types of cellular contacts in which mediation could occur. The partnership between two cells of one organism or different organisms is based on a balanced exchange of metabolites. Cell-cell recognition is of several types: between sexual cells, between somatic cells, and between non-specialized cells (Roshchina, 1999a). There are two types contacts between cells: (1) structured contacts having gap junctions or non-gap junctions between the cells of the same oganisms and (2) non-structured contacts formed at the joining points of two cells from different organisms (Fig. 61). Structured junctions are physical linkages between the cells formed by transmembrane protein channels and are called gap junctions. These structures permit the diffusion-mediated transfer of low molecular weight (< 1.7 kDa) cytoplasmic components, chemical signals, and metabolic or energy precursors. Similar contacts in plants occur through the plasmadesmata. The basic unit of the gap junction is called connexon and is composed of six polypeptide chains that form a single channel of 2 nm diameter. The head-to-head association of two connexons between contacting cells forms a gap junction. In a non-gap junction, the contacts may be through synapsis between nervous and lymphoid cells within the animal. The communication between cells, forming synapsis, occurs

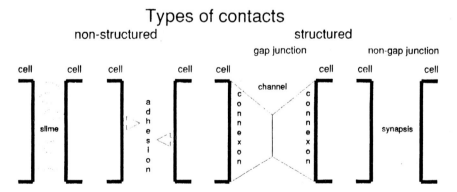

Fig. 61. Types of cell-cell contact in plants (Roshchina, 1999a)

through special chemical signals called transmitters, e.g. acetylcholine and biogenic amines, which move in the synaptic space from one plasmatic membrane to another. We do not know yet whether similar types of contact exist between different organisms. Generally two unicellular organisms form an unstructured linkage. It can be a simple combination based on adhesion or an indirect communication through the slime, released by a receiving cell surface, e.g., olfactory signals in insects and lower organisms or in tentacle hairs of carnivorous plants.

Independent on the type of contact, the chemical signal received by the acceptor cell surface is spread along the organism. Coordination and synchrony in a number of cellular activities in multicellular plants and animals occur through the transfer of low-molecular-weight signaling molecules via plasmodesmata in plant cells and gap junctions in mammalian cells (Roshchina, 1999a).

4.1.1. Intracellular and Intercellular Chemical Signaling in Plants

Signaling with participation of neurotransmitters may occur at contacts between plant cells and between organelles within the same cell. Two experimental models (Roshchina, 2000) could be considered for the analysis of the process:

neurotransmitters at contacts

Model	Model
Within cell	*Between cells*
↓	↓
cytoplasm-organelle system (chloroplast)	pollen-pistil system

Model of intracellular signaling. Based on the concept of universal mechanism of intracellular chemical signalization from plasmalemma to organelles, the reactions of chloroplasts on mediators are considered (Roshchina, 1989a, 1990a). According to the symbiotic theory of plastid origin (Whatley *et al.*, 1979), the chloroplast is the descendant of cyanobacteria that has been caught by the eukaryotic cell by means of endocytosis, which led to the symbiosis of two earlier independent organisms. Naturally, sensory features of chloroplasts should manifest within cell-like reactions of unicellular organism to external chemical irritator. This hypothesis is confirmed by the plastid structure. Like an individual cell, the chloroplast is surrounded by an envelope consisting of two different membranes. The interior of the chloroplast is filled with thylakoids, which are seen as plane disk-like vesicles surrounded by a third type of membrane, named thylakoid membrane. The external membrane of the plastid envelope is highly permeable, whereas the internal one is less permeable, and special transport proteins are inserted in this membrane. There is a narrow intermembraneous space between internal and external membranes of the chloroplast envelope. The internal membrane of this envelope surrounds a large central part, the stroma, which contains a large amount of solubilized enzymes. Thylakoids are immersed in the stroma. Unlike thylakoid membranes, the internal membrane of the chloroplast envelope has no pigments and the electron carriers. On the contrary, thylakoid membranes include the reaction centers of photosystems, components of electron transport chain and ATP synthesis. Internal cavities of thylakoids communicate with each other and form a third internal compartment of chloroplasts, the thylakoid space. Thus, there are three types of membranes in chloroplasts—three zones of receiving of intracellular signals.

The reception of chemical signals by the envelope of chloroplast or thylakoid membrane can be indicated by changes, in ion permeability or in photochemical activity respectively, that permit the characterization of chloroplasts as biological structures, subject to universal lows of chemical signalization (Roshchina, 1989a).

In order to confirm this hypothesis, the influence of acetylcholine and biogenic amines on photochemical activity and ion permeability of pea chloroplasts was studied (Chapters 2 and 3). The possible participation of these substances in intracellular signaling is confirmed by some data, such as the effects of low concentrations (10^{-10}–10^{-7} M) of exogenous mediators, the presence of endogenous acetylcholine and biogenic amines in chloroplasts, and the finding of cholinesterase in plastids.

Model of intercellular signaling. Intercellular signaling is considered in the pollen-pistil system (Roshchina, 1999b). At fertilization, the interaction between pollen grains and a pistil stigma includes several events: (1) recognition of specific signal stimulus in plant excretions by cell-receptor; (2) spreading of chemical information within cell-acceptor; (3) formation of characteristic response of cell-acceptor, germination of the pollen tube. The signal stimulus is recognized at the contact of pistil with the pollen surface. The chemosignaling substance of pistil is recepted by the external and internal pollen cover (exine and intine) and a secretion located within the cover is released just after it is moistened. The exine is a rigid polymer complex structure composed of carotenoid and/or phenol residues, but sensitive to various chemical stimulus. If pollen is on the pistil of own species and receives the stimulus (perhaps neurotransmitter), the physiological response is seen as pollen germination, fertilization, and formation of fruits and seeds (Roshchina and Melnikova, 1998a, b; Roshchina et al., 1998a, b, c). Autofluorescence of pollen and pistil, dealing with changes in the state of components under excitation with ultraviolet light 360–380 nm, serves as sensory reaction during pollen-pollen and pollen-pistil interactions (Roshchina et al., 1996, 1997, 1998a). The response to the chemosignal is quick, within several minutes. Autofluorescence is connected with the state of surface components; redox reactions with neurotransmitter could be estimated by the recording of the fluorescence spectra of pollen and pistil (Roshchina and Melnikova, 1998a, b). Pollen secretion is a primary medium for a chemosignal spreading from pollen surface to plasmalemma (Roshchina et al., 1998b). Secretion of the recipient pollen cover may serve as a recognition and transportation liquid, like olfactory slime in animals for chemosignal. The secretion may contain neurotransmitters acetylcholine and histamine (Marquardt and Vogg, 1952), catecholamines (Fig. 62) (Roshchina and Melnikova, 1999), and cholinesterase (Bednarska, 1992; Roshchina et al., 1994). In male cells of animals (spermatozoa), acetylcholine (Nelson, 1978; Clegg, 1983), catecholamines, and serotonin (Young and Laing, 1990) are also found and participate in fertilization as hormones.

Among the contacting liquid ingredients in pollen secretion are proteins (peroxidase, esterases, etc.) and lipids (Stanley and Linskens, 1974). Proteins 87, 75, 58–55, 40–45, and 28 kDa, perhaps participating in the recognition of chemical substances released by cells of pistil stigma, are found in pollen secretion from *Betula verrucosa, Papaver orientale, Petunia hybrida,* and *Hippeastrum hybridum* by polyacrylamide gel electrophoresis with sodium dodecyl sulfate (Roshchina et al., 1998b, c). In secretion of *H. hybridum* pollen, high-molecular-weight proteins were completely absent, excluding protein 90 kDa (Roshchina et al., 1998b, c). In olfactory slime, proteins 28 kDa with a glutathione peroxidase activity and 40–45 kDa are supposed to participate in redox reactions and recognition. Similar proteins are also found in pollen excretions studied. The occurrence of acetylcholinesterase in pollen is demonstrated one of the possible sensors for natural substances with an anticholinesterase activity or participation of the enzyme in recognition of excretions from both competitive pollen species and pistil. Moreover, cholinesterase inhibitors applied on the pistil stigma promoted fruit development (Roshchina and Melnikova, 1998a, b). This is a first stage in chemical interaction that can influence the triggering or depression of pollen germination.

Free radicals (superoxide radical and others) and hydrogen peroxide are formed on the pollen surface because of interaction with strong oxidants as natural catecholamines dopamine and noradrenaline on contact with oxygen of air (Roshchina et al., 1998a; Roshchina and Melnikova, 1998a, b). Reactive oxygen intermediates can serve as mediators, activating the transcription factor. Superoxide radical is converted into hydrogen peroxide by superoxide dismutase, and the enzyme is often found in the extracellular space (Roshchina, 1996, 1999b). In high concentrations

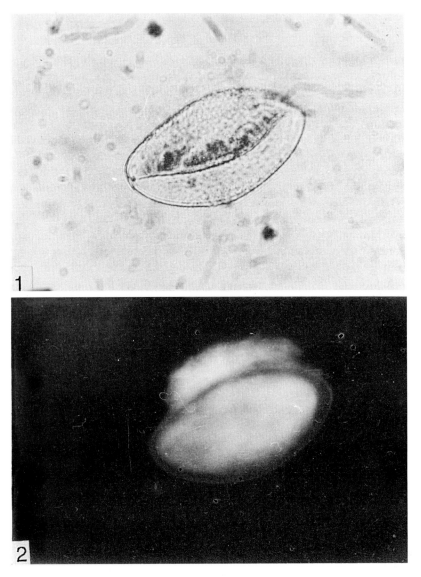

Fig. 62. Pollen of *Hippeastrum* in transmitting white light (1) or in UV light (2) after treatment with 1% glyoxalic acid (reagent on the catecholamine presence), × 260 (Roshchina and Melnikova, 1999). The excretions were clearly seen under UV light.

(> 10^{-3} M), superoxide radical and peroxides are known as factors of damage. Noradrenaline decreased the index of pollen germination, perhaps because of those factors. Reactive oxygen forms can initiate systems of secondary messengers (see Chapter 3, Section 3.2), and the transformed signal spreads into the cell. The free radical chain can be ruptured by antioxidants of the cell-acceptor excretions or antioxidant enzymes such as peroxidase or superoxide dismutase. The first enzyme decomposes H_2O_2 and other peroxides and so prevents the free radical from spreading. Endogenous peroxidase decreased the index of pollen germination (Roshchina and Melnikova, 1998b), but to a lesser degree than superoxide radical generator noradrenaline. In fruits of *H. hybridum* in which pistil was treated with peroxidase (0.6 mg/ml), there were no seeds. Unlike peroxidase, which depresses pollen germination, superoxide dismutase stimulated the process during the first 2 hours after the moistening, which indicated the participation of superoxide radical in inhibition of pollen tube formation.

The examples of intercellular signaling with participation of neurotransmitters in allelopathic relations (chemical relations between different plant species) may also be observed in such systems as pollen-pollen interactions on the surface of pistil (Roshchina and Melnikova, 1998a, b) and root-root relations of various species, because these plant parts concentrate acetylcholine (see Table 1) and acetylcholinesterase (see Table 18), as well as diamine oxidases (Angelini *et al.*, 1990). The cell that senses the secretions of positively or negatively influencing plant species may recognize the neurotransmitters released (Roshchina, 1994, 1999a). The intercellular relations between plants and fungi are another example of the participation of neuromediators in cell-cell recognition (Tretyn and Kendrick, 1991). Both fungi and plants contain acetylcholine and cholinesterase (see Chapters 1 and 3), therefore the mediation is possible.

4.1.2. Electric Potentials in Plants and Possible Role of Neuromediators

Acetylcholine and biogenic amines in animals are known to mediate by two modes: either by opening of ion channels, which leads to the changes of biopotential up to the arising of action potential, or by initiation of adenylate cyclase or guanylate cyclase systems resulting in the formation of intracellular secondary messengers cAMP or cGMP. Let us consider which of these mechanisms could have analogs in plants.

There is now evidence for the participation of mediators in the transmission of chemical signal between animal cells in synaptic contacts. The spreading of the excitation wave as electric impulse along the animal cell is also well known. A similar phenomenon was found in plants by Gunar and Sinyukhin (1959, 1962). The rate of the spreading of the excitation wave in various plant species is usually about 30–40 cm/min, whereas in the conducting system of gourd *Cucurbita pepo* it reaches 40–80 cm/min (Gunar and Sinyukhin, 1963). Moreover, the action potential was registered in reproductive organs of *Ipomoea* flower, when pollen was added to the pistil stigma (Sinyukhin and Britikov, 1967), and at fertilization of ferns and mosses (Sinyukhin, 1973). Now action potential is demonstrated for many plants (Pickard, 1973; Opritov et al., 1991).

Prosser (1986) proposed a universal molecular mechanism of mediator action in animal and plant cells, which consists in the regulation of ion fluxes. The changes of membrane potential are determinated by the shift in ion permeability of membranes by opening or closing of ionic channels (Satter and Moran, 1988). These events are also connected with mechanism of arising and spreading of the action potential in animals, in which membranes have Na^+/K^+ channels controlled by acetylcholine and Ca^{2+} channels mainly dependent on biogenic amines.

The existence of K^+ and Ca^{2+} channels in plasmalemma of plant cells as well as Cl^- channels has been demonstrated by several laboratories (Alexandrov et al., 1976; Yurin et al., 1979a, b; Tester, 1990). H^+ channels are known to be in cell types of membranes, and it is interesting to note that in membranes of mitochondria and chloroplasts they are composed by subunits of CF_o-factor of ATP synthetase (Skulachev, 1989). The existence of K^+ channels has been demonstrated on thylakoids of spinach chloroplasts that have been incorporated in the lipid bilayer (Tester and Blatt, 1989) and similar univalent cation channels incorporated in the chloroplast envelope of *Nitella* by patch-clamp method (Pottosin, 1992). In plasmalemma and vacuolar membrane tonoplast of algae, belonging to Characeae, voltage-dependent calcium channels and Cl^- channels are included in electrical excitation and play the same role as Na^+ channels in nerve cells (Lunevsky et al., 1983). According to Hartmann and Gupta (1989), the ion basis of action potential in plant cells is given by K^+, Cl^-, and Ca^{2+} fluxes through the respective channels. The water transport into and out of the cells, in particular in guard cells of stomata (Madhavan et al., 1995), as well as water transport in conducting systems of multicellular plant organisms sensitive to neuromediators (Zholkevich, 1981; Zholkevich et al., 1990), is regulated by ion fluxes and the state of channel-forming elements. The sensory mechanism in guard cells results in electrical signal by light activation of electrogenic pumps and potassium channels in their plasmalemma (Serrano and Zeiger, 1989). In plant membranes water permeability is regulated by water-channel-forming proteins, named aquaporins (Schaffner, 1998). Swelling of *Hippeastrum hybridum* pollen after wetting was stimulated two-fold with 10^{-5} M acetylcholine and decreased in the presence of d-tubocurarine (Roshchina, unpublished data). Therefore, acetylcholine can control the water exchange of pollen via similar aquaporin pores.

In plants the spreading of the action potential is believed to be associated with calcium and potassium channels and in some cases with chlorine channels as well (Beilby, 1984; Tester, 1990). This mechanism is described as follows:

Mediator → receptor → opening of ion channels controlled by receptor
↓
depolarization of membrane
↓
spreading of the action potential
opening of potential (voltage)-dependent ion channels
↓
depolarization of membrane
↓
opening of potential (voltage)-dependent ion channels

It is believed that there are two types of ion channels: channels controlled by receptors and potential (voltage)-dependent channels. The interaction of mediator with receptor leads to the opening of ion channels controlled by receptor. This in turn induces the drop in membrane potential at the part of membrane lower than a certain value (threshold) and opening potential-dependent ion channels. The depolarization resulting from opening of potential-dependent channels leads to the opening of new potential-dependent channels and so on. By such means the action potential spreads. Cells in a multicellular plant organism communicate via cytoplasmic streams passing through cellular pores from cell to cell and so connecting all cells in a common system. Such a symplastic system of organization promotes not only the movement of ions and metabolites, but also the spreading of electric impulses that transmit information over a distance. These electron impulses are transmitted between cells of filamentous cyanobacteria (Levin et al., 1982) and stem cells of higher plants (Opritov et al., 1991). However, the mechanism of spreading of the action potential in plants is still unclear. Since cells in a multicellular plant communicate via cytoplasmic streams passing from one cell to another, there is no need to excrete chemical mediators in intercellular spaces as happens in synapses of animals. Thus, the excitation is transmitted within intercellular space and throughout the plasmadesmata to other cells, spreading the irritation along the plant organism.

Acetylcholine and biogenic amines play an important role in electric events in animals, inducing the changes in cell membrane potential and the arising and spreading of the excitation wave, known as action potential (McQueen, 1987). The possible participation of acetylcholine in similar reactions in plants is confirmed by some studies. Acetylcholine induces the drop in membrane potential in root cells of higher plants (Jaffe, 1970; Slezak, 1984) and cells of algae *Nitella* (Yurin et al., 1979a, b). In the dark, on incubation of etiolated hypocotyls with acetylcholine (10^{-6} M),

K⁺ uptake from external medium occurred (Hartmann, 1977). Similar concentration of the neurotransmitter can regulate K⁺ uptake by stomata (Madhavan et al., 1995). Acetylcholine decreases membrane potential of chloroplasts in light, measuring as light-induced absorbance changes at 520 nm (Roshchina and Mukhin, 1985a). Adrenaline and noradrenaline are also capable of changing membrane potential of cells, as has been shown for algae *Nitella* (Oniani et al., 1974, 1977a, b). Membrane potential decreases with the increase in its ion permeability. Acetylcholine (10^{-5} M) can induce the increase in potassium (Yurin et al., 1979b) and H⁺ conductivity of plasmalemma (Jaffe, 1970; Slezak, 1984). Lower concentrations (10^{-9}–10^{-7} M) of acetylcholine stimulate the efflux of ions Na⁺ and K⁺ from intact chloroplasts (Roshchina, 1987; Roshchina and Mukhin, 1987a) and the uptake of H⁺ ions independently on radiation by both intact plastids and thylakoids (Roshchina and Mukhin, 1987a). Unlike acetylcholine, adrenaline and noradrenaline (10^{-9}–10^{-7} M) have no action on Na⁺/K⁺ flux, but induce weak efflux of Ca^{2+} and Mg^{2+} ions from intact chloroplasts (Roshchina, 1989a, b).

The transmission of external irritation from plasmalemma to other organelles and between organelles with participation of mediators is imagined to be quite probable if we consider the following. The rest potential of cell is found as the difference between concentrations of ions in the external medium and within the cell, while the rest potential of any organelle is the difference between cytoplasm and the surface of organelle. The separation of the surface charges of membrane, which is connected with asymmetry of the localization of charged groups in membranous proteins, is also important in this phenomenon. Negative charge of the surface of both plasmalemma and organelles is determined by all these factors. Approximate values of membrane potential for intact leaf of *Peperomia metallica* (family Piperaceae) are from –60 to –130 mV (Bulychev et al., 1971); for cell of green algae *Nitella* it is –136 mV (Yurin et al., 1979a, b); for chloroplast inside cell of *P. metallica* it is +15 to –60 mV in the dark and to –80 mV in light (Bulychev et al., 1971), for mitochondria of wheat cells it is –132 mV in the dark and –157 mV in light (Valles et al., 1984). Negative charge promotes the interaction of cations, including acetylcholine and biogenic amines, with plasmalemma and membranes of organelles. In model experiments with isolated intact pea chloroplasts and thylakoids the drop in membrane potential under exogenous acetylcholine was demonstrated (Roshchina and Mukhin, 1985a). These changes are due to efflux of Na⁺ and K⁺ ions from intact chloroplasts in the surrounding medium (Roshchina, 1987a). Within pea and spinach chloroplasts concentrations of these ions are higher than in cytoplasm (Robinson and Downton, 1984), but the envelope of plastid is not permeable to them and only opening of ion channels, when acetylcholine acts so, perhaps to promote a passive ion transport according to their concentration gradient. The opening of ion channels in the presence of acetylcholine leads to the uptake of H^+ ions by both inact organelles and thylakoids (Roshchina and Mukhin, 1987a). When pH is compared in cytoplasm (7.0), stroma of chloroplasts (7.8–8.0), and thylakoids (5.9) (Barber, 1986), it is evident that the movement of protons in intact plastids occurs according to the concentration gradient, whereas in thylakoids it occurs against the gradient, perhaps by antiflux to K^+ ions that have gone out. Thus, from the example with chloroplasts it is seen that acetylcholine localized in cytoplasm and stroma of plastids can induce the opening of ion channels and decrease in membrane potential, which, if it spreads along the membrane, itself represents intracellular action potential. An analogous event induced by opening of Ca^{2+} channels under treatment by catechloamines and serotonin is found in chloroplasts (Roshchina, 1989a, b). However, the existence of intercellular or intracellular contacts, to which mediators could be released, as is observed for synapses, has not been established yet in plants. In connection with this problem, there is an interesting interpretation by Skulachev (1989), made on the base of data gathered in his laboratory and dealing with intramitochondrial contacts in muscle cells. Such contacts, as he supposes, could be similar to synapses.

A concept of non-synaptic binding of neurotransmitters has also been developed in neurology (Vizi and Lendvai, 1999). It is similar to the above-mentioned descriptions in plants.

4.1.3. Neuromediators as Possible Triggers of Secondary Messenger Systems

The presence of systems of secondary messengers, including cAMP, cGMP, adenylate cyclase and guanylate cyclase, inositol triphosphate, Ca, and related components, in plants has been considered in Chapter 3. In animal cells these systems participate in a regulation of organism after the binding of mediator with appropriate receptor on external surface of plasmatic membrane. The main role of mediator in the this process is the activation of protein kinases needed for protein phosphorylation by cAMP or cGMP. Components of cholinergic and adrenergic system of regulation as well as components of secondary messenger systems are found in plant cells and even in individual organelles (see Chapter 3). Therefore, a similar mechanism of triggering of cascade reactions with participation of secondary messengers is also proposed. Lack of synapses in plants leads to a hypothesis of the participation of the compounds in intracellular mediation in their organisms. Thus, the principle of intercellular signalization by mediators and secondary messengers can be applied to

intracellular interactions between organelles. It may be supposed that cellular organelles (chloroplasts and mitochondria) originated from independent organisms with complex internal organization and retained their structures after their inclusion into the eukaryotic cell. Besides, adenylate cyclase, guanylate cyclase, and protein kinase are found in plasmalemma, tonoplast (Ladror and Zielinskii, 1989), membranes of many other organelles, envelope of chloroplasts, and thylakoids (Brown and Newton, 1981; Barber, 1986; Brown et al., 1989). At the transmission of external signals three steps of intracellular signaling can be distinguished:
(1) external factor → plasmalemma → secondary messengers;
(2) secondary messengers → organelle → mediators within organelle;
(3) mediator within organelle → internal structure of organelle.

Intracellular mediators are transported within secretory vesicles. Receiving signal from plasmalemma or other organelles, these vesicles are either built into or fused to the surface membranes of organelles and liberate mediator. Then the mediator reacts with appropriate receptors and thus triggers systems of secondary messengers. This mechanism can be illustrated by an example of transmission of signal from plasmalemma to chloroplasts (Figs. 63 and 64). The diagrams are based on experimental data on the presence of acetylcholine, catecholamines, adenylate cyclase, guanylate cyclase, and protein kinase in chloroplasts (see Chapters 1 and 3), stimulation of ATP synthesis in chloroplasts by low concentrations of acetylcholine and biogenic amines (Chapter 2), changes in Na^+/K^+ permeability of chloroplasts induced by acetylcholine, and Ca^{2+}/Mg^{2+} permeability under the influence of catecholamines and serotonin (see Chapter 2). Besides, the character of the action of agonists and antagonists of acetylcholine on membranes of chloroplasts (Chapter 3) confirms the possibility of the existence of membrane receptors for this mediator. Mediators, being in cytoplasm, can interact with proposed receptors of plastid envelope. This results in changes in permeability to Na^+/K^+ (when acetylcholine acts) or Ca^{2+}/Mg^{2+} (when catecholamines or serotonin act) and triggering of secondary messenger systems by activation of adenylate cyclase on the internal side of the chloroplast envelope. Inside the chloroplast, acetylcholine and biogenic amines can bind with receptors of thylakoid membranes, which also induces the activation of secondary messenger systems. The pathways of chemosignaling with neurotransmitters via secondary messengers may be alternate and/or crossed.

4.2. NON-MEDIATORY FUNCTIONS

The non-mediatory functions of acetylcholine and biogenic amines, namely regulatory and protective functions, are considered.

Physiological Role of Neuromediators in Plants

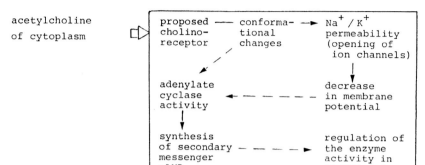

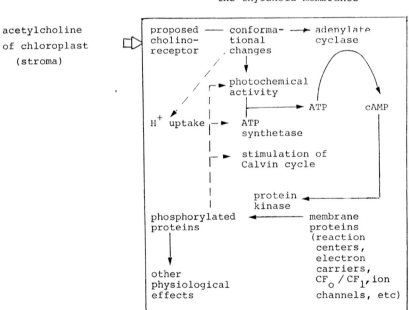

Fig. 63. Regulation of chloroplast processes by acetylcholine (Roshchina, 1990c, d)

194 *Neurotransmitters in Plant Life*

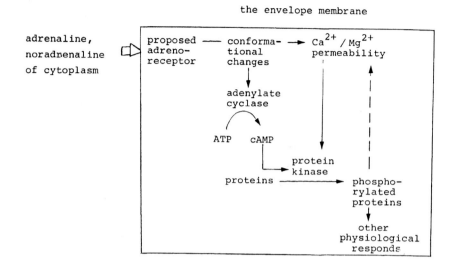

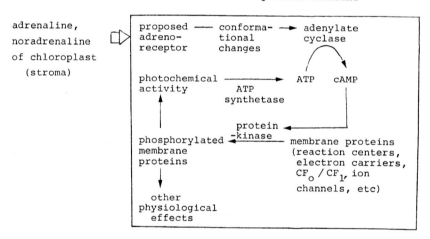

Fig. 64. Regulation of chloroplast processes by catecholamines (Roshchina, 1990c, d).

4.2.1. Mediators as Regulators of Energetic and Metabolic Processes

Acetylcholine and biogenic amines can regulate energetic and metabolic processes in animal cells (Shostakovskaya and Babskii, 1984; Kondrashova and Doliba, 1989). In particular, adrenaline and acetylcholine stimulate the substrate phosphorylation in animal tissues and oxidative phosphorylation in mitochondria isolated from them (Shostakovskaya and Babskii, 1984; Kondrashova and Doliba, 1989). Acetylcholine, dopamine, adrenaline, and noradrenaline also promote photophosphorylation in isolated chloroplasts (Roshchina and Mukhin, 1987a; Roshchina, 1989a, b). The formation of ATP is a source of energy for many cellular reactions as well as the substrate for the synthesis of secondary messenger cAMP. ATP also is needed for active ion transport through membranes. In animal physiology, ATP also serves as local mediator (Alberts et al., 1994). Thus, one of the functions of acetylcholine and biogenic amines consists in stimulation of energetic processes as well. They often serve as cations for maintaining physiological pH (Smith, 1981). The accumulation of adrenaline and noradrenaline in blood and myocardium of animals is usually connected with stress situation and characterizes the reaction to the irritation (Gurin et al., 1989). Acetylcholine-induced release of ATP from animal cells (Rojas et al., 1985) and the product of neuromediator hydrolysis choline acted on Na^+ and K^+ interaction with Na^+/K^+-ATPase (Robinson and Pratap, 1991).

Acetylcholinesterase, enzyme of hydrolysis of acetylcholine, is also supposed to play a role in energetic processes. The possible universal role of acetylcholine hydrolysis for the vital biological systems with participation of acetylcholinesterase is discussed from the position of energy transformation in such processes as synaptic transfer, muscle contraction, and photosynthesis (Borodyuk, 1990). One reason for this opinion is an accumulation of H^+ protons at the hydrolysis of acetylcholine that influences proton pumps of the membranes.

In higher concentrations dopamine, adrenaline, and noradrenaline participate in redox reactions of chloroplasts (Roshchina, 1989c, d, 1990b). As seen from standard potentials of catecholamines and electron carriers as well as of artificial donors and acceptors of electrons (Table 39), dopamine can be a most effective electron donor in electron transport

Table 39. Standard potentials of catecholamines and electron donors and acceptors in photosynthesis

Compound	E, W	Compound	E, W
Adrenaline red/ox	+0.39	Potassium ferricyanide red/ox	+0.42
Dopamine red/ox	+0.126	Dichlorophenol indophenol	+0.217
Ascorbinic acid/ dihydroascorbinic acid	–0.054	Cytochrome f red/ox	+0.365
$NADP^+$/NADPH	–0.32	Plastocyanin red/ox	+0.375

chain between photosystems 2 and 1. Paired with ascorbate it is able to reduce cytochrome f and plastocyanin (Roshchina, 1990b). Since noradrenaline and adrenaline have potentials that are approximately equal to these electron carriers, the mediators are not effective electron donors for the proteins, although they reduce isolated and purified cytochrome f and plastocyanin in model experiments (Roshchina, 1990b).

Acetylcholine and biogenic amines can regulate enzymic activity or be involved in metabolism as substrates. For instance, histamine inhibited lysozyme activity in latex of *Asclepias syriaca* (Lynn, 1989), and noradrenaline depressed the activity of pea ribulose bisphospate carboxylase by 10–20% (Roshchina and Karpilova, 1993). Enzymic activity may be indirectly regulated via synthesis of secondary messengers cAMP and cGMP, which phosphorylate key proteins of cells (see section 3.2). Based on the fact that acetylcholine and biogenic amines are contained in secretions of some animal and plant species and serve as defensive agents, the mediators have been considered as possible final products of metabolism (Mukhelson and Zeimal, 1973; Fodor, 1980). However, the finding of active enzymes such as cholinesterase (Chapter 3) and cholinoxidase (Chapter 1) in plants permits us to suppose the active participation of acetylcholine in metabolism as substrate for the formation of choline (see Chapter 1). It is rapidly incorporated in biosynthesis of betaine and sinapinecholine. Catecholamines, serotonin and histamine are even more active in metabolism as precursors for many reactions of biosynthesis of taxon-specific alkaloids (Schütte, 1989; 1991). Besides alkaloids, many plants produce highly toxic stressory amines. Catecholamines serve as precursor for the formation of phenolic compounds such as vanillic acid, vanillic aldehyde, and dioximandalic acid (Chapter 1). It should be noted that in some plants alkaloids that are blockers of cholinoreceptor or agonists of acetylcholine originate from dopamine, whereas adrenoblockers arise after condensation of this catecholamine with serotonin alkaloids. Among the substances are atropine from *Atropa belladonna*, arecoline from *Areca catechu*, d-tubocurarine from *Chondrodendron tomentosum*, capsaicin from *Capsicum annum*, and reserpine from *Rauwolfia serpentina*. There are also alkaloids that have anticholinesterase activity, such as physostigmine (eserine) from *Physostigma venenosum*. Participation of catecholamines in synthesis of similar products is an example of a reverse connection when stimulators of the processes are precursors in the formation of inhibitors of the same reactions.

4.2.2. Role of Neuromediators in Growth and Morphogenetic Reactions

Adrenaline, noradrenaline, and serotonin in animals may function as hormones. In this case synthesized in one organ, they move along the organism to cell targets in which they induce an effect. As mediators, they act in the same site in which they are synthesized. The hormonal role of acetylcholine is also hypothesized (Roshchina, 1991a), although it is not confirmed by many animal physiologists. In plants, growth stimulation or depression is observed only at high concentrations of acetylcholine (10^{-4}–10^{-3} M). In comparison with the action of phytohormones, such as indole acetic acid or gibberellin, these effects are not quite clear. Serotonin is supposed to play a more important role in plant growth regulation (Grosse, 1982; Regula et al., 1989; Roshchina, 1991a, b), because changes in growth induced by the substance are more significant. Serotonin (Regula, 1970; Lembeck and Skofitsch, 1984) and histamine (Haartmann et al., 1966) are found in marked amounts in seeds and may regulate their germination. There are data showing that acetylcholine reverses growth inhibition in wheat (Dekhuijzen, 1973) and hyacinth dolichos *Dolichos lablab* (Tung and Raghavan, 1969) by retardant chloroethyltrimethlyl ammonium.

The action of mediators on morphogenetic reactions is poorly studied as a whole. Attention should be given to acetylcholine participation in processes of photomorphogenesis of plants (see Chapter 2). First of all, red light increases the level of endogenous acetylcholine in plant tissues (Jaffe, 1970; 1972a, b), which influences morphogenesis (Hartmann, 1971; Hartmann and Kilbinger, 1974a). Acetylcholine imitates the action of red light on some phytochrome-controlled processes (Table 40).

The comparison of effects induced by acetylcholine or red light shows the stimulation of morphogenetic, energetic, and bioelectric reactions, dependent on phytochrome in both cases. Far red light shows an opposite effect. However, there are processes controlled by phytochrome, but not regulated by acetylcholine, unlike red light (Satter et al., 1972). All these data are not enough for a final conclusion that mediators play a hormonal role. This is perhaps a subject for future studies.

4.2.3. Protective Role of Neuromediators

Most adaptive reactions exist for the stabilization and defense of cellular structures by the production of appropriate chemical compounds. Acetylcholine, biogenic amines, and their derivatives can serve as stabilizers and protectors. In particular, glycine betaine (a product in acetylcholine metabolism) is accumulated in conditions of salinity, whereas histamine is accumulated during drought (Chapter 1). Choline and betaine are accumulated under stress. Their level is supposed to be an indicator of the stress (Varsheney et al., 1988). Dopamine is found in certain sorts of sugar beet and makes them tolerant to injury by fungus *Cercospora beticola* (Gardner et al., 1967). The mediator contained in *Beta vulgaris* is oxidized

Table 40. Effects of acetylcholine and red light on the phytochrome-controlled processes

Plant	Process	Action of acetylcholine	Action of red light	Reference
Phaseolus aureus (secondary roots)	Oxygen uptake	Increase	Increase	Jaffe, 1972a, b
Phaseolus aureus (secondary roots)	Formation of ADP and P	Increase	Increase	Yunghans and Jaffe, 1972
Phaseolus aureus (secondary roots)	Efflux of H^+ ions	Increase	Increase	Jaffe, 1970
Echinochloa crusgalli	Seed germination	Increase	Increase	Holm and Miller, 1972
Brassica kaber	Seed germination	Increase	Increase	Holm and Miller, 1972
Agropyron repens	Seed germination	Increase	Increase	Holm and Miller, 1972
Setaria viridis	Seed germination	Increase	Increase	Holm and Miller, 1972
Chenopodium album	Seed germination	Increase	Increase	Holm and Miller, 1972
Spinacia oleracea	Bioelectric reactions	Increase	Increase	Greppin et al., 1973
Trichoderma viride (fungi)	Sporulation	Increase	Increase	Gressel et al., 1971

and prevents the invasion of sugar beet leaf spot *Cercospora beticola*. The resistance of the plant to the fungus is explained by this process. When cactus *Carneginea gigantea* is wounded the defensive callus arises, and a main phenolic component of its cortical tissue (pulpy cortex) is dopamine (Steelink et al., 1967). The concentration of the substance increases at the site of wounding from 0.3–0.4% per gram of fresh mass in unwounded parts to 0.58–0.64% per gram of fresh mass in wounded ones. Mechanical damage, such as by rock, birds, wind, and vandalism, perhaps leads to the oxidation of dopamine because red pigment (adrenochrome) is seen at the site of the injury. Total concentration of dopamine at wounding can reach 1% of common pulp of callus.

Serotonin and its derviatives are also known protectors. Serotonin inhibits growth of tumors in potato tubers (Regula et al., 1988, 1989) and can protect against ultraviolet inactivation (Fraikin, 1987) and X-rays (Lozeron and Maggiora, 1965). The mechanism of this defense is still unclear. However, it has been shown that serotonin is easily oxidized in light, and products of its oxidation may stabilize membranes. The mechanism of photoprotection in cell of yeast *Saccharomyces cerevisiae* consists in a binding of serotonin with DNA that leads to the increase in survival of cells by 50% in comparsion with control (Strakhovskaya *et al.*, 1982). Serotonin stimulates DNA synthesis (Seuwen *et al.*, 1988). Under 400 Rad dose of X-radiation serotonin is able to protect completely the capacity to form secondary roots in broad bean *Vicia faba* (Lozeron and Maggiora, 1965). In animals acetylcholine and catecholamines can also function as radioprotectors (Suvorov and Shashkov, 1975). In addition, catecholamines have antiviral and antitumor functions (Potopalskii *et al.*, 1989). Antioxidant features of mediators appear to be a base for the protective characteristics, as has been shown for catecholamines, serotonin, and histamine in model experiments on the fluorescence of phycoerythrin (DeLange and Glazer, 1989). Under γ-radiation of banana fruits their polyphenol oxidase preferably and at highest rate (10–50 times) oxidizes noradrenaline and dopamine in comparison with other phenols (Thomas and Nair, 1971). Studies of protective functions of biogenic amines can give a key to the understanding of stressor mechanisms in plants that is important for agricultural practice and environmental protection (Artyukhov, 1973; 1983; Villalobos-Peetrini and Laguarda-Figueras, 1967).

4.3. EVOLUTION OF REGULATORY SYSTEM WITH PARTICIPATION OF NEUROMEDIATORS

The similarity and universality of basic mechanisms of the development and receptor mechanisms of protozoa and invertebrates are discussed (Csaba and Muller, 1996). Plant systems are also included in the discussion.

During the course of evolution of living cells, intracellular systems of regulation must first have arisen on the levels of enzymes, membranes, and genes, and they were based on the universal principal called receptor-conformational (Polevoi, 1975, 1985). In all cases any protein molecule—enzyme, receptor, or regulator—binds with a ligand specific for a given protein. As a result of this interaction the protein changes its configuration. In multicomponent complexes of enzymes, genes, and membranes, conformational changes of receptor molecules are cooperatively transmitted on the whole complex, acting on its functional activity. External stimuli from surrounding media or from other cells are received by special receptors of plasmatic membranes of cells. As a result, the cell comes to the excited or functionally active state. On the molecular level the base of irritation is the lability of protein molecules and their capacity to change their configuration and characteristics after interaction with chemical and physical factors. The primary and simplest indicator of irritation in animal and plant cells is hyperpolarization or depolarization of plasmalemma. Cellular specialization during evolution from unicellular to multicellular organisms leads to the formation of specialized cells, receiving signals from external media and then transmitting them to other kinds of specialized cells via the same means. Here, the external signal is transformed into other kinds of signals, hormonal or electric. This signal is translated by cells belonging to informational channels, in particular conducting bundles of phloem. As soon as this signal reaches a competent cell, it triggers its functional activity which is a response of the whole organism to external stimulus.

It is possible that over the course of evolution principles of intracellular regulation on molecular and membrane levels are preserved in any cell of a multicellular organism. They are also the basis of formation of special intercellular mechanism of signaling and regulation in animals and plants. One of the principles realized as electric contacts between cells via spreading action potentials is evident for all kingdoms (Opritov et al., 1991). As for animals with nervous system, the participation of chemical mediators or transmitters in synapses is known in the above-mentioned processes. However, these substances are also present in unicellular organism (Tsavkelova et al., 2000). Perhaps, initially intracellular regulators or local mediators have evolved to acquire functions of intercellular mediators or, like adrenaline and serotonin, functions of hormones as well. Intracellular and intercellular systems of regulation function in connection with each other.

We do not know exactly how the receptor in membranes evolved. Analysis of primary structures of receptor proteins such as rhodopsin, β-adrenoreceptor, muscarinic and nicotinic cholinoreceptors, and bacteriorhodopsin by computer method showed (Tzendina et al., 1988) that the first three proteins are 15–20% homologous parts with identical sequence of amino acid residues. It is proposed that a conservative domain, consisting of certain amino acids (GTP-binding site), exists in primary structures and is preserved independently on the evolutionary ladder, whereas other amino acids residues were replaced during phylogenesis without loss of functional activity, although they have acquired higher specialization. In particular, rhodopsin is a photoreceptor, while β-adrenoreceptor and muscarinic cholinoreceptor are chemoreceptors. Thus, receptor proteins have a common precursor and may have undergone divergent evolution from a unicellular organism more than one billion years ago. As confirmation of this hypothesis, it has been shown that yeast cells synthesize α-subunits of cholinoreceptor from *Torpedo californica* (Fujita et al., 1986).

The evolution of both effectors (hormones and mediators) and receptors is discussed below (Pertseva, 1988, 1990). Some possibilities are proposed:
(1) Receptor protein can have non-receptor functions in cells and only during evolution acquire receptor features.
(2) Proreceptors arose during the earliest stage of evolution and quickly formed structures, analogous functionally to ones known now-a-days, i.e., to receptors.

(3) Receptors and hormones arose independently of each other, and their capacity for interaction has evolved accidentally.

The above-mentioned data confirm preferentially the second hypothesis. Evolutionary stability of the receptor is evidence of the conservatism and universality of membrane parts or domains responsible for receptor features. Perhaps the function of effectors-mediators and hormones themselves changed in phylogenesis from local intracellular to intercellular, as well as their affinity to receptor. Throughout evolution there has been simultaneous perfection of complementation of effector and receptor either by modification or by structural packing of some parts of receptor protein (Hall, 1992; Hamilton, 1992).

Evidently, cholinoreceptors of mammals are a result of the evolution of membranous receptors; however, their functional analog, i.e., structures capable of regulating ion channel or triggering systems of synthesis of secondary messengers (Levitzki, 1988), are present perhaps in any living cell. In unicellular organisms receptor structures are not highly specialized. However, Na^+, K^+, Ca^{2+}, and Cl^- channels, the opening of which lead to membrane depolarization and arising of action potential, are found here (Saimi et al., 1988). Saimi and coworkers (1988) have shown that, like nervous and muscle cells, the simplest unicellular animal *Paramecium* has ion channels, partcipating in the formation of action potential and controlling the behavior of this organism. Analogous data have been received for yeast *Saccharomyces cerevisiae* cells. The authors came to the conclusion that common mechanisms controlled the functioning of Na^+, K^+, and Ca^{2+} channels in both unicellular and multicellular organisms, perhaps with the participation of receptors. Besides, chloroplasts and mitochondria undoubtedly also have structures similar to channels and certain types of receptors, which are connected with their movement and functioning in cells. The existence of Na^+ and K^+ channels sensitive to acetylcholine and Ca^{2+} channels sensitive to catecholamines and serotonin has been confirmed for intact chloroplasts (Roshchina, 1987a, 1989a, b, c). The generation of action potential in cells of moss *Anthoceros* appears to be connected with chloroplasts (Niyazova and Bulychev, 1989). In maize mitochondria, there are potential-dependent channels, mainly ion-selective, which open still at potential shifts from 0 to 30 mW (Smack and Colombini, 1985).

According to M.N. Pertseva (1989), ion channels are the most ancient systems, representing short variant of nicotinic cholinoreceptor where the receptor and ion channel are combined in a single molecule. Moreover, archaic receptors could switch on the relevant reaction without secondary messengers. This could be due to a receptor molecule such as an insulinic molecule, that has protein kinase activity and directly phosphorylates proteins (Pertseva, 1989). On the other hand, adenylate cyclase system has arisen in bacteria and other unicellular organisms

(Levitzki, 1988; Pertseva, 1989), as well as regulation of transmembranous signalization via phosphorylation of receptor (Sibley *et al.*, 1987). Phosphorylation of proteins in cell is important for transport of substances, including mediators, by endocytosis, where receptors, binding ligands, move to special "clathrinic" regions of plasmalemma. Clathrinic parts are invaginated, catching the molecule and forming a vesicle coated by clathrin.

Cholino-, adreno-, serotonino-, and other types of receptors also exists in primitive animals and are somewhat connected with their development and morphogenesis (Buznikov, 1987). However, their main function is to receive a signal from external media (for unicellular organisms) or from other cells (for multicellular organisms), and then transmit it to the reaction and adaptation systems, ultimately regulating the growth and development.

The evolution of effector-mediator functions can be imagined only approximately. Acetylcholine (Stephenson and Rowatt, 1947) and serotonin (Strakhovskaya *et al.*, 1982) are found in bacteria and yeast cells, which indicates their ancient origin.

Acetylcholine appears to play the role of intracellular mediator between organelles and plasmalemma in unicellular forms (Roshchina, 1989a, c) as well as regulators of energetic and metabolic processes (Roshchina and Mukhin, 1985a, b) and serves as a source of carbon (Goldstein and Goldstein, 1953). Only when multicellular organisms arose and specialization of organs and tissues began, acetylcholine and biogenic amines, in addition to their former functions, start to serve as intercellular mediators or hormones in animals (Sastry and Sadavongvivad, 1979). The function of acetylcholine and biogenic amines as intercellular mediators in plants is not yet established as well as their hormonal role. Perhaps their common biological functions as intracellular mediators and regulators of energetic and metabolic processes have been preserved in any cell. Especially for the plant as a whole, the role of acetylcholine and biogenic amines may be significant in specialized nitrogen exchange as a source of nitrogen for the synthesis of secondary metabolites, including stabilizers of membranes at stresses and toxins needed for defense of the organism. This becomes clear if we keep in mind that amines themselves are widespread in plants and often serve as precursors to toxic alkaloids. Considerable amounts of acetylcholine are accumulated in light, and its synthesis depends on the light quality (see Chapter 1), which indicates that photosynthesis participates in this process. It is natural that in animals, lacking chloroplasts and chlorophyll, this phenomenon is not found. Moreover, acetylcholine accumulation is higher under red light than under white light. But the process decreases when the plant undergoes far red radiation, which indicates the participation of phytochrome in regulation of the formation of acetylcholine. The significance of photosynthesis and phytochrome-dependent reactions as factors needed for the increase in acetylcholine concentration is a determinant of the specific functions that were later acquired by plants when they evolved from heterotrophic to autotrophic nutrition. Perhaps this is also connected with the peculiarities of plant nitrogen exchange (see above). Other evolutionary factors should be kept in mind, such as different physiological activity of organs and concentration of active matter. According to Vdovichenko (1966), the effects of acetylcholine on physiological processes in animals are variable and depend on its concentration as well as individual features of the objects studied. The same can be supposed for plants. The author feels that in objects that are tolerant to acetylcholine, there are adaptations preserving cells against an excess of this bioactive agent. The regulatory role of acetylcholine in respiration and photosynthesis can be important when the functional state of plant tissue is disturbed.

The evolution of functions of tryptophan derivatives is discussed by V.V. Polevoi (1985), who supposed that at the divergence of animal and plant kingdoms, their common precursor tryptophan was converted to indole acetic acid in plants and to serotonin in animals. Thus, indole acetic acid is only a plant hormone that has an analog in both plants and animals. However, experiments in which serotonin has been found in both unicellular and multicellular organisms contradict this hypothesis. In animals serotonin functions as local hormone or tissue hormone, and it is a neurotransmitter only in animals with a nervous system. Its role in plants is still unclear, but experimental data demonstrate its regulatory role in the growth processes (Chapter 2). Besides, in plants direct biochemical transformations of serotonin to indole acetic acid and back are also proposed.

Enzymes of synthesis and catabolism of acetylcholine and biogenic amines have undergone evolution as well. It is known that choline acetyltransferase and cholinesterases are found in all living cells. Choline acetyltransferase acetylates not only choline, but other compounds too (Sastry and Sadavongvivad, 1979). It is logical to suppose that initially the function of the enzyme was not so specialized. In unicellular organisms acetylcholine, choline acetyltransferase, and cholinesterase are often concentrated in a flagellum; that is a reason for possible regulation of motility by cholinergic system as an element of contractile structures. The characteristics of cholineacetylase of plants are studied to lesser degree than those of animals. Plant enzymes synthesize acetylcholine at slower rates than animal enzymes, although for common nettle they are of the same order as for animals (Smallman and Manekjee, 1979). Cells of some bacteria contain active choline acetyltransferase (Girvin and Stevenson 1954). For instance, in *Lactobacillus plantarum* the enzyme produces up to 1 nmole of acetylcholine per min per mg of dry cells (White and Cavallito, 1970).

Evolution of cholinesterases can be explained on the basis that all the enzymes have the capacity to hydrolyze not only acetylcholine and some other cholinic esters, but peptidase and arylacylamidase activities as well, i.e., they have universal characteristics as hydrolases (Balasubramanian, 1984; Greenfield, 1984). High specialization and, preferably, cholinesterase activity are common only in highly organized animals. In sessile mollusks and plants the enzymes, even purified by affinity chromatography, have lower cholinesterase activity (Sastry and Sadavongvivad, 1979; Solyakov et al., 1989). Unlike animal cholinesterase, bacterial acetylcholinesterase reacts very slowly with such organophosphates as diisopropyl phosphofluoridate and, on the contrary, very quickly with tabun. It is sensitive to neostigmine and tolerant to physostigmine (Sastry and Sadavongvivad, 1979). Higher resistance to physostigmine is also characteristic for cholinesterases from plants (Vackova et al., 1984) and some planaria, unicellular flagellata (Sastry and Sadavongvivad, 1979). Sensitivity to physostigmine is observed in unicellular protozoans such as *Paramecium*. In highly organized animals and plants the main characteristics of cholinesterases are similar, except rates of acetylcholine hydrolysis, which is 10–100 times higher, and correspondingly K_m for animal enzymes is lower. It appears to connect with high specialization of the organisms (Chapter 3). More developed organisms have expressed substrate specificity of cholinesterase. However, in leguminous and some other species of plants highly specific acetylcholinesterase is present, as is observed in nerve cells. Unlike in highly specialized animals, pseudocholinesterases or butyrylcholinesterases are not yet found in plants. Only intermediate types between acetylcholine and pseudocholinesterases are found in roots of plants such as *Cicer arietinum* (Gupta and Maheshwari, 1980) as well as in lower animals.

There is practically no information on evolution of the enzymes participating in synthesis and catabolism of biogenic amines, barring the data of Ehrismann and coworkers (1948), who showed the presence of histidine decarboxylase and histaminase in bacteria. In comparison with other systems, cholinergic mechanism of cell regulation has a special position in evolution. In animals it is supposed to be a main channel of signal control (Denisenko, 1980) and functions in accordance with other biochemical systems: adrenergic, serotoninergic, histaminergic, and γ-aminobutyric acid systems. Acetylcholine can not only mediate impulses, but also liberate other mediators and hormones. Thus, the increase in membrane permeability under acetylcholine induces the release of noradrenaline. The blockade of cholinoreceptors promotes the liberation of catecholamines, which are responsible for effects of central cholinolytics. Many monoamines, such as histamine, serotonin, and catecholamines, stimulate some processes in secretory cells. Analogous interaction of similar systems of regulation is possible for plants as well. However, special experimental data are needed to prove this.

4.4. PERSPECTIVES IN PRACTICAL USE OF NEUROTRANSMITTERS IN PLANTS

Until now we have given no attention to possible applications of the presence of neurotransmitters in plants. Below we briefly consider such applications.

4.4.1. Medicine and Pharmacology

Plants with a high content of certain neurotransmitters may be used in medicine and pharmacology as natural components of drugs. A list of plant species enriched in some neurotransmitters is given in Appendix 1. These are both medicinal and food plants. Usually acetylcholine is found in blood clotting plant species or in species stimulating cardiac activity, but these peculiarities are not mentioned by specialists. Catecholamines regulate blood pressure and vascular contraction, as well as participate in brain disorders, particularly when they interact with copper ions. They are contained in species of some Cactaceae and Mimosaceae. Their deficit leads to motile disturbances after cerebral stroke and intoxications, Parkinson's disease, and psychic disorders such as schizophrenia. Neurotransmitters regulates embroyonal developments (Buznikov et al., 1996) and microbial invasion (Lyte, 1993). Allergic reactions to some plant species can be explained by the presence of high concentrations of histamine in the leaves, stems, and pollen (Table 9).

4.4.2. Agriculture and Protection of the Environment

As biosensors, plant cholinergic and aminoenergic systems could be used for the analysis of agricultural practice and environmental monitoring both in laboratory and in field conditions. The action of many pesticides is based on the depression of pest cholinesterases or weed cholinesterases in plant agricultural communities. Cholinesterase activity may be inhibited by various toxins and pesticides (see Chapter 3) and used as an indicator of environment pollution. New pesticides can be screened and plant sensitivity to them assessed by means of analysis of cholinesterase in plants belonging to families Fabaceae or Solanaceae (Table 18). Some retardants could be also tested. The search for male or female sterility or self-incompatibility among various clones of plants, based on their cholinesterase activity, is useful for genetic and breeding laboratories. Some toxins (perhaps pesticides too) may prevent normal fertilization, acting on the plant receptor systems of neurotransmitters, and their testing on the model pistil-pollen system is also recommended (Roshchina and Melnikova, 1998a, b). Moreover, acetylcholine and biogenic amines are involved in pollen allelopathy, when pollen grains of foreign species compete with pollen of own species added on the pistil stigma through the release of inhibitory or stimulatory substances (Roshchina and Melnikova, 1998a, 1999; Roshchina, 1999a, b, 2001a). Similar testing on the antineurotransmitters agents of pollen is very important for an understanding of the composition of artificial agricultural and forest phytocenosis in order to limit the breeding of incompatible or weed-like species in the plant communities.

A special aspect is the use of biogenic amines for the stimulation of germination of seeds and pollen with low viability (Roshchina, 1992, 1994; Roshchina and Melnikova, 1998a, b). This is important for the analysis of valuable or simple examplars of decorative plants grown in hothouse (Roshchina et al., 2001).

The accumulation of biogenic amines in plants is also correlated with stress, therefore their content can be a test of the stress conditions of growth. Such tests should also be used in environmental monitoring. The correlation between particular types of stress (e.g., water deficit, heavy metals, salt accumulation, air pollutants) and the enhanced synthesis of specific neurotransmitters must be investigated.

CONCLUSION

Neurotransmitters have multiple functions in plants, which differ from known synaptic functions in animals. In plants neurotransmitters may regulate electric events, growth and morphogenesis, and fertilization, protect against stress factors, and serve other functions. Within the plant cell and between plant cells some constant or non-constant contacts are functionally similar to synapsis and, perhaps, more ancient than synaptic contacts may be. Moreover, modern nervous structures are supposed to have evolved from such contacts. Molecular and cellular mechanisms of the neurotransmitters' reception, synthesis, and metabolism may be the same in microbial, fungal, plant, and animal cells. The main differences are in multicellular specialized structures of the various living organisms. Neurotransmitter occurrence in plants has applications in medicine, pharmacology, agriculture, environmental monitoring, and many other fields.

Conclusion

The presence of classic neurotransmitters of synapses such as acetylcholine, catecholamines, serotonin, and histamine in plants, their significant physiological activity, and the finding of components of cholinergic and aminergic systems of regulation in plant membranes make quite real the idea of the universal principles of signalization and transmission of information in the form of electric and chemical signals in all living organisms. The differences lie mainly in particular mechanisms of intercellular signaling in multicellular animals and plants, which consist in specialization, structural organization, and peculiarities of energetic and metabolic exchange. There is a large gap in our knowledge of transmission of the information between plant cells. The participation of chemical mediators as generators of changes in membrane potential and their role in the arising of action potentials in plants is only postulated, based on what is known about synaptic cells of animals so far. On the cellular level the presence of acetylcholine and biogenic amines in unicellular and multicellular organisms, including plants, which lack synaptic contacts, can be explained by their dual role: as regulators of intracellular metabolic exchange and as local intracellular mediators. These intracellular effectors interact with organelles in a response to the external signal that the plasmalemma receives from the environment surrounding the cell.

Reactions of acetylcholine and biogenic amines, consisting in changes of membrane permeability for Na^+, K^+, Ca^{2+} and/or an induction of synthesis of secondary messengers cAMP and cGMP, inositol triphosphate, Ca^{2+}, as well as enzyme adenylate cyclase in plant cells and its organelles, confirm the possibility of local mediation, which may be imagined as shown in Fig. 65: The sequence of events represented in this figure is possible first of all for mitochondria and chloroplasts, which were perhaps independent organisms at the beginning of evolution and have preserved characteristic principles of structural organization. Information transmission with participation of acetylcholine and biogenic amines from plasmalemma to organelles or between organelles is similar to the classic principle of intercellular transmission of excitation in synapses. This seems

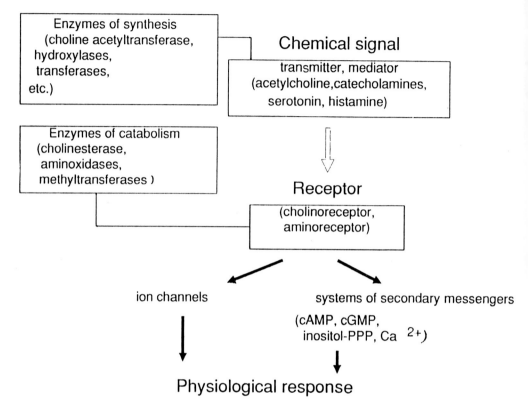

Fig. 65. Cholinergic and aminoenergic system regulation (Roshchina, 1999a)

more probable because acetylcholine and biogenic amines act as mediators and regulators in low concentrations (10^{-9}–10^{-7} M).

The accumulation of acetylcholine and biogenic amines in plants to high concentrations is connected with their other functions. It is primarily due to stress conditions. Accumulation of catecholamines leads to redox reactions with production of toxic substances as a result of their oxidation (aminochromes) or to synthesis of toxic alkaloids (d-tubocurarine, berberastine, etc.) when the mediators are used as precursors. The increase in acetylcholine concentrations depresses the activity of enzymes, in particular cholinesterases. Under unfavorable factors of external media (UV, γ and X-radiation) the protective role of serotonin is observed. Analogous function is proposed for histamine in conditions of salinity, although there are few studies dealing with this question. However, accumulation of mediators not only indicates stress, but also can be a taxonomic sign, for instance of family Urticaceae. In plants belonging to this group, localization of mediators in stinging hairs is a mode of defense against mechanical injury by animals. Thus, the roles of acetylcholine and biogenic amines vary from common biological roles (regulatory, signaling, mediatory) to special ones.

Analysis of the information presented in this book shows that experimental studies of acetylcholine and biogenic amines in plants and their role in functioning of plant organisms were mainly of phenomenological character. This is a necessary step in the development of any new problem. Systematization of received data, the comparison of data with known data for animals, and the inclusion of knowledge from adjacent fields of common cellular biochemistry permit us to make a conclusion about the state of the problem and the perspectives of future work. First of all, while continuing to accumulate experimental data about the nature of action of acetylcholine and biogenic amines on various processes in plant cells, we need to search for approaches to deciphering this mechanism and analyzing the events in the complex system of operative regulation in plants.

The presence of acetylcholine and biogenic amines, some components of cholinergic regulatory system such as choline acetyltransferase and cholinesterase, and their physiological activity are now established. The following key problems need to be better studied:
(1) sites of synthesis and compartmentation of mediators within plant cell;
(2) transport of these compounds within plant cell and between cells (plant-plant cells, plant-microbe cells, and plant-animal cells);
(3) mechanism of the transmitter reception by plant membranes; and
(4) role of acetylcholine and biogenic amines in electric events, including the arising of action potential and opening and closing of ion channels.

This is a look at the problem as a whole. However, the problem of the functions of acetylcholine and biogenic amines in plants is large and has several aspects, each of which represents an independent direction of scientific investigation. Among them are the finding, isolation, and purification of cholinesterases and choline acetyltransferases. Although the question is more studied than others, nevertheless this is only the first step for plant physiologists and biochemists. For in-depth analysis of active centers and protein structures, the participation of molecular biologists and geneticists is needed. There are few works dealing with the separate stages of biosynthesis and catabolism of acetylcholine and biogenic amines, and the position of these processes in a whole system of nitrogen exchange and secondary metabolism. As for the plants themselves, the interactions of mediators with phytohormones and inhibitors of growth are especially important. Conceptual and methodological approaches to studies on plants are also needed, since the existing approaches are derived from animal physiology, which is characterized by higher sensitivity to acetylcholine and biogenic amines.

Notwithstanding the reality that fundamental studies of mechanisms of sensory reactions in plants are still at the initial stages, the perspectives for practical use of existing data need to be discussed. First of all, attention should be given to plant injuries on treatment by insecticides. The occurrence of cholinesterase in plants has not so far been taken into account, although enzyme depression is one of the possible causes of crop damage by some insecticides. Sensory capacities of plants (cholinesterase activity, receptor systems, etc.) can be used in laboratory tests on pesticides and for examination of new modes of plant protection. Stress-induced accumulation of acetylcholine and catecholamines can be tested in the same way as for medicinal tests. The simplicity and accessibility of plant material for analysis makes the separated plant organ, cells, and organelles more suitable models for the working out of biosensors in biochemical diagnostics. Data about the content and metabolism of mediators in plants have a practical significance for medicine and pharmacology as a base for the production of new drugs of plant origin.

PERSPECTIVE DIRECTIONS IN THE FUTURE STUDIES

1. Role in intermembrane contacts in plants on cellular and molecular level;
2. Mechanisms of interaction and recognition of the signal transfer with participation of the neurotransmitters systems.
3. Multiplicity of the functions of the neurotransmitter systems in plant cell;
4. Evolution of the neurotransmitter functions.

APPLIED DIRECTIONS

1. Usage of plants enriched in neurotransmitters in pharmacology
2. Usage of plants as tests on antineurotransmitter poisons and pesticides in toxicology
3. Analysis of biological activity of pesticides on plant objects.

Appendix 1

PHARMACOLOGICAL EFFECTS OF NEUROTRANSMITTERS

Acetylcholine (0.05-0.15 g per day) decreases arterial pressure, acting as vasodilator, retards the cardiac rhythm and induces the stenosis of pupils, enhances the contraction of smooth musculature in internal organs, as well as increases the secretion of sudoriferous, lacrimal and bronchial glands. Unlike acetylcholine catecholamines (25-100 mg per day), stimulating, mainly α- and β-receptors, increase blood pressure. Dopamine and noradrenaline are used at shock states and at acute cardiac insufficiency while adrenaline stops the bronchial asthma and other acute allergic reactions, against glaucoma, and as vasoconstrictive and antiinflammatory agents in otorhinolaryngologic and ophthalmic practices. Serotonin (5-10 mg per day) is useful against hemorrhagin syndromes, and increases the capillary stability and decreases the hemorrhage at anemia. It acts on blood vessels and as a hormone. Histamine is known as a component of drugs against polyartria and rheumatism, and in small doses ($\sim 10^{-7}$ M) in order to prevent acute allergic reactions. It excites a secretion of gastric glands and spastic contraction of enteric musculature. Perspectives for medicine is in a usage of plants enriched in the neurotransmitters.

REFERENCES

Mashkovskii, M.D. (1993). *Medicinal Drugs*. Moscow: Meditsina. 12 edition. 2 volumes, 736 and 740 pp.

Medicinal Drugs used in Medicinal Practice in USSR (1991). Ed. M.A. Kluev, Moscow: Meditsina. 512 pp.

List of plant species contained high amount of neurotransmitters and known as medicinal (*) or food (**) sources

Acetylcholine	Dopamine	Noradrenaline	Serotonin	Histamine
Asteraceae *Helianthus annuus*** (stems, shoots)	**Cactaceae** *Carnegiea gigantea*** (leaves)	**Fabaceae** *Phaseolus aureus*** (roots, leaves)	**Ananaceae** *Ananas commosus*** (fruits)	**Chenopodiaceae** *Beta vulgaris*** (all parts)
Cucurbitaceae *Cucurbita pepo*** (stems)	**Musaceae** *Musa sp.*** (fruits)	*Pisum sativum*** (all parts)	**Elaegnaceae** *Hippophae rhamnoides** and ** (fruits)	*Spinacia oleracea*** (leaves)
Euphorbiaceae *Codiaeum variegatum** (leaves)		*Robinia pseudoacacia** (leaves, fruits)	**Fabaceae** *Mucuna pruriens** (fruits)	**Euphorbiaceae** *Jatropha urens** (leaves) **Geraniaceae** *Erodium cicutarium**
Funariaceae *Funaria hygrometrica** (callus)		**Mimosaceae** *Albizzia julibrissin** (leaves, seeds)	**Juglandaceae** *Juglans nigra*** (fruits, leaves)	**Labiatae** *Lamium album** (leaves)
Geraniaceae *Erodium cicutarium** *Geranium thunbergii**				**Loranthaceae** *Viscum abietis** *Viscum austriacum**
Gramineae *Avena sativa*** (green seedlings)		**Musaceae** *Musa sp.*** (fruits)	*Juglans zegia*** (fruits, leaves)	**Malvaceae** *Gossipium sp.** (shoots, fruits)
*Stipa tenasissima*** (leaves)		**Portulacaceae** *Portulaca oleracea*** (shoots)	**Musaceae** *Musa sp.*** (fruits)	**Mimosaceae** *Mimosa sp** (leaves, seeds)
Loranthaceae *Viscum abietis* *Viscum austriacum*				
Moraceae *Artocarpus champeden** (fruits)*			**Rosaceae** *Prunus domestica*** (fruits, leaves)	**Papaveraceae** *Chelidonium majus** (leaves)
*Artocarpus integra*** (fruits)			**Saxifragaceae** *Grossularia reclinata* (L.) Mill. (fruits)**	**Urticaceae** *Laportea moroides* (leaves, stems)
Rosaceae *Crataegus oxyacantha** (leaves, flowers, fruits)			**Solanaceae** *Lycopersicon esculentum*** (fruits)	*Urtica dioica* (leaves, stems)
Scrophulariaceae *Digitalis ferruginea** (leaves)				
Urticaceae *Laportea moroides* (leaves, stems) *Urtica dioica* (leaves, stems)				

List of plants with anticholinesterase activity

Plant	Active matter
Anabasis aphylla	Anabasine, alkaloid
Berberis sp.	Berberine
Brucea antidysenterica	Brucine, alkaloid
Chelidonium album	Sanguirithrine (sangvinarine and chelerythrine, mixture of alkaloids)
Citisus laburnum	Citisine, alkaloid
Ephedra equisetina, *E. monosperma*	Ephedrine, alkaloid
Evoidia rutaecarpa Bentham	Dehydroevodiamine
Galanthus woronowi A. Los	Galanthamine, alkaloid
Galega officinalis	Petanin, alkaloid
Gossipium hirsutum	Gossipol, phenolic alcohol
Linaria vulgaris Mill	Peganin, alkaloid
Magnolia grandiflora L.	Magnolin, alkaloid
Papaver somniferum	Morpholine, morphine, alkaloids
Salsola richteri	Salsoline, alkaloid
Solanum tuberosum L.	α-Chaconine, glycoalkaloid
Solanum dulcamara	α-Solanine, glycoalkaloid
Strychnos nux-vomica	Strychnine; alkaloid
Thermopsis lanceolata	Citisine, alkaloid

List of plants enriched in neuromediators and demonstrated antiradiation activity (*) or adrenoblocking or cholinolitic (antagonist) activity (**)

Plant	Neurotransmitter or antagonist
*Arnica montana***	Arnifolin
*Gnaphalium uliginosum***	Arnifolin
*Galanthus woronowi***	Galanthamine
*Hippophae rhamnoides**	Serotonin
Rawwolfia serpentina	Yohimbine
*Urtica dioica**	Acetylcholine, catecholamines, serotonin

SOURCES

Khaitbaev, A.Kh., Tilyabaev, Z., Achilova, G.Sh., Khaitbaev, Kh.Kh., Auelbekov, S.A. (1995) Synthesis and biological activity of some gossypol derivatives. *Chemistry of Natural Compounds* N 1: 44-49.

Kuzin A.M., Kopylov, V.A., and Revin, A.F. (1994) New approaches to a problem of the enhanced viability of γ-irradiated animals. *Doklady of Russian Academy of Sciences* 336: 829-830

Ivanchenko, V.A., Grodziskii, A.M., Cherevchenko, T.M., Lebeda, A.F., Makarchuk, N.M., and Snezhko, V.V. (1989) *Phytoergonomica*. Kiev, Naukova Dumka 296 s.

Lovenberg W. (1974). Psycho and vasoactive compounds in Food Substances. *Journal of Agricultural and Food Chemistry* 22: 23-26.

Makhlayuk, V.P. (1967). *Medicinal Plants in Folk Medicine*. Privolzhskoe Izdatestvo: Saratov. 360 pp.

Mashkovskii, M.D. (1993). *Medicinal Drugs*. Moscow: Meditsina. 12 edition. 2 volumes, 736 and 740 pp.

Medicinal Drugs used in Medicinal Practice in USSR (1991). Ed. M.A. Kluev. Moscow: Meditsina. 512 pp.

Mizina, T.Y. and Sitnikova, S.G. (1999) Antiradiation activity of juice concentrated from *Hippophae rhamnoides* L. *fruits, Rastitelnye Resursy* (Russia) 35: 85-92.

Nishimoto, N., Inoue, J., Ogava, S., and Takemoto, T. (1980) Studies on the active constituents of *geranii herba*. *Shoyakugaku zasshi* 34: 122-126.

Nishimoto, N., Inoue, J., Ogava, S., and Takemoto, T. (1980) Studies on the active constituents of geranii herba. *Shoyakugaku zasshi* 34: 127-130.

Pearce, F.L. (1991). Biological effects of histamine: an overview. *Agents Action* 33: 4-7.

Rastitelnie Resursy USSR, (Ed. Sokolov, P.D). Flower plants, their chemical composition and usage. Familes Rutaceae-Elaeagnaceae. Leningard: Nauka, 1988. 357 pp.

Rino, S.M., and Rothschild, A.M. (1976). Estudo sobre a natureza e o local de sintese de principios farmacologicamente ativos presentes nos espinhos de vegetais *Cnidosculus*. *Ciencia e Cultura* 28: 588 (Supplement).

Sokolov, S.A. and Zamotaev, I.P. (1990). *The Dictionary on the Medicinal Plants*. Moscow: Meditsina, 464 pp.

Slorach, S.A. (1991). Histamine in food. In: *Histamine and Histamine Antagonists* (Ed. H.van der Goot). Berlin, Heidelberg: Springer-Verlag, pp. 515–519.

Tilyabaev, E., Kushiev, Kh.Kh., Abdullaev L.K. and Dalimov, D.N. (1995) Alkaloids and their phosphorylated derivatives as regulators of catalytic activity of insect cholinesterases. *Chemistry of Natural Compounds* N 2: 183-187.

Turova, A.D. (1974) *Medicinal Plants of USSR and their Usage*. Moscow: Meditsina, 424 pp.

Appendix 2

METHODS FOR PLANT ASSAYS

1. Cholinesterase

1.1. Cholinesterase in flowers and pollen

Objects of the study were petals, sepals, anthers, and pistils from flower buds. Water extracts from unhomogenized or homogenized tissues were used for the analysis of cholinesterase activity. Unhomogenized fresh samples were exposed for 1 hour at 20°C with 0.01 M K/Na phosphate buffer pH 7.4 (w/w 1:5), then filtered through paper filter. In some cases tissues (anthers of *Hippeastrum* and *Gladiolus*) were homogenized in 20 mM K/Na phosphate buffer pH 7.4 with an electric homogenizer, using a teflon-glass system, at 0°C for 60 s and then filtered through nylon gauze (Semenova and Roshchina, 1993). Cholinesterase activity was measured according to a modified Ellman method developed for crude preparations (Gorun *et al.*, 1978), using Ellman reagent 5,5"-dithio-bis (p-nitrobenzoic acid), which interacts with thiocholine, forming yellow product (maximum of absorbance at 412 nm).

For the experiment with pollen, another method of cholinesterase determination with an analog of the Ellman reagent on the thiol groups 2,2-dithio-bis-(p-phenyleneazo)-bis-(1-oxy-8-chlorine-3,6)-disulfur acid in the form of sodium salt (Roshchina *et al.*, 1994) was also used (Fig. 1). When the red reagent first used for the coloration of polyacrylamide gel electrophoregrams with cholinesterase (Roshchina and Alexsandrova, 1991) interacts with thiocholine, a blue product is formed (maximum of absorbance at 620 nm). The reaction medium contained in 0.2 ml: 50 µl of water extract of protein, 149 µl 0.1 M K-phosphate or Na/K-phosphate buffer pH 7.4, 1 µl of a substrate for 20 µl of homogenate, 10 µl of the substrate solution, and 170 µl of 20 mM Na/K-phosphate buffer (pH 7.4) for homogenized and unhomogenized tissues of sepals, petals, anthers, and pistils and unhomogenated pollen grains.

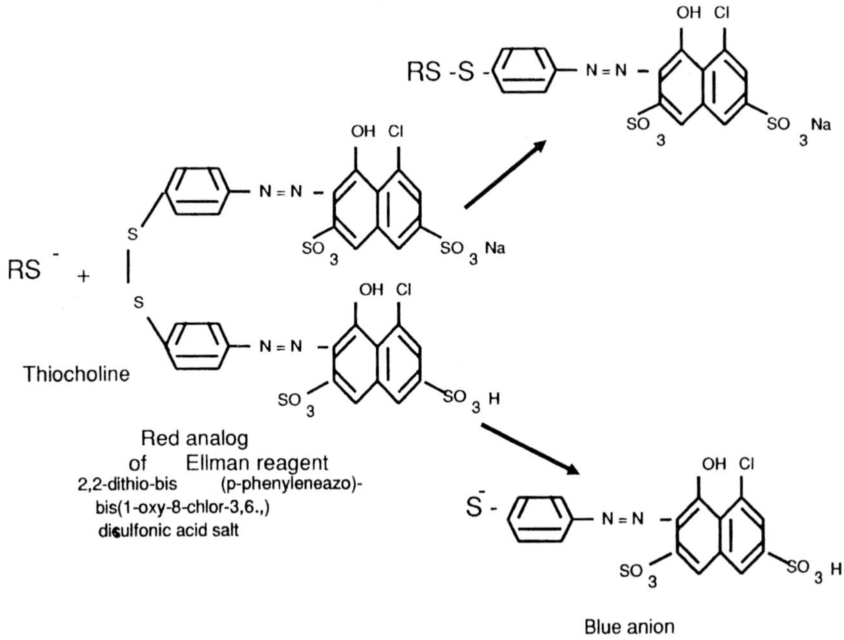

Fig. 1. The reaction thiocompounds with red analog of Ellman reagent DTPDD

In the experiments with inhibitors, instead of the 10 µl of buffer, 10 µl of inhibitor solution was added to the mixture 10–20 min or 1 hour before the substrate was added. The enzyme-substrate reaction proceeded for 60 min at 25°C. Simultaneously, mixtures containing 20 µl of protein probes and 180 µl of buffer were used as reference solutions, measuring of optical density of the samples. To measure a spontaneous hydrolysis of substrates, mixtures containing 10 µl of substrate and 190 µl of phosphate buffer were used. The reaction was stopped by adding 1.8 ml of solution containing 12.4 mg DTNB (Sigma, USA), 120 ml 96% ethanol, 50 ml 0.1 M Na/K-phosphate buffer (pH 7.4), and 50 ml distilled water. When Ellman analog for the reaction stop was used this mixture contained 6 mg red analog of Ellman reagent (DTPDD), 120 ml 96% ethanol, 50 ml 0.1 M Na/K-phosphate buffer (pH 7.4), and 50 ml distilled water. The optical density of the product reaction between thiocholine and Ellman reagent or its red analog were measured with a Perkin-Elmer model spectrophotometer or Specord M-40 (Jena) spectrophotometer at 412 nm or 620 nm respectively. The activity of cholinesterase was expressed in terms of µmoles of substrate hydrolyzed by 1 g (kg) of the tissue (fresh weight) in 1 hour. Acetylthiocholine iodide, propionylthiocholine iodide, butyrylthiocholine iodide, and benzoylthiocholine iodide (synthesized at the Institute of Elementoorganic Chemistry, Russian Academy of

Sciences) were used as substrates. The rates of hydrolysis were plotted against substrate concentrations (pS dependence). Michaelis-Menten constants (K_m) and maximal rates of hydrolysis (V_{max}) were calculated using Lineweaver-Burk graphs and Cornish-Bowden statistical method.

The activity of organophosphorous inhibitors was evaluated by the enzyme-inhibitor interaction rate constant (k_2, M^{-1} min^{-1}), calculated according to the formula

$$k_2 = \frac{2.3}{t[I]} \log \frac{A_o}{A_i}$$

where A_i is the rate of substrate hydrolysis in the presence of inhibitor; A_o is a rate of substrate hydrolysis without inhibitor; [I] is the inhibitor concentration, M; and t is the time of pretreatment with the inhibitor before the substrate was added, min.

The following organophosphorus inhibitors were used: iso-OMPA (Serva), a specific butyrylcholinesterase inhibitor and specific acetylcholinesterase inhibitor GD-42 and its uncharged analog GD-7 (both from the Institute of Elementoorganic Chemistry, Russian Academy of Sciences). For the reversible inhibitors physostigmine and neostigmine the concentration of inhibitor causing 50% inhibition of enzyme activity (I_{50}, M) was calculated. Physostigmine salicylate, physostigmine sulfate, and neostigmine (all Sigma) were used.

The localization of cholinesterase in pollen was determined by histochemical coloration by red analog of Ellman reagent (see above) by the following means. The intact pollen was exposed in 10^{-3} M solution of acetylthiocholine in 0.1 M N-/K-phosphate buffer at pH 7.4 for 1 hour on the subject glass in a petri dish. When physostigmine or neostigmine was used as inhibitor, the sample of pollen was first exposed in 10^{-4} M solutions of the inhibitors, and after that the substrate was added. Then the probe was slightly air-dried. After that, red color analog of Ellman reagent (6 mg in 200 ml of the same buffer) was added to the reaction media, and the coloration was observed under the Fluoval microscope (Karl Zeiss) in white or ultraviolet light. An example is seen in Fig. 2. Thiocholine reacts with the red analog of Ellman reagent, forming a blue coloration. After incubation with acetylthiocholine, DTPDD stained the exine of pollen blue, while the inner part of the pollen grain has the red initial color of the reagent (Fig. 2). These reactions changed the fluorescence emitted by the pollen; its maximum shifted from 480 nm (untreated pollen) to 510 nm (blue-stained pollen zone) and a new maximum appeared at 640–650 nm. In addition, the fluorescence of the blue-stained zone of pollen grains at 510 nm increased. This effect can be used for qualitative and semi-quantitative analysis for the enzyme; that is important for acetylcholine indication, where cholinesterase serves as a marker.

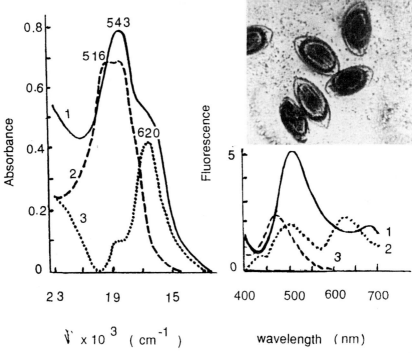

Fig. 2. The histochemical reaction on the cholinesterase activity in pollen, left. The absorbance spectra of the color-generated solutions in the reaction to cholinesterase activity. 1—DTPDD + 10^{-3} M thiocholine; 2—DTPDD ; 3—difference spectrum 1 minus 2, showing the arising of blue component right, below. The fluorescence spectra of *Hippeastrum* pollen (after Roshchina et al., 1994). 1—pollen, its edge after 1 h of incubation with acetylthiocholine followed by DTPDD; blue color is seen; 2—(in central part of the pollen after the same procedure, red color of the reagent DTPDD is retained), 3—untreated pollen grains.

1.2 Purification of leaf and chloroplast cholinesterases

Cholinesterases from leaves were isolated in the following way. Frozen leaves (1 kg) were pounded with liquid nitrogen. The proteins were extracted by 1500 to 2000 cm^3 of 0.01 M K-phosphate buffer pH 7.0 + 4.5% $(NH_2)_2SO_4$ during 2–3 hours at 4°C. After centrifugation for 1 hour at 12,000–15000 × g a pellet was dissolved in 3.0–5.0 cm^3 of the same buffer. The proteins were unsalted on the column with Sephadex G-50 (100 × 2.5 cm) and purified on the column (7 × 2 cm) of DEAE-cellulose Servacel-32 (Reanal, Hungary) equilibrated by 0.01 M K-phosphate buffer, pH 7.0. The enzymes with cholinesterase activity were eluted from DEAE-cellulose by 0.05–0.1 M K-phosphate buffer, pH 7.0. Then proteins were purified on Sepharose CL-6B (60 × 2 cm) equilibrated with 0.01 M K-phosphate buffer (pH 7.0). The enzymes were eluted by the same buffer.

Cholinesterases from intact chloroplasts were first isolated from 150-200 g leaves (Roshchina, 1988a, b; 1990b) and purified by two methods of purification. The main procedure included fractionation of proteins by 80% $(NH_4)_2SO_4$ after extraction from chloroplasts, unsalting on Sephadex G-25, and purification on Sepharose CL-6B and Sephadex G-200 (Roshchina, 1988a, b). Unlike the method described by Roshchina (1988a, b), after fractionation of enzymes by ammonium sulfate we applied enzymes with salt on the column of Toyopearl HW-55 (27 × 1 cm) equilibrated with 0.01 M K-phosphate buffer pH 7.0 (Roschina, 1990a). The proteins were eluted by the same buffer. If needed, the enzymes were unsalted on Sepharose CL-6B (60 × 1 cm).

2. Histochemical Assay of Catecholamines in Pollen

The induction fluorescence of pollen treated with 1% glyoxalic acid is considered to be associated with catecholamines noradrenaline, dopamine, or adrenaline (Roshchina and Melnikova, 1999). The microspectrofluorimetric assay of catecholamines was applied to the pollen grains of *Hippeastrum*. After treatment of intact microspores with 1% glyoxalic acid and excitation by UV light (360–380 nm), the surfaces of pollen fluoresce as well as the secretion around it (Fig. 62). Owing to the light emission the secretions become visible. The fluorescence spectra of the system has a large maximum at 550 nm, unlike untreated pollen with maximum at 475 nm (Fig. 3). The intensity of the fluorescence is enhanced 2–3 times in comparison with untreated pollen grains. Noradrenaline and adrenaline as markers of catecholamines themselves have no such emission in the visible region of spectra when not treated with 1% glyoxalic acid, but after such treatment a maximum of fluorescence at 550 nm is seen. This shows the presence of catecholamines on the surface of pollen grains and in the secretory liquid. A small contribution of other phenol-containing substances in the fluorescence intensity is also not excluded at 520-530 nm.

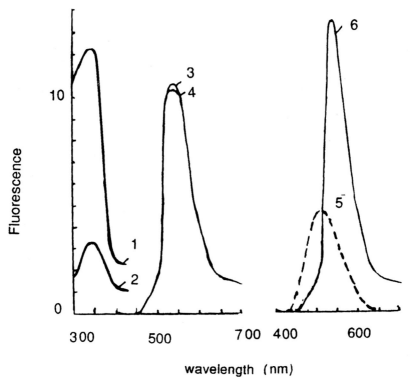

Fig. 3. The fluorescence spectra of *Hippeastrum* pollen treated by 1% glyoxalic acid. 1–4, solutions; 5–6, pollen grains. 1, 10^{-5} M adrenaline hydrochloride; 2, 1% glyoxalic acid; 3, 10^{-5} M adrenaline + 1% glyoxalic acid; 4, 10^{-5} M noradrenaline hydrotartrate + 1% glyoxalic acid; 5, untreated pollen of *Hippeastrum*; 6, pollen of *Hippeastrum* treated by 1% glyoxalic acid, λ of excitation: 1.2–280 nm; 3 to 6, 360–380 nm.

Bibliography

Ackermann, D., and List, P.H. (1957). Über das Vorcommen beträchtlicher Mengen Histamin in der niederen Tierwelt. *Hoppe-Seyler's Zeitschrift für Physiologische Chemie* 308: 274–276.
Aebi, A. (1956). Über die Isolierung von Casimiroedin aus den Samen von *Casimiroa edulis* La Llave et Lejarza. *Helvetica Chemica Acta* 39: 1495–1500.
Ahmad, S.A., and Zaman, A. (1973). Chemical constituents of *Glochidon venulatum*, *Maesa indica* and *Rhamnus triquetra*. *Phytochemistry* 12: 1826.
Alberts, B., Bray, D., Lewis, J., Raff, M., Roberts, K. and Watson, J.D. (1994). *Molecular Biology of the Cell*. 3rd edition. New York and London: Garland Publishing Inc. 1294 pp.
Aleksandrov, A.A., Berestovskii, G.N., Volkova, S.P., Vostrikov, I.Ya., Zherelova, O.M., Kravchik, S., and Lunevskii, V.Z. (1976). Reconstruction of single calcium-sodium cellular channel in lipid bilayer. *Doklady AN SSSR* 227: 732–726.
Amlaiky, N., and Caron, M.G. (1986). Identification of the D_2-dopamine receptor binding subunit in several mammalian tissues and species by photoaffinity labeling. *Journal of Neurochemistry* 47: 196–204.
Anderson, L.A., Homeyer, B.C., Phillipson, J.D., and Roberts, M.F. (1983). Dopamine and cryptopine production by cell suspension cultures of *Papaver somniferum*. *Journal of Pharmacy and Pharmacology* 35: 21 pp.
Andreo, C.S., Orellano, E.G., and Niemeyer, H.M. (1984). Uncoupling of spinach thylakoids by gramine. *Zeitschrift für Naturforschung* 39c: 746–748.
Angelini, R., and Federico, R. (1986). Occurrence of diamine oxidase in the apoplast of Leguminosae seedlings. In: *Biomedical Studies of Natiral Polyamines* (Eds. C.M. Caldarera, C. Clo and C. Guarnieri). pp. 183–189. Bologna: CLUEB.
Angelini, R., Manes, F. and Federico, R. (1990). Spatial and functional correlation between diamine-oxidase and peroxidase activities and their dependence upon de-etiolation and wounding in chick-pea stems. *Planta* 182: 89-96.
Antweiler, H., and Pallade, S. (1972). Über den Acetylcholingehalt von *Stica tenacissima* Linne (Espartoqras). *Archives für Toxikologie* 29: 117–127.
Appel, W., and Werle, E. (1959). Nachweis von Histamin, N-Dimethylhistamin, N-Acetylhistamin und Acetylcholin in *Spinacia oleracea*. *Arzneimittelforschung* 9: 22–26.
Applewhite, P.B. (1973). Serotonin and norepinephrine in plant tissues. *Phytochemistry* 12: 191–192.
Appleyard, M.E. (1992). Secreted acetylcholinesterase: non-classical aspects of a classical enzyme. *Trends in Neurobiology* 15: 485–490.
Arkhipova, L.V., Tretyak, T.M., and Ozolin O.N. (1988). The influence of catecholamines and serotonin on RNA-synthesizing capacity of isolated nucleus and chromatin of brain and rat liver. *Biochemistry (USSR)* 53: 1078–1081.
Artyukhov, V.G. (1978). Effect of UV-radiation in the presence of serotonin on optical properties of cytochrome C solution. *Biofizika* 38: 580–583.

Artyukhov, V.G. (1983). On the protective action of serotonin according to oxyhaemoglobin at ultraviolet irradiation of the solution mixture. *Biologicheskie Nauki:* 29–34.

Asada, K., Kiso, K., and Yoshikava, K. (1974). Univalent reduction of molecular oxygen by spinach chloroplasts on illumination. *Journal of Biological Chemistry* 249: 2175–2181.

Askar, A., Rubach, K., and Schormüller, J. (1972). Dünnschichtchromatographische Trennung der in Bananen vorkommenden Amin-Fraktion. *Chemiche Microbiologie und Technologie Lebenshem* 1: 187–190.

Atta-ur-Rahman and Basha, A. (1999). *Indole Alkaloids.* Berks: Harwood Academic Publishers. 324 pp.

Augurell, S., Bruhn, J.G., and Sheth, K. (1972). Structure and biosynthesis of alkaloids in *Carnegiea gigantea.* Part 1. In: *Biochemie und Physiologie der Alkaloide* (Eds. K. Mothes, K. Schreiber, H.R. Schütte). Berlin: Akademie-Verlag (Int. Symposium Halle (Saale) 25–28 June 1969), pp. 275–278.

Augustinsson, K.B. (1963). Classification and comparative enzymology of the cholinesterases and methods for their determination. In: *Cholinesterase and Anticholinesterase Agents* (Ed. G.B. Koelle). *Handbuch der Experimentale Pharmacologie.* Berlin, New York: Springer-Verlag, 15: 89–128.

Ayer, W.A., and Browne, L.M. (1970). Alkaloids of *Shepherdia argentea* and *Shepherdia canadensis. Canadian Journal of Chemistry* 48: 1980–1984.

Babakov, A.V., and Abramycheva, N.Yu. (1989). GTP-binding protein in plasmic membrane of higher plants. *Biological Membranes (USSR)* 6: 262–266.

Babskii, A.M., and Shostakovskaya, I.V. (1984). High sensibility of mitochondrial processes to adrenaline. *Bulletin of Experimental Biology and Medicine (USSR)* 43: 286–288.

Babskii, A.M., Kondrashova, M.N., and Shostakovskaya, I.V. (1985). Action and post-action of adrenaline on the respiration of mitochondria in rat liver. *Fiziologicheskii Zhurnal (USSR)* 31: 301–306.

Bahre, R. (1975). Zur Regulation des Protonemawachstums von *Athyrium filix-femina* (L.) Roth. II. Acetylcholin als Äthylenantagoinst. *Zeitschrift für Pflanzenphysiologie* 76: 248–251.

Bahre, R. (1977). Zur Regulation des Protonemawachstums von *Athyrium filix-femina* (L.) Roth. IV. Wirking cholinerger Substanzen in Gegenwart eines Antiauxins (PCIB). *Zeitschrift für Pflanzenphysiologie* 81: 278–282.

Bajjalieh, S.M., and Scheller, R.H. (1995). The biochemistry of neurotransmitter secretion. *Journal of Biological Chemistry* 270: 1971–1974.

Balasubramanian, A.S. (1984). Have cholinesterase more than one function? *Trends in Neuroscience* 7: 467–468.

Balasubramanian, A.S., and Bhanumathy, C.D. (1993). Noncholinergic functions of cholinesterases. *FASEB Journal* 7: 1354–1358.

Bandyopadhyay, R. (1984). Inhibition of acetylcholine esterase by permethrin and its reversion by acetycholine. *Indian Journal of Experimental Biology* 20: 488–491.

Bansal, V.S., and Majerus, P.W. (1990). Phosphatidylinositol-derived precursors and signals. *Annual Review of Cell Biology* 6: 41–67.

Barber, J. (1986). Surface electrical changes and protein phosphorylation. *Encyclopedia of Plant Physiology*, Vol. 19. pp. 653-664. Photosynthesis III (Eds. L.A. Staehelin and C.J. Arntzen). Berlin: Springer-Verlag.

Barber, J (1976). Ionic regulation in intact chloroplasts and its effect on primary photosynthetic processes. In: *Intact Chloroplast.* Ed. J. Barber. Amsterdam: Elsevier, pp. 89–134.

Barger, G., and Dale, H.H. (1910). A third active principle in ergot extracts. *Proceedings of Chemical Society* 26: 128–129.

Barlow, R.B., and Dixon, R.O.D. (1973). Cholineacetyltransferase in the nettle *Urtica dioica* L. *Biochemical Journal (London)* 132: 15–18.

Barnard, E.A. (1992). Receptor classes and the transmitter-gated ion channels. *Trends in Biochemical Sciences* 17: 368–374.

Barr, R., Troxel, K.S., and Crane, F.L. (1983). A calcium-selective site in photosystem II of spinach chloroplasts. *Plant Physiology* 73: 309–315.
Barrantes, F.J. Ed. (1998). *The Nicotinic Acetylcholine Receptor. Current Views and Future Trends.* Berlin: Springer-Verlag, 226 pp.
Baumgarten, H.G., and Göthert, M. (Eds) (1997). *Serotoninergic Neurons and 5-HT Receptors in CNS.* Berlin, Heidelberg: Springer-Verlag. 767 pp.
Bednarska, E. (1992). The localization of nonspecific esterase and cholinesterase activity in germinating pollen and in pollen tube of *Vicia faba* L. The effect of actinomycin D and cycloheximide. *Biologia Plantarum* 34: 229–240.
Bednarska, E., and Tretyn, A. (1989). Ultrastructural localization of acetylcholinesterase activity in the stigma of *Pharbitis nil* L. *Cellular Biology International Reports* 13: 275–281.
Beilby, M.J. (1984). Calcium and plant action potentials. *Plant Cell and Environment* 7: 415–421.
Benovic, J.L., Shorr, R.G.L., Caron, M.G., and Lefkowitz, R.J. (1984). The mammalian β_2-adrenergic receptor: purification and characterization: *Biochemistry* 23: 4510–4518.
Bergmann, L., Grosse, W., and Ruppel, H.G. (1970). Die Bildung von Serotonin *in Juglans regia* L. *Planta* 94: 47–59.
Berlin, J., Kuzovkina, I.N, Rügenhagen, C., Fecker, L., Commandeur, U., and Wray, V. (1992). Hairy root cultures of *Peganum harmala*. 2. Characterization of cell-lines and effect of culture conditions on the accumulation of Beta-carboline alkaloids and serotonin. *Zeitschrift für Naturforschung* 47 C: 222-230.
Berlin, J., Mollenschott, C., Sasse, F., Witte, L., Piehl, G.W., and Büntemeyer, H. (1987) Restoration of serotonin biosynthesis in cell-suspension cultures of *Peganum harmala* by by selection for 4.methyltryptophan-tolerant cell-lines. *Journal of Plant Physiology* 131: 225-236.
Berlin, J., Rügenhagen, C., Greidziak, N., Kuzovkina, I.N., Witte, L., and Wray, V. (1993). Biosynthesis of serotonin and β-carboline alkaloids in hairy root cultures of *Peganum harmala*. *Phytochemistry* 33: 593–599.
Bernard, C. (1878). *Life Phenomena Common to Animals and Plants.* (Russian Translation). Sanct-Peterburg: Bilibin Publ Press. 316 pp.
Berridge, M.J. (1984). Inositol triphosphate and diacylglycerol as secondary messengers. *Biochemical Journal* 220: 345–360.
Berridge, M.J. (1987). Inositol triphosphate and diacylglycerol: Two intergrating second messengers. *Annual Review of Biochemistry* 56: 159–193.
Berridge, M.J., and Irvine, R.F. (1984) Inositol triphosphate, a novel second messenger in cellular signal transduction. *Nature* 312: 315-321.
Biro, R.L. (1978). *Choline Acetylation and Its Phytochrome Control in Etiolated Pea Extracts.* Master's Thesis, Ohio University, 24 pp.
Blusztajn, J.K., Liscovitch, M., Richardson, U.I., and Wurtman, R.J. (1987). Phosphatidylcholine as a precursor of choline for acetylcholine synthesis. In: *Cellular and Molecular Basis of Cholinergic Function* (Eds. M.J. Dawdall and J.N. Hawthorne). Chichester, Weinheim: Ellis Horwood Ltd and VCH Verlagsgesselschaft, pp. 341–346.
Borodyuk, N.R. (1990). Role of the acetylcholine hydrolysis for the vital biological systems. *Vestnik Selkhozyastvennoi Nauki* 6: 87–95.
Boruttau, H., and Cappenberg, H. (1921). Beitrage zur Kennthis der Wirksmen Bestandteile des Hirtentaschelkrautes (Herba *Capsella Bursa Pastoris*). *Archives der Pharmacologie* 259: 33–52.
Bose, J.C. (1926). *The Nervous Mechanism of Plants.* London, New York: Langmans, Green and Co., 224 pp.
Bossen, M., Dassen, H.H.D., Kendrick, R.E., and Vredenberg, W.J. (1988). The role of calcium ions in phytochrome-controlled swelling of etiolated wheat (*Triticum aestivum* L.) protoplasts. *Planta* 174:. 94–100.

Bossen, M., Tretyn, A., Kendrick, R.E., and Vredenberg, W.J. (1991). Comparison between swelling of etoilated wheat (*Triticum aestivum* L.) protoplasts induced by phytochrome and α-naphthaleneacetic acid, benzylaminopurine, gibberellic acid, abscisic acid and acetylcholine. *Journal of Plant Physiology* 137: 706–710.

Bowden, K., Brown, B.G., and Batty, J.E. (1954). 5-hydroxytryptamine: its occurrence in cowhage. *Nature* 174: 925–926.

Bravo, F., Somavilla, C., Burdaspal, P., and Inigo, B. (1986). Histaminogenesis. V Contenido en histamina en vinos con poblaciones microbianas residuales. *Alimentaria* 23: 6–70.

Brestkin, A.P., Brick, I.L., and Grigoreva, G.M. (1973). Comparative pharmacology of cholinesterases. In: *Comparative Pharmacology* (Ed. M.I. Michelson), Vol. 1, Oxford, New York: Pergamon Press, pp. 241–344.

Brewin, N.J., and Northcote, D.H. (1973). Partial purification of a cyclic AMP a non-dialysable inhibitor. *Biochimica et Biophysica Acta* 320: 104–122.

Brown, E.G., and Newton, R.P. (1981). Cyclic AMP and higher plants. *Phytochemistry* 20: 2453–2463.

Brown, E.G., Newton, R.P., Evans, D.F., Walton, T.J., Younis, L.M., and Vaughan, J.M. (1989). Influence of light on cyclic nucleotide metabolism in plants, effect of dibutyryl cyclic nucleotides on chloroplast components. *Phytochemistry* 28: 2559–2563.

Brownlee, C., Taylor, A.P., and Roberts, S. (1991). Ionic and molecular mechanisms of intracellular signalling in plants. *Journal of Marine Biology Association of UK, JMBA* 71: 930–932.

Bruce, D.W. (1960). Serotonin in pineapple. *Nature* 188: 147.

Brzin, M., Sketelj, J., and Klinar, B. (1983). Cholinesterases. In: *Handbook of Neurochemistry*. Vol. 4, Enzyme in the Nervous System. New York, London: Plenum Press, pp. 251–292.

Buelow, W. and Gisvold, O. (1944). A phytochemical investigations of *Hermidium alipes*. *Journal of the American Pharmaceutical Association* 33: 270–274.

Bulard, C., and Leopold, M.A. (1958). 5-Hydroxytryptamine chez les plantes superieures. *Comptes Rendus de l'Academie de Science* 247: 1382–1384.

Bulychev, G.A., Andrianov, B.K., Kurella, G.A., and Litvin, F.F. (1971). Transmembrane potenital of cell and chloroplast of higher terrestrial plants. *Fiziologia rastenii (USSR)* 18: 248–256.

Bürcky, K., and Kauss, H. (1974). Veränderung in Gehalt an ATP and ADP in Wurzelsptzen der Mungobohne nach Hellrotbelichtung. *Zeitschrift für Pflazenphysiologie* 73: 184–186.

Burgun, C., de Munoz, D., and Aunis, D. (1985). Osmotic fragility of chromaffin granules prepared under isoosmotic or hyperosmotic conditions and localization of acetylcholinesterase. *Biochimica et Biophysica Acta* 839: 219–227.

Buznikov, G.A. (1967). *Low Molecular Regulators of Embryonal Development*. Moscow: Nauka, 265 pp.

Buznikov, G.A. (1987). *Neurotransmitters in Embryogenesis*. Moscow: Nauka, 232 pp.

Buznikov, G.A. (1989). Transmitters in early embryogenesis: New data. *Ontogenez* (Ontogenesis, USSR) 20: 637–646.

Buznikov, G.A. (1990). *Neurotransmitters in Embryogenesis*. London: Harwood, 526 pp.

Buznikov, G.A., and Turpaev, T.M. (1988). Specific features in intra and intercellular signalling systems as revealed in the biological models 'oocyte–early embryo-neuron'. In *Intracellular Signalling* (Eds. P.G. Kostyuk and M.A. Ostrovsky). Moscow: Nauka, pp. 7–15.

Buznikov, G.A., Shmukler, Yu.B., and Lauder, J.M. (1996). From oocyte to neuron: do neurotransmitters function in the same way throughout development? *Cellular and Molecular Neurobiology* 16: 532–559.

Bygdeman, S. (1960). Noradrenaline, 3-hydroxytyramine and related compounds in *Solanum tuberosum*. *Arkiv. för Kemi* 16: 247–253.

Carricarte, V.C., Bianchini, G.M., Muschietti, J.P., Tellez-Inon, M.T., Perticari, A., Torres, N., and Flawia, M.M. (1988). Adenylate cyclase activity in a higher plant alfalfa (*Medicago sativa*). *Biochemical Journal* 249: 807–811.

Cash, J.N., Sistrunk, W.A., and Stutte, C.A. (1976). Characteristics of Concord grape polyphenoloxidase involved in juice color loss. *Journal of Food Science* 41: 1398.

Cerione, R.A., Codina, J., Benovic, J.L., Lefkowitz, R.J., Birnbaumer, L., and Caron, M.G. (1984). The mammalian β_2-adrenergic receptor: Reconstitution of functional interactions between pure receptor and pure stimulatory nucleotide binding protein of the adenylate cyclase system. *Biochemistry* 23: 4519–4525.

Ceha, L.J., Presperin, C., Young, E., Allswede, M., and Erickson, T. (1997). Anticholinergic toxicity from nightshade berry poisoning responsive to physostigmine. *Journal of Emergence Medicine* 15: 65–69.

Chadwick, L.E. (1963). Actions on insects and other invertebrates. In: *Cholinesterases and Anticholinesterases Agents* (Ed. Koelle, G.B.). Berlin: Springer-Verlag, pp. 741–798.

Changeux, J.P., and Revah, F. (1987). The acetylcholine receptor molecule: allosteric sites and the ion channel. *Trends in Neurosciences* 10: 245–250.

Changeux, J.P., Devillers-Thiery, A., and Chemouilli, P. (1984). Acetylcholine receptor: An allosteric protein. *Science* 225: 1335–1345.

Changeux, J.P., Giraudat, J., and Dennis, M. (1987). The nicotinic acetylcholine receptor: molecular architecture of a ligand-regulated ion channel. *Trends in Pharmacological Sciences* 8: 459–465.

Che, F.S., Cho, C., Hyeon, S.B., Isogai, A., and Suzuki, A. (1990). Metabolism of choline chloride and its analogs in wheat seedlings. *Plant and Cell Physiology* 31: 45–50.

Chet, I., Henis, Y., and Mitchell, R. (1973). Effect of biogenic amines and cannabinoids on bacterial chemotaxis. *Journal of Bacteriology* 115: 1215–1218.

Chew, W.L. (1965). *Laportea* and allied genera (Urticaceae). *Gardens Bulletin of Singapore* 21: 195–208.

Chhabra, N., and Malik, C.P. (1978). Influence of spectral quality of light on pollen tube elongation in *Arachis hypogaea*. *Annals of Botany* 42: 1109–1117.

Chiang, G.G., Wooten, D.C. and Dilley, R.A. (1992). Calcium-dependent interaction of clorpromazine with the chloroplast 8-kilodalton CF_0 protein and calcium gating of H^+ fluxes between thylakoid membrane domains and the lumen. *Biochemistry* 31: 5808–5819.

Chkhve Tkhesop (1987). *Medicinal Plants*. Moscow: Meditsina, 608 pp.

Clegg, E.D. (1983). Mechanisms of mammalian sperm capacitation. In: *Mechanism and Control of Aninmal Fertilization* (Ed. J.F. Hartmann). New York, London: Academic Press, pp. 177–211.

Coleman, J., Evans, D., and Hawes, C. (1988). Plant coated vesicles. *Plant Cell and Environment* 11: 669–684.

Collier, H.O.J., and Chesher, G.B. (1956). Identification of 5-hydroxytryptamine in the sting of the nettle (*Urtica dioica*). *Britain Journal of Pharmacology* 11: 186–189.

Cooper, J.R., Bloom, F.E., and Roth, R.H. (1996). *Neuropharmacology*. Oxford: Oxford University Press.

Cordeiro, R.S.B., Aragao, J.B., and Morhy, L. (1983). The presence of histamine in *Cnidosculus oligandrus* (Euphorbiaceae). *Anais da Academia Brasiliera de Ciencias* 55: 123–128.

Cormier, M.J. (1983). Calmodulin: The regulation of cellular function. In: *Calcium in Biology* (Ed. T.G. Spiro). New York etc.: J. Wiley Intersci. Publ., pp. 53–106.

Cote, G.G., Depass, A.L., Quarmby, L.M., Tate, B.F., Morse M.J., Satter, R.L., and Crain, R.C. (1989). Separation and characterization of inositol phospholipids from the pulvini of *Samanea saman*. *Plant Physiology* 90: 1422-1428.

Crane, E. (1955). *Honey*. New York: Crane, Russak and Company.

Crane, E. (1975). The flowers honey comes from. In: *Honey. A Comprehensive Survey* (Ed. E. Crane). pp. 3-76.. New York: Crane, Russak and Company.

Cross, G.A. (1990). Glycolipid anchoring of plasma membrane proteins. *Annual Reviews in Cell Biology* 6: 1-39.

Coughlan, S.J., and Hind, G. (1986). Purification and characterization of a membrane-bound protein kinase from spinach thylakoids. *Journal of Biological Chemistry* 261: 11378–11385.

Csaba, G., and Pal, K. (1982). Effect of insulin triodothyronine and serotonin on plant seed development. *Protoplasma* 110: 20–22.

Csaba G., and Muller W.E.G (1996). *Signalling Mechanisms in Protozoa and Invertebrates*. Berlin: Springer-Verlag. 212 pp.

Cubero, A., and Malborn, C.C. (1984). The fat cell β-adrenergic receptor. Purification and characterization of a mammalian β_1-adrenergic receptor. *Journal of Biological Chemistry* 259: 1344–1350.

Dai, Y.R., Michaels, P.J., and Flores, H.E. (1993). Stimulation of ethylene production by catecholamines and phenylethylamine in potato cell suspension cultures. *Plant Growth Regulation* 12: 219–222.

Dannenburg, W.N., and Liverman, J.L. (1957). Conversion of tryptophan 2-C^{14} to indoleacetic acid by watermelon tissues slices. *Plant Physiology* 32: 263–269.

Dasgupta, B. (1966). Chemical investigations of *Alangium lamarckii*. II. Isolation of choline from the leaves. *Experientia* 22: 287–288.

Deacon, W., and Marsh, H.V. (1971). Properties of an enzyme from bananas (*Musa sapientum*) which hydroxylates tyramine to dopamine. *Phytochemistry* 10: 2915–2924.

Defilippi, P., Tarone, G., Gismondi, A., and Santoni, A. (1997). *Signal Transduction by Integrins*. New York, Berlin: Springer-Verlag, 188 pp.

Dekhuijzen, H.M. (1973). The effect of acetylcholine on the growth and on growth inhibition by CCC in wheat seedlings. *Planta* 111: 149–156.

De Lange, R.J., and Glazer, A.N. (1989). Phycoerythrin fluorescence-based assay for peroxy radicals: a screen for biologically relevant protective agents. *Analytical Biochemistry* 177: 300–306.

Delhaize, E., Dilworth, M.J., and Webb, J. (1986). The effects of copper nutrition and developmental state on the biosynthesis of diamine oxidase in clover leaves. *Plant Physiology* 82: 1126–1131.

de Melo, A.C., Perec, C.Y., and Rubio, M.C. (1978). Acetylcholine-like activity in the fruit of the black night shade (Solanaceae). *Acta Physiologica Latinoamericana* 28: 171–178.

Denisenko, P.P. (1980). *Role of Cholinereactive Systems in Regulatory Processes*. Moscow: Meditsina. 296 pp.

de Robertis, E. (1967). Ultrastructure and cytochemistry of the synaptic region. *Science* 156: 907–914.

Dettbarn, W.D. (1962). Acetylcholinesterase activity in *Nitella*. *Nature* 194: 1175–1176.

Devasankaraiah, G., Hanin, I., Haranath, P.S.R.K., and Ramanamurthy, P.S.V. (1974). Cholinomimetic effects of aqueous extracts from *Carum copticum* seeds. *Britain Journal of Pharmacology* 52: 613–614.

Dhalla, N.S., Gupta, K.C., Sastry, M.S., and Malhotra, C.L. (1961). Chemical composition of the fruit of *Momordica charantia* Linn. *Indian Journal of Pharmacy* 23: 128–129.

Dieter, P. (1984). Calmodulin and calmodulin-mediated processes in plants. *Plant, Cell and Environment* 7: 371–380.

Dobbins, M. (1968). *Trichocereus pachanoi*, a mescaline cactus used in folk healing in Peru. *Economical Botany* 22: 191–194.

Doi, T., and Ohmori, H. (1993). Acetylcholine increases intracellular Ca^{2+} concentration and hyperpolarizes the guinea-pig outer hair cell. *Hearing Research* 67: 179–188.

Doliba, H.M. (1986). Cholinergic regulation of respiration, oxidative phosphorylation and calcium volume of liver mitochondria. Various aspects of analysis of biological systems. *Doklady of Moscow Society of Naturalists. Common Biology*. Moscow: Nauka, pp. 94–95.

Doman, N.G., and Fedenko, E.P. (1976). Biological role of cyclic AMP. *Uspekhi Biologicheskoi Khimii* (USSR) 17: 63–101.
Drøbak, B.K. (1993). Plant phosphoinositides and ultracellular signaling. *Plant Physiology* 102: 705–709.
Drøbak, B.K. and Ferguson, I.B. (1985). Release of Ca^{2+} from hypocotyl microsomes by inositol-1,4,5-triphosphate. *Biochem. Biophys. Res. Commun.* 130: 1241-1246.
Durand, E., Ellington, E.V., Feng, P.C., Haynes, L.J., Magnus, K.E., and Philip, N. (1962). Simple hypotensive and hypertensive principles from some West Indian medicinal plants. *Journal of Pharmacy and Pharmacology* 14: 562–566.
Earnst, D.Y., and Turek, F.W. (1983). Role of acetylcholine in mediating effects of light on reproduction. *Science* 219: 77–79.
Eglen, R.M. (Ed.) (1997) *Muscarinic Receptor Subtypes in Smooth Muscle*. Boca Raton: CRC Press. 210 pp.
Ehrismann, O., and Werle, E. (1948). Über Histidindecarboxylase, Histidindehydrase und Histaminase der Bakterien. *Biochemische Zeitschrift* 318: 560-578.
Eichorn, M., Hermann, G., and Augsten, H. (1987). Rapid effects of isolated phytochrome upon the photosynthetic electron transport. *Abstracts of the 14th Int. Botanical Congress*. West Berlin: Berlin-Dahlem Botanical Museum. 481 pp.
Eijk, J.L. (1957). Fytochemisch oandersoak von *Erodium cicutarium*. II. *Pharm. Weekbl. Ned.* 92: 581–587.
Einspahr, R.J. (1990). Transmembrane signaling via phosphophatidylinositol 4,5-bisphosphate hydrolysis in plants. *Plants Physiology* 93: 361–366.
Einspahr, R.J. (1992). Inositol phospholipids in plant cell signalling. *Trends in Genetics* 8: 228.
Eisenberg, R.C. (1990). Channels as enzymes. *Journal of Membrane Biology* 115: 1–12.
Elazar, Z., Kanety, H., and David, C. (1988). Purification of the D_2 dopamine receptor from bovine striatum. *Biochemistry and Biophysics Research Communications* 156: 602–609.
Ellman, G.L., Courtney, K.D., Andres, V., and Featherstone, R.M. (1961). A new and rapid colorimetric determination of acetylcholinesterase activity. *Biochemcial Pharmacology* 7: 88–95.
Emmelin, N., and Feldberg, W. (1947). The mechanism of the sting of the common nettle (*Urtica urens*). *Journal of Physiology* 106: 440–455.
Emmelin, N., and Feldberg. W. (1949). Distribution of acetylcholine and histamine in nettle plants. *New Phytologist* 48: 143–148.
Endress, R. (1977). Einfluss moeglicher Phosphodiesterase Inhibitoren and cAMP auf die Betacyanin-Akkumulation. *Phytochemistry* 16: 1549–1554.
Endress, R., Jager, A., and Kreis, W. (1984). Catecholamine biosynthesis depends on the dark in betacyanin-forming *Portulaca* callus. *Journal of Plant Physiology* 115: 291–295.
Ernst, M., and Hartmann, E. (1980). Biochemical characterization of an acetylcholine-hydrolyzing enzyme from bean seedling. *Plant Physiology* 65: 447–450.
Erspamer, V. (1961). Recent research in the field of 5-hydroxytryptamine and related indolealkylamines. In: *Fortshritte der Arzneimittelforschung. Progress in Drug Research* (Ed. E. Jucker), Vol. 3. Basel, Stuttgart: Birkhaser Verlag, pp. 151–367.
Evans, M.L. (1972). Promotion of cell elongation in *Avena* coleoptiles by acetylcholine. *Plant Physiology* 50: 414–416.
Ewins, A.J. (1914). Acetylcholine, a new active principle of ergot. *Biochemical Journal* 8: 44–49.
Fairely-Grenot, K., and Assmann, S.M. (1991). Evidence for G-protein regulation of inward K^+-channel current in guard cells of fava bean. *The Plant Cell* 3: 1037–1044.
Faugeras, G.M. (1967). Presence d'amino-phenols sympathomimetiques chez les aconits. Distribution de la dopamine (3-hydroxy-tyramine), de la tyramine et de la noradrenaline chez l'*Aconitum napellus* L. et l'*Aconitum paniculatum* Lam. *Plant Medicinales et Phytotherapie* 1: 87–98.

Faugeras, G.M., Debelmas, J., and Paris, R.R.M. (1967). Sur la presence et la repartition d'aminophenols (tyramine, 3-hydroxy-tyramine ou dopamine, et nor-adrenaline), chez l'Aconit Napel: *Aconitum napellus* L. *Comptes Rendus de l'Academie de Science*, Paris Ser. D 264: 1864–1867.

Faust, M.A., and Doestsch, R.N. (1971). Effect of drugs that alter excitable membranes on the motility of *Rhodospirillum rubrum* and *Thiospirillum rubrum* and *Thiospirillum jenense*. *Canadian Journal of Microbiology* 17: 191–196.

Federico, R. and Angelini, R. (1986) Occurrence of diamine oxidase in the apoplast of pea epicotyls. *Planta* 167: 300-303.

Federico, R. and Angelini, R. (1988) Distribution of polyamines and their related catabolic enzyme in etiolated and light-grown Leguminosae seedlings. *Planta* 173: 317-321.

Federico, R., Alisi, C., and Angelini, R. (1988). On the occurrence of oxidoreductases in the apoplast of Leguminosae and Gramineae and their significance in the study of plasmalemma membrane-bound redox activities. In: *Plasma Membrane Oxidoreductase in Control of Animal and Plant Growth* (Eds. F.L. Crane, D.J. Morre and H. Low). pp. 333-337. New York, Plenum Press.

Fellows, L.E., and Bell, E.A. (1970). 5–Hydroxy-L-tryptophan, 5-hydroxytryptamine and L-tryptophan-5-hydroxylase in *Griffonia simplicifolia*. *Phytochemistry* 9: 2389–2396.

Fellows, L.E., and Bell, E.A. (1971). Indole metabolism in *Piptadenia peregrina*. *Phytochemistry* 10: 2083–2091.

Feng, P.C., Haynes, L.I., and Magnus, K.E. (1961). High concentration of (–) noradrenaline in *Portulaca oleracea* L. *Nature* 191: 1108.

Ferret, B., Sorrentino, G., Hattori, H., Kanfer, J., and Massarelli, R. (1987). Membrane-bound choline acetyltransferase. In: *Cellular and Molecular Basis of Cholinergic Function*. (Eds. M.J. Dawdall and J.N. Hawthorne). Chichester, Weinheim: Ellis Horwood and VCH Verlagsgesellschaft, pp. 361–367.

Fielder, U., Hildebrand, G., and Neu, R. (1953). Weitere Inhaltstoffe des Weissdorns: Der Nachweis von Cholin und Acetylcholin. *Arzneimittelforschung* 3: 436–437.

Field, R.A., Haines, A.H., Chrystal, E.J.T., and Luczniak, M.C. (1991). Histidines, histamines and imidazoles as glycosidase inhibitors. *Biochemical Journal* 274: 885–889.

Finberg, J.P.M., and Youdim, M.B.H. (1983). Monoaminooxidases. In: *Handbook of Neurochemistry* (Ed. A. Lajtha), Vol. 4, Enzymes in the nervous system., New York and London: Plenum Press, pp. 293–313.

Firn, R.D. (1987). Too many binding proteins, not enough receptors? In: *Plant Hormone Receptors*. (Ed. D. Klambt). pp. 1-11. Berlin, Heidelberg: Springer-Verlag.

Fitch, W.M. (1963a). Studies on a cholinesterase of *Pseudomonas fluorescens*. I. Enzyme induction and the metabolism of acetylcholine. *Biochemistry* 2: 1217–1221.

Fitch, W.M. (1963b). Studies on a cholinesterase of *Pseudomonas fluorescens*. II. Purification and properties. *Biochemistry* 2: 1221–1227.

Fitzgerald, J.S. (1964). Alkaloids of the Australian Leguminosae, IV. Cinnamoylhistamine, the alkaloid of *Acacia argentea* and *A. polystacha*. *Australian Journal of Chemistry* 17: 375–378.

Fluck, R.A., and Jaffe, M.J. (1974a). The acetylcholine system in plants. In: *Current Advances in Plant Sciences* (Ed. E. Smith), Vol. 5. Oxford: Science Engineering, Medical and Data Ltd., pp. 1–22.

Fluck, R.A., and Jaffe, M.J. (1974b). The distribution of cholinesterases in plant species. *Phytochemistry* 13: 2475–2480.

Fluck, R.A., and Jaffe, M.J. (1974c). Cholinesterases from plant tissues VI. Distribution and subcellular localization in *Phaseolus aureus* Roxb. *Plant Physiology* 53: 752–758.

Fluck, R.A., and Jaffe, M.J. (1974d). Cholinesterases from plant tissues. VI. Cholinesterase is not pectin esterase. *Plant Physiology* 54: 797–798.

Fluck, R.A., and Jaffe, M.J. (1975). Cholinesterases from plant tissues VI. Preliminary characterization on enzymes from *Solanum melongena* L. and *Zea mays* L. *Biochimica et Biophysica Acta* 410: 130–134.
Fluck, R.A., and Jaffe, M.J. (1976). The acetylcholine system in plants. In: *Commentaries in Plant Science* (Ed. H. Smith). Oxford: Pergamon Press, pp. 119–136.
Fodor, G.B. (1980). Alkaloids derived from phenylalanine and tyrosine. In: *Secondary Plant Products. Encyclopedia of Plant Physiology* (Eds. E.A. Bell and B.V. Charlwood), Vol. 8. New Ser. Berlin, New York, Heidelberg: Springer-Verlag, pp. 92–127.
Fonteless, M.C., Freire, J.G.L., and Rao, V.S.N. (1993) Identification of acetylcholine-like activity in the fruits of *Solanum ambrosiacum*. *Fitoterapia* 64: 404-406.
Fournier, D., Mutero, A., and Duri, R. (1992). Drosophila acetylcholinesterase. Expression of a functional precursor in *Xenopus* oocytes. *European Journal of Biochemsitry* 203: 513–519.
Fowler, H.D. (1962). Histamine in grasses and clovers. *Nature* (London) 193: 582–583.
Foy, J.M., and Parratt, J.R. (1960). A note on the presence of noradrenaline and 5-hydroxytryptamine in plantain (*Musa sapientum* var. *paradisiaca*). *Journal of Pharmacy and Pharmacology* 12: 360–364.
Foy, J.M., Parratt, J.R. (1961). 5-Hydroxytryptamine in pineapples. *Journal of Pharmacy and Pharmacology* 13: 382–383.
Fraikin, G.Ya. (1987). Some problems of modern ultra-violet photobiology. *Soviet Plant Physiology* 34: 712-719.
Franklin, F.C.H., Atwal, K.K., Ride, J.P., and Franklin-Tong, V.E. (1994). Towards the elicidation of the mechanisms of pollen tube inhibition during the self-incompatibility response in *Papaver rhoeas*. In: *Molecular and Cellular Aspects of Plant Reproduction*. (Eds. R.J. Scott and A.D. Stead). pp. 173-190. Cambridge, Cambridge University Press.
Frazer, A., Maayani, S., and Wolfe, B.B. (1990) Subtypes of receptors for serotonin. *Annual Review of Pharmacology and Toxicology*. 30: 307-348.
Frenzel, T., Beale, J.M., Kobayashi M., Zenk, M.H., and Floss, H.G. (1988). Stereochemistry of enzymatic formation of the berberine bridge in protoberberine alkaloids. *Journal of American Chemistry Society* 110: 7878-7882.
Fujita, N., Nelson, N., Fox, T.D., Claudio, T., Lindström, J., Riezman, H., and Hess, G.P. (1986). Biosynthesis of the *Torpedo californica* acetylcholine receptor α-subunit in yeast. *Science* 231: 1284–1287.
Gallon, J.R., and Butt, V.S. (1971). L-Tyrosine decarboxylase from barley roots. *Biochemical Journal* 123: 5–6.
Ganellin, C.R., Fkyerat, A., Hosseini, S.K., Khalaf, Y.S., Pizipitsi, A., Tertiuk, W., Arrang, J.M., Garbarg, M., Ligneau, X., and Schwartz, J.C. (1995). Structure-activity studies with histamine H_3-receptor ligands. *Ars Pharmaceutica* 36: 445–454.
Gardner, R.L., Kerst, A.F., Wilson, D.M., and Payne, M.G. (1967). *Beta vulgaris* L. the characterization of three polyphenols isolated from the leaves. *Phytochemistry* 6: 417–422.
Gasparov, V.S., Nalbandyan, R.M., and Buniatian, H.Ch. (1979). Interaction of neurocuprein with adrenaline. *FEBS Letters* 97: 37–39.
Gear, J.I. and Spenser, I.D. (1961). Biosynthesis of hydrastine. *Nature* 191: 1393-1395.
Gear, J.I. and Spenser, I.D. (1963). The biosynthesis of hydrastine and berberine. *Canadian Journal of Chemistry* 41: 783-803.
Gewitz, H.S., and Volker, V. (1961). 3-Hydroxytyramin' ein biologischer Oxydations katalysator der Photosynthese. *Zeitschrift für Naturforschung* 16B: 559–560.
Gibson, R.A., Barret, G., and Wightman, F. (1972). Biosynthesis and metabolism of indol-3yl-acetic acid. III. Partial purification and properties of a tryptamine-forming L-tryptophan decarboxylase from tomato shoots. *Journal of Experimental Botany* 23: 775–786.
Ginanatasio, M., Mandato, E. and Maccia, V. (1974). Content of cyclic 3',5'-AMP and cyclic AMP phosphodiesterase in dormant and activated tissues of Jerusalem artichoke tubers. *Biochem Biophys. Res. Commun.* 57: 365-371.

Girvin, G.T., and Stevenson, J.W. (1954). Cell free 'choline acetylase' from *Lactobacillus plantarum*. *Canadian Journal of Biochemistry and Physiology* 32: 131–146.

Goldstein, D.B., and Goldstein, A. (1953). An adaptive bacterial cholinesterase from a *Pseudomonas* species. *Journal of General Microbiology* 8: 8–17.

Goldschmidt, E.L. and Burkert, H. (1955). Die Hydrolyse des cholinergischen Honigwirkstoffes und anderer Cholinerster mittels Cholinesterasen und deren Hemmung im Honig. *Hoppe-Seyler's Zeitschrift für Physiologie und Chemische* 301: 78-89.

Golikov, C.N., Dolgo-Saburov. V.B., Eliev, N.R., and Kuleshov, V.I. (1985). *Cholinergic Regulation of Biochemical Cellular Systems*. Moscow: Meditsina, 223 pp.

Goot, H. van der, Bast, A., and Timmerman, H. (1991). Structural requirements for histamine H_2 agonists and H_2 antagonists. In: *Histamine and Histamine Antagonists* (Ed. H. van der Goot). Berlin, Heidelberg: Springer-Verlag, pp. 573–583.

Gorska-Brylass, A. and Smolinski, D.J. (1992). Ultrastructural localization of acetylcholinesterase activity in stomata of *Marchantia polymorpha* L. *Electron Microscopy* 3: 439–440.

Gorun, V., Proinov, I., Baltescu, V., Balaban, G., and Barru, O. (1978). Modified Ellman procedure for assay of cholinesterases in crude enzymatic preparations. *Analytical Biochemistry* 86: 324–326.

Govindappa, T., Govardham, L., Jyothy, P.S., and Veerabhadrappa, P.S. (1987). Purification and characterization of acetylcholinesterase isozymes from the latex of *Synadenium grantii* Hook, 'f'. *Indian Journal of Biochemistry and Biophysics* 24: 209–217.

Grawe, W., and Strack, D. (1968). Partial purification and some properties of 1-sinapoylglucose: choline sinapoyltransferase ('sinapine synthase') from seeds of *Raphanus sativus* L. and *Sinapis alba* L. *Zeitschrift für Naturforschung* 41: 28–33.

Green, S., Mazur, A. and Shorr, E. (1956). Mechanism of the catalytic oxidation of adrenaline by ferritin. *Journal of Biological Chemistry* 220: 237-255.

Greenfield, S. (1984). Acetylcholinesterase may have novel function in the brain. *Trends in Neuroscience* 7: 364–368.

Greppin, H., Horwitz, B.A., and Horwitz, L.P. (1973). Light-stimulated bioelectric response of spinach leaves and photoperiodic induction. *Zeitschrift für Pflanzenphysiologie* 68: 336–345.

Gressel, J., Strausbauch, L., and Galun, E. (1971). Photomimetic effect of acetylcholine on morphogenesis in *Trichoderma*. *Nature* 232: 648–649.

Grosse, W. (1982). Function of serotonin in seeds of walnuts. *Phytochemistry* 21: 819–822.

Grosse, W. (1983). Biosynthesis of serotonin and auxin in seeds of *Juglans regia*. In: *Progress in Tryptophan and Serotonin Research* (Eds. H. Schlossberger, W. Kochen, B. Linzen and H. Steinhart). Berlin, New York: de Gruyter, pp. 803–806 (Proc: 4th Meet. of Int. Stud. Group to Tryptophan Research ISTRY. Martinsried, Federal Republic of Germany, April 19–22, 1983).

Grosse, W., and Artigas, F. (1983). Incorporation of N-ammonia into serotonin in cotyledons of maturing walnuts. *Zeitschrift für Naturforschung* 38c: 1057–1058.

Grosse, W., and Klapheck, S.E. (1979). Enzymatische Untersuchungen uber die Biosynthese von Serotonin und ihre Regulation in Samen von *Juglans regia* L. *Zeitschrift für Pflanzenphysiologie* 93: 359–363.

Grosse, W., Karisch, M., and Schroder, P. (1983). Serotonin biosynthesis and its regulation in seeds of *Juglans regia* L. *Zeitschrift für Pflanzenphysiologie* 110: 221–229.

Guggenheim, M. (1958). Die biogenen Amine in der Pflanzenweit. In: *Handbuch für Pflanzenphysiologie* (Ed. W. Ruhland). Berlin: Springer-Verlag, pp. 889–988.

Gunar, I.I., and Sinyukhin, A.M. (1959). Electrophysiological characteristic of plant irritation. *Izvestiya of Timiryazev Agricultural Academy of Sciences (USSR)* 29: 7–22.

Gunar, I.I., and Sinyukhin, A.M. (1962). The propagating excitation wave in higher plants. *Doklady AN SSSR* 142: 954–956.

Gunar, I. I. and Sinyukhin, A.M. (1963). The functional significance of exitation currents in the changes of gas exchange in higher plants. *Soviet Plant Physiology* 10: 265-275.
Gupta, A., and Gupta, R. (1997). A survey of plants for presence of cholinesterase activity. *Phytochemistry* 46: 827–831.
Gupta, R. (1983). Studies on Cholinesterase in Bengal Gram and Probable Role of Acetylcholine in Plants. Ph.D. Thesis, University of Delhi, 24 pp.
Gupta, R., and Maheshwari, S.C. (1980). Preliminary characterization of a cholinesterase from roots of Bengal gram—*Cicer arietinum* L. *Plant and Cell Physiology* 21: 1675–1679.
Gupta, A., Vijayaraghavan, M.R., and Gupta, R. (1998). The presence of cholinesterase in marine algae. *Phytochemistry* 49: 1875–1877.
Gurin, V.N., Lapsha, V.I., and Streletskaya, L.G. (1989). Changes in the catecholamine content and energy metabolism in the rat myocardium during cold and emotional stress. *Sechenov Physiological Journal of the USSR* 75: 542–547.
Guye, M.G. (1989). Phospholipid, sterol composition and ethylene production in relation to choline-induced chill-tolerance in mung bean (*Vigna radiata* L. Wilcz.) during a chill-warm cycle. *Journal of Experimental Botany* 40: 369–374.
Guye, M.G., Vigh, L., and Wilson, J.M. (1987). Choline induced chill-tolerance in mung bean (*Vigna radiata* L. Wilcz.). *Plant Science* 53: 223–228.
Haartmann, U., Kahlson G., and Steinhardt, C. (1966). Histamine formation in germinating seeds. *Life Sciences* 5: 1–9.
Hadacova, V., Hofman, J., Almeida, R.M., Vackova K., Kutacek, M., and Klozova, E. (1981). Cholinesterases and choline acetyltransferase in the seeds of *Allium altaicum* (Pall.) Reyse. *Biologia Plantarum* 23: 220–227.
Hadacova, V., Vackova, K., Klozova, E., Kutacek, M. and Pitterova, K. (1983). Cholinesterase activity in some species of the *Allium* genus. *Biologia Plantarum* 25: 209–215.
Haga, T., and Berstein, G. (Eds.) (2000) *G Protein-coupled Receptors*. Boca Raton: CRC Press. 423 pp.
Hanson, A.D., and Grumet, R. (1985). Betaine accumulation: metabolic pathways and genetics. In: *Cellular and Molecular Biology of Plant Stress* (Eds. J.L. Key and T. Kosuge), Vol. 22. UCLA Symposia on Molecular and Cellular Biology, New Series. New York: Alan R. Liss, pp. 72–82.
Hanson, A.D., Brouguisse, R., Lerma, C., Weigel, P., and Weretilnyk, E.A. (1988). Enzymes and genes of glycine betaine synthesis. *Abstracts of 14th International Congress of Biochemistry.*, July 10-15, 1988. Prague: Videopress. 1: 50.
Harrison, W.H., Whisler, W.W., and Ko, S. (1967). Detection and study by fluorescence spectrometry of stereospecificity in mushroom tyrosine-catalyzed oxidations. A proposal a copper-containing reaction control site. *Journal of Biological Chemistry* 242: 1660–1667.
Hartmann, E. (1971). Über den Nachweis eines Neurohormones beim Laubmosscallus und seine Beeinflussung durch das Phytochrom. *Planta* 101: 159–165.
Hartmann, E. (1975). Influence of light on the bioelectric potential of the bean (*Phaseolus vulgaris*) hypocotyl hook. *Physiologia Plantarum* 33: 266–275.
Hartmann, E. (1977). Influence of acetylcholine and light on the bioelectrical potential of bean (*Phaseolus vulgaris* L.) hypocotyl hook. *Plant and Cell Physiology* 18: 1203–1207.
Hartmann, E. (1978). Uptake of acetylcholine by bean hypocotyl hooks. *Zeitschrift für Pflanzenphysiologie* 86: 303–311.
Hartmann, E. (1979). Attempts to demonstrate incorporation of labelled precursors into acetylcholine by *Phaseolus vulgaris* seedlings. *Phytochemistry* 18: 1643–1646.
Hartmann, E., and Gupta, R. (1989). Acetycholine as a signaling system in plants. In: *Second Messengers in Plant Growth and Development* (Eds. W.F. Boss and D. I. Morve). New York: Allan Liss, pp. 257–287.

Hartmann, E., and Kilbinger, H. (1974a). Gas-liquid chromatographic determination of light-dependent acetylcholine concentrations in moss callus. *Biochemical Journal* 137: 249–252.

Hartmann, E., and Kilbinger, H. (1974b). Occurrence of light-dependent acetylcholine concentrations in higher plants. *Experientia* 30: 1387–1388.

Hartmann, E., and Schleicher, W. (1977). Isolierung und Characterisierung einer Cholinkinase aus *Phaseolus aureus* L. keimlingen. *Zeitschrit für Pflanzenphysiologie* 83: 69-80.

Hartmann, E., Grasmusck, I., Lehrback, N., and Muller, R. (1980). The influence of acetylcholine and choline on the incorporation of phosphate into phospholipides of etiolated bean hypocotyl hooks. *Zeitschrift für Pflanzenphysiologie* 97: 377–389.

Hassler, J., Macko, V., and Novacky, A. (1969). Separation of plant diamine oxidase by disc electrophoresis in polyacrylamide gel. *Biologia* 24: 223–228.

Heacock, R.A., and Powell, W.S. (1973). Adrenochrome and related compounds. In: *Progress in Medicinal Chemistry* (Eds. G.P. Ellis and G.B. West), Vol. 9. Amsterdam, London, New York: North-Holland Publishing Company, pp. 275–340.

Hegnauer, R. (1964). *Chemotaxonomie der Pflanzen*, 1st ed., Vol. 3. Basel: Birkhauser, p. 362.

Hegnauer, R. (1990). *Chemotaxonomie der Pflanzen*, Bd. 109. Basel: Birkhauser, p. 362.

Heirman, P. (1939). Recherches sur l'adrenoxine. 1. Variabilite des proprietes physiologiques et biochimiques du jus de presse du champignon de couche. *Archives of International Physiology and Biochemistry* 49: 449–454.

Heise, K.P., and Treede, H.J. (1987). Regulation of acetylcoenzyme A synthesis in chloroplasts. In: *Metabolism, Structure and Function of Plant Lipids*. Proc. 7th Int Symp. Plant Lipids, Davis, Califorina (July 27 to August 1, 1986). (eds.) P.K. Stumpf, J.B. Mudd, and W.W.D. Nes), Plenum Press, New York, London: pp. 505–507.

Hikino, H., Ogata, M., and Konno, C. (1983). Structure of ferruloylhistamine, a hypotensive principle of *Ephedra* roots. *Planta Medica* 48: 108–110.

Hill, J.M., and Mann, P.J.G. (1964). Further properties of the diamine oxidase of pea seedlings. *Biochemcial Journal* 91: 171–182.

Hill, R.K. (1984). The imidazole alkaloids. In: *Alkaloids: Chemical and Biological Perspectives* (Ed. S.W. Pelletier), Vol. 2. New York, Chichester, Brisbane, Toronto, Singapore: J. Wiley and Sons, pp. 49–104.

Hock, K.F. (1983). Untersuchungen zur Kompartimentierung der Enzyme des Pflanzlichen Acetylcholine Metabolismus mit Hilfe Pflanzlicher Protoplasten. Ph.D. Thesis, University of Mainz, 24 pp.

Hofer, P., and Fringeli, U.P. (1981). Acetylcholinesterase kinetics. *Biophysical Structure and Mechanisms* 8: 45–59.

Hoffmann, N. (1982). Lokalisierung, Reiningung and biochemische Charakterisierung der Cholinacetyltransferase aus Hypocotylhaken von *Phaseolus vulgaris* L. Master's thesis, University of Mainz.

Hoffmann, A. von. (1969). Mexikaniscer zauberdrogen und ihre wirkstoffe. *Planta Medica* 132: 341–352.

Holm, R.E., and Miller, M.R. (1972). Hormonal control of weed seed germination. *Weed Science* 20: 209–219.

Holtz, P., and Janisch, H. (1937). Über das Vorkommen von Azetylcholin, Histamin und Adenosin in Pflanzen. *Archives Experimental Pathology Pharmacology* 187: 336–343.

Hong, Ma (1993). Protein phosphorylation in plants: enzymes, substrated and regulators. *Trends in Genetics* 9: 228–230.

Horn, A.S. and Rodgers, J.R. (1980). 2-Amino-6,7-dihydroxytetrahydronaphthalene and the receptor site preferred conformation of dopamine—a commentary. *Journal of Pharmacy and Pharmacology* 32: 521-524.

Horton, E.W., and Felippe, G.M. (1973). An acetylcholine-like substance in *Porophyllum lanceolatum*. *Biologia Plantarum* 15: 150–151.

Hoshino, T. (1979). Simulation of acetylcholine action by β-indole acetic acid in inducing diuranol change of floral response to under continuous light in *Lemna gibba* G3. *Plant and Cell Physiology* 20: 43–50.
Hoshino, T. (1983). Effects of acetylcholine on the growth of the *Vigna* seedlings. *Plant and Cell Physiology* 24: 551–556.
Hoshino, T., and Oata, Y. (1978). The occurrence of acetylcholine in *Lemna gibba* G-3. *Plant and Cell Physiology* 19: 769–776.
Hosoi, K. (1974). Purification and some properties of L-tyrosine carboxylase from barley roots. *Plant and Cell Physiology* 15: 429–440.
Hosoi, K., Yoshida, S., and Hasegawa, M. (1970). L-Tyrosine carboxy-lyase of barley roots. *Plant and Cell Physiology* 11: 899–906.
Hourmant, A., Rapt, F., Morzadec, J.M., Feray, A., and Caroff, J. (1998). Involvement of catecholic compounds in morphogenesis of in vitro potato plants. Effect of methylglyoxal-bis (guanylhydrzone). *Journal of Plant Physiology* 152: 64–69.
Hyeon, S.B., Cho, C., Che, F.S., Tsukamoto, C., Tanaka, A., Furushima, M., and Suzuki, A. (1987). Effects of choline chloride and its analogues on photosynthesis in wheat protoplasts. *Agricultural and Biological Chemistry* 51: 917–919.
Irvine, R.E., Letcher, A.I., and Dawson, R.M. (1980). Phosphatidylionsitol phosphodiesterase in higher plants. *Biochemical Journal* 192: 279-283.
Irvine, R.F., Letcher, A.J., Stephens, L.R., and Musgrave, A. (1992). Inositol polyphosphate metabolism and inositol lipids in a green algae *Chlamydomonas eugametos*. *Biochemical Journal* 281: 261–266.
Iwasa, K., Kamigauchi, M., Saiki, K., Takao, N., and Wiegrebe, W. (1993). O-Methylating enzymes of dopamine and dopamine derived tetrahydroisoquinoline, salsolinol. *Phytochemistry* 32: 1443–1448.
Iwasa, K., Kondon, Y., Kamigauchi, M., Lee, D.U., Tanner, U., and Wiegrebe, W. (1995). Effects of pyrogallol on O-methylation of dopamine and salsolinol. *Phytochemistry* 38: 335–342.
Izotova, T.E., Yusupova, I.K. and Kiselev, F.M. (1973) The effect of rigor, a organophosphate, insecticide on the development of pollen in winter wheat. In: *Sexual Process and Embryogenesis of Plants*. (Ed. N.A. Tsitsin). pp. 84-85. Moscow: Central Botanical Garden.
Jacobsohn, K., and Azevedo, M.D. (1962). A propos de la cholinesterase vegetale. *Comptes Rendus de la Societe de Biologie* 156: 202–204.
Jaffe, M.J. (1970). Evidence for the regulation of phytochrome mediated processes in bean roots by the neurohumor, acetylcholine. *Plant Physiology* 46: 768–777.
Jaffe, M.J. (1972a). Acetylcholine as a native metabolic regulator of phytochrome mediated processes in bean roots. In: *Recent Advances in Phytochemistry* (Eds. V.C. Runeckles and T.C. Tso), Vol. 5, New York: Academic Press, pp. 81–104.
Jaffe, M.J. (1972b). Acetylcholine as a native metabolic regulator of phytochrome mediated processes in bean roots. Proceedings of 8th Annual Symposium of Phytochemical Society N. America, *Advances in Phytochemistry* 7: 81–104.
Jaffe, M.J. (1976). Phytochrome-controlled acetylcholine synthesis at the endoplasmatic reticulum. In: *Light and Plant Development* (Ed. H. Smith). London: Butterworth, pp. 333–344.
Jaffe, M.J., and Thoma, L. (1973). Rapid phytochrome-mediated changes in the uptake by bean roots of sodium acetate (1-^{14}C) and their modification by cholinergic drugs. *Planta* 113, 283–291.
Janakidevi, K., Dewey, V.C., and Kidder, G.W. (1966a). Serotonin in Protozoa. *Archives of Biochemistry and Biophysical* 113: 758–759.
Janakidevi, K., Dewey, V.C., and Kidder, G.W. (1966b). The biosynthesis of catecholamines in two genera of Protozoa. *Journal of Biological Chemistry* 241: 2576–2578.

Janistin, B. (1972). Indol-3-essigsäure-induzierte Nukleotida bgabe bei gleichzeiting erhohter Adenosin-3',5'-monophosphor-saure (c-AMP)-Synthese in Maiskoleoptilzylinden. *Zeitschrift für Naturforschung* 27b: 273–276.

Janistyn, B.C. (1989). AMP promoted protein phosphorylation of dialysed coconut milk. *Phytochemistry* 28: 329–331.

Jansen, E.F., Nutting, M.D.F., and Balls, A.K. (1948). The reversible inhibition of acetylesterase by diisopropyl fluorophosphate and tetraethyl pyrophosphate. *Journal of Biological Chemistry* 175: 975.

Jansen, E.F., Jang, R., and Macdonnell, L.R. (1974). *Citrus* acetylesterases. *Archives of Biochemistry and Biophysics* 15: 415–431.

Jewett, S.L., Eddy, L.J., and Hochstein, P. (1989). Is the autooxidation of catecholamines involved in ischemia-reperfusion injury? *Free Radical Biology and Medicine* 6: 185-188.

Johns, S.R., and Lamberton, J.A. (1966). New histamine alkaloids from *Glochidon* species. *Chemical Communications* 10: 312–313.

Johns, S.R., and Lamberton, J.A. (1967). New imidazole alkaloids from *Glochidon* species (Family Euphorbiaceae). *Australian Journal of Chemistry* 20: 555–560.

Johns, S.R., Lamberton, J.A., Loder, J.W., Redicliffe, A.H., and Siouimis, A.A. (1969). Alkaloids of *Argyrodendron peralatum* (Sterculiaceae): identification of N^{α}- cinnamoylhistamine. *Australian Journal of Chemistry* 22: 1309–1310.

Jones, H., and Halliwell, B. (1984). Calcium ion and calmodulin in pea chloroplasts as a function of plant age. *Photobiochemistry and Photobiophysics* 7: 293–297.

Jones, R.S., and Stutte, C.A. (1984). The acetylcholine-ethylene connection. *Proceedings of Plant Growth Regulators Society of America* 11: 46–51.

Jones, R.A., and Stutte, C.A. (1986). Acetylcholine and red-light influence on ethylene evolution from soybean leaf tissues. *Annals of Botany* 57: 897–900.

Jones, R.S., and Stutte, C.A. (1988). Inhibition of ACC-mediated ethylene evolution from soybean leaf tissue by acetylcholine. *Phyton* 48: 107–113.

Jürisson, S. (1971). Hiirekorva *Capsella bursa-pastoris* (L.) Med toimeaine maaramine. *Tartu Riiklikku Tlikooli Toimetised (Transactions of the Tartu State University, Farmatsia)* 26: 71–79.

Kamaravel, M., Sundaran, R., and Rao, D.S. (1979). Studies on active chemical principles responsible for nyctinastic behavior of *Samanea saman*. *Indian Journal of Biochemistry and Biophysics. Suppl.* 16: 89–90.

Kamisaka, S. (1979). Catecholamine stimulation of the gibberellin action that induced lettuce hypocotyl elongation. *Plant and Cell Physiology* 20: 1199–1207.

Kamisaka, S., and Shibata, K. (1982). Identification in lettuce seedlings of catecholamine active synergistically enchancing the gibberellin effect on lettuce hypocotyl elongation. *Plant Growth Regulation* 1: 3–10.

Kandeler, R. (1972). Die Wirkung von Acetylcholin auf die photoperiodische Steurung der Blutenbildung bei Lemnaceen. *Zeitschrift für Pflanzenphysiologie* 67: 86–92.

Kandle (1961). Cited in Werle *et al.* (1961).

Karnovsky, M.J., and Roots, L.A. (1964). A direct-coloring thiocholine method for cholines terases. *Journal of Histochemistry and Cytochemistry* 12: 219–221.

Kasemir, H., and Mohr, H. (1972). Involvement of acetylcholine in phytochrome-mediated processes. *Plant Physiology* 49: 453–454.

Kasturi, R. (1978). De novo synthesis of acetylcholinesterase in roots of *Pisum sativum*. *Phytochemistry* 17: 647–649.

Kasturi, R. (1979). Influence of light, phytohormone and acetylcholine on the de novo synthesis of acetylcholinesterase in roots of *Pisum sativum*. *Indian Journal of Biochemistry and Biophysics* 16: Supplement. 14–17.

Kasturi, R., and Vasantharajan, V.N. (1976). Properties of acetylcholineesterase from *Pisum sativum*. *Phytochemistry* 15: 1345–1347.

Katz, M.M. and Lavretskaya, E.F. (1986). *Receptors of Brain Biogenic Amines: Structure, Mechanisms of Function and Interaction with Physiologically Active Substances.* (Ser. Itogi Nauki i Tekniki: Bioorganic Chemistry). Moscow: VINITI, 228 pp.

Kawasaki, S., and Takeuchi, J. (1989). Senescence-induced thylakoid-bound diisopropyl fluorophosphate-binding protein in spinach. Induction pattern, localization and some properties. *Plant Physiology* 90: 338–344.

Kawashima, K., Watanabe, N., Oohata, H., Fujimoto, K., Suzuki, T., Ishizaki, Y., Morita, I., and Murota, S. (1990). Synthesis and release of acetylcholine by cultivated bovine arterial endothelial cells. *Neuroscience Letters* 19: 156–158.

Kefeli, V.I., Kutacek, M., Vackova, K., Makhachkova, I., Zmrgal, Z., Vlasov, P.V., Guskov, A.V., and Shapkin, V.I. (1977). Derivatives of sinapic acid in seedlings of kohlrabi *Brassica oleracea*. *Fiziologia Rastenii (USSR)* 24: 1200–1205.

Keil, W., and Portner, F. (1934). Zur Chemie und Pharmakologie vergorener Nahrungsmittel. *Biochemiche Zeitschrift* 276: 61–65.

Keil, W., and Portner, F. (1935). Zur Chemie und Pharmakologie vergorener Nahrungsmittel. *Biochemiche Zeitschrift* 280: 61–64.

Kelly, R.B. (1993). Storage and release of neurotransmitters. *Cell* 72: Neuron 10, (suppl.), 43–53.

Kenten, R.H., and Mann, P.J.G. (1952). The oxidation of amines by extracts of pea seedlings. *Biochemical Journal* 50: 360–369.

Kessler, B., and Levinstein, R. (1974). Adenosine 3,5-cyclic monophosphate in higher plants: assay, distribution and age-dependency. *Biochimica et Biophysica Acta* 343: 156–166.

Kesy, J., Tretyn, A., Lukasiewicz, H., and Kopcewicz, J. (1991). Acetylcholinesterase from oat seedlings. 1. Preliminary biochemical characterization of the enzyme. *Biologia Plantarum* 33: 131–143.

Khurana, J.P., Tamot, B.K., Maheshwari, N., and Maheshwari, S.C. (1987). Role of catecholamines in promotion of flowering in a short-day duckweed *Lemna paucicostata* 6746. *Plant Physiology* 87: 10–12.

Kim, H.Y., Kim, T.I., Kim, H.K., and Chae, Q. (1990). The effect of phytochrome action on the activity of cytosolic cholinesterase in oat cells. *Biochem. Biophys. Res. Commun.* 169: 159–164.

Kimbrough, T.D., Reynolds, J.D., Humphereys, K.J., and Weekley, L.B. (1987). Diurnal changes in tissue leaf levels of tryptophan, tyrosine and amine metabolites in *Sedum morganianum* and *Sedum pachyphyllum*. *Biochemie und Physiologie der Pflanzen* 182: 67–72.

Kirberger, E., and Braun, L. (1961). Über das Vorkommen von 5-Hydroxytriptamin in der Walnus (*Juglans regia*). *Biochimica et Biophysica Acta* 49: 391–393.

Kirshner, R.L., White, J.M., and Pike, C.S. (1975). Contol of bean bud ATP levels by regulatory molecules and phytochrome. *Physiologia Plantarum* 34: 373–377.

Klapheck, S., and Grosse, W. (1980). Ammonia detoxification by formation of serotonin in seeds of *Juglans regia* L. II. De novo synthesis of aromatic amino acid decarboxylase during seed maturation. *Zeitschrift für Pflanzenphysiologie* 100: 325–331.

Klegeris, A., Korkina, L.G., and Greenfield, S.A. (1995). A possible interaction between acetylcholinesterase and dopamine molecules during autoxidation of the amine. *Free Radical Biology and Medicine* 18: 223–230.

Kondrashova, M.N., and Doliba, N.M. (1979). Polarographic observation of substrate-level phosphorylation and its stimulation by acetylcholine. *FEBS Letters* 243: 153–155.

Kopcewicz, J., and Cymerski, M. (1980). On the involvement of acetylcholine in phytochrome-mediated changes of gibberellin content in scots pine seedlings. *Bulleten Academie Poloniae Science* 28: 247–252.

Kopcewicz, J., Cymerski, M., and Porazinski, Z. (1977). Influence of red and far red irradiation on the acetylcholine and gibberellin content in scots pine seedlings. *Bulleten Academie Poloniae Science* 25: 114–117.

Kopcewicz, J., Cymerski, M, and Porazinski, Z. (1979). The effect of acetylcholine on gibberellin content in scots pine seedlings. *Bulleten Academie Poloniae Science* 27: 295–302.

Korobova, L.N., Beletskii, Yu.D., and Karnaukhova, T.B. (1988). The experiment of the selection of salt-tolerant forms of sunflower among the selection material based on the content of histamine in seeds. *Fiziologiya and Biokhimiya Kulturnikh Rastenii (USSR)* 20: 403–406.

Korolev, A.V. and Zholkevich, V.N. (1990). The influence of metabolic regulators on the root pumping activity. *Doklady of Russian Academy of Sciences* 310: 507–511.

Koshland, D.E. (1981). Biochemistry of sensing and adaptation in a simple bacterial system. *Annual Review of Biochemistry* 50: 765–782.

Koshtoyantz, Ch.S. (1963). *Problems of Enzymochemistry of the Processes of Excitation and Depression and Evolution of the Function of Nervous System.* 17th Bakh Lection (USSR), AN SSSR Pub. House, Moscow, 31 pp.

Kostir, J., Klenha, J., and Jiracek, V. (1965). Vliv cholinu a acetylcholinu na Kliceni semen hospodarsky dulezitych plodin. *Journal Rostlinna Vyroba.* Rocnik 11 (28): 1239–1280.

Kovaleva, L.V. and Roshchina, V.V. (1997). Does cholinesterase participate in the intercellular interaction in pollen-pistil system? *Biologia Plantarum* 39: 207–213.

Krzaczek, T. (1977) Badania farmakologiczne podgatunkov *Viscum album* L. *Ann. UMCSD* 33: 281-291.

Kesy, J., Tretyn, A., Lukasiewicz, H., and Kopcewicz, J. (1991). Acetylcholinesterase from oat seedlings. I. Preliminary biochemical characterisation of the enzyme. *Biologia Plantarum* 33: 131–143.

Kuhn, D.M., and Lovenberg, W. (1983). Hydroxylases In: *Handbook of Neurochemistry* (Ed. A. Laytha), Vol. 4, Enzymes in the Nervous System. New York, London: Plenum Press, pp. 133–150.

Kumar, M., and Flurkey, W.H. (1991). Activity, isoenzymes and purity of mushroom tyrosinase in commercial preparations. *Phytochemistry* 30: 3899–3902.

Kumaravel, M., Sundaran, R., and Rao, D.S. (1979). Studies on active chemical principles responsible for nyctinastic behavoir of *Samanea saman*. *Indian Journal of Biochemistry and Biophysics.* Supplement. 16: 89–90.

Kurosawa, M. (1991). Histamine. *Pharmaceutical Medicine* 9: 35–39.

Kurchii, V.M. and Kurchii, B.A. (2000). Decomposition of acetylcholine with ethylene formation in vitro. Possible free radical mechanisms of acetylcholine action. *The Ukrainian Biochemical Journal* 72: 69–72.

Kutachek, M., Pospishil, F., de Almeida, R.M., Vackova, K., and Mikhailovskii, N. (1981). The derivatives of oxicinnamic acid in plant of family Brassicaceae and the participation of cholinesterase in their metabolism. Biological role of the studied systems. In: *Plant Growth and Differentiation* (Ed. V.I. Kefeli). Moscow: Nauka, pp. 133–150.

Kutchan, T.M., Rush, M., and Coscia, C.J. (1986). Subcellular localization of alkaloids and dopamine in different vacuolar compartment of *Papaver bracteatum*. *Plant Physiology* 81: 161–166.

Kutcher, F. (1910). Die physiologische Wirkung einer Secalebase und des Imidazolylathyamin. *Zentralblatt der Physiologie* 24: 163–165.

Kwasniewski, V. (1958). Untersuchungen uber die nichtalkaloidischen Inhaltstoffe des Schoollkrauts (*Chelidonium majus* L.). *Pharmazie* 13: 363–364.

Kwasniewski, V. (1959). Zur Kenntnis der Inhaltstoffe der Bluten der Taubnessel. *Planta Medica* 7: 35–40.

Lachmann, K.U., and Grosse, W. (1990). The effect of phytohormones and sucrose supply on the induction of tryptophan decarboxylase in developing embryos of *Juglans regia* L. *Zeitschrift für Naturforschung* 45c: 785–792.

Ladeira, A.M., Felippe, G.M., and Dietrich, S.M.C. (1982a). Detection of acetylcholine and effects of exogenous ACh on *Rumex* and *Cucumis* photoblasticity. *Revta Brasilian Botanica* 5: 17–19.

Ladeira, A.M., Dietrich, S.M.C., and Felippe, G.M. (1982b). Acetylcholine and flowering of photoperiodic plants. *Revta Brasilian Botanica* 5: 21–24.
Ladror, U.S., and Zielinski, R.E. (1989). Protein kinase activities in tonoplast and plasmalemma membranes from corn roots. *Plant Physiology* 89: 151–158.
Larsen, P.O., Pedersen, E., Sorensen, H., and Sorup, P. (1973). Tyrosine 0-glucoside and dopamine 3-0-glucoside in seeds of *Entada pursaetha. Phytochemistry* 12: 2243–2247.
Lauffenburger, D.A. (1991). Models for receptor-mediated cell phenomena: adhesion and migration. *Annual Review of Biophysics and Biophysical Chemistry* 20: 387–414.
Lawson, V.R., Brady, R.M., Campbell, A., Knox, B.G., Knox, G.D., and Walls, R.L. (1978). Interaction of acetylcholine chloride with IAA, GA_3, and red light in the growth of excised apical coleoptile segments. *Bulletin of Torrey Botanical Club* 105: 187–191.
Lees, G.L., and Thompson, J.E. (1975). The effect of germination on the subcellular distribution of cholinesterase in cotyledons of *Phaseolus vulgaris. Physiologia Plantarum* 34: 230–237.
Lees, G.L., Lahue, R., and Thompson, J.E. (1978). Changes in the acetylcholine titre of senescing cotyledons. *Jounral of Experimental Botany* 29: 1117–1124.
Leete, E., and Murrill, S.J.B. (1964). The incorporation of dopamine into chelidonine and morphine. *Tetrahedron Letters* 3: 147–151.
Legendre, L., Derckel, J.P., Wrisez, F., Correze, C., Audran, J.C., Haye, B., and Lambert, B. (1997). Evidence for the existence of cAMP in lily plant flower tissues. *Phytochemistry* 44: 769–774.
Lembeck, F., and Skofitsch, G. (1984). Distribution of serotonin in *Juglans regia* seeds during ontogenetic development and germination. *Zeitschrift für Pflanzenphysiologie* 114: 349–353.
Lerma, C., Hanson, A.D., and Rhodes, D. (1988). Oxygen-18 and Deuterium labeling studies of choline oxidation by spinach and sugar beet. *Plant Physiology* 88: 695–702.
Levin, S.A., Potapova, T.V., Skulachev, V.P., and Chailakhyan, L.M. (1982). The propagation of electric potential changes in filamentous cyanobacteria. *Biofizika (USSR)* 27: 280–284.
Levitzki, A. (1988). Transmembrane signalling to adenylate cyclase in mammalian cells and in *Saccharomyces cerevisiae. Trends in Biochemical Science* 13: 298–301.
Lew, J.Y., Meller, E., and Goldstein, M. (1985). Photoaffinity labelling and purification of solubilized D_2 dopamine receptors. *European Journal of Pharmacology* 113: 145–146.
Lin, Lu (1993). Shengmingde huaxue. *Chemistry Life* 13: 19–20.
Lin, P.C., and Varmer, J.E. (1972). Cyclic nucleotide phosphodiesterase in pea seedlings. *Biochimica et Biophysica Acta* 276: 454–474.
Lin, R.C.Y. (1955). Presence of acetylcholine in the Malayan jackfruit, *Artocarpus integra. Britain Journal of Pharmacology* 10: 247–253.
Lin, R.C.Y. (1957). Distribution of acetylcholine in the Malayan jack-fruit plant, *Artocarpus integra. Britain Journal of Pharmacology* 12: 256–269.
Linscheid, M., Wendisch, D., and Strack, D. (1980). The structures of sinapic acid esters and their metabolism in cotyledons of *Raphanus sativus. Zeitschrift für Naturforschung* 35c: 907–914.
List, P.H., and Reinhard, C. (1962). Pilzinhaltsstoffe. 11. Eine schwefelhaltige Diaminodicarbon-saure im Hexenei von *Phallus impudicus* L. *Archives der Pharmacognosie* 295: 564–571.
Lloyd, G.R., and Nicholis, P.J. (1964). The presence of histamine in the cotton plant. *Journal of Physiology* (London) 172: 56–57.
Lloyd, G.R., and Nicholis, P.J. (1965). Formation of histamine in the cotton plant. *Nature* 206: 298–299.
Loewi, O. (1937). Strychninerregung und Acetylcholingehalt des Zentralnervensystems. *Naturwissenschaften* 25: 526–529.
Löffler, S., Stadler, R., and Zenk, M.H. (1987). Biosynthesis of morphinandienone alkaloids in opium poppy. *Abstracts 14th Int. Bot. Congress*, 24 July to 1 August 1987, Berlin-Dahlem Museum, Berlin, p. 51.

Lookado, S.E. and Pollard, A.J. (1991). Chemical contents of stinging trichomes of *Cnidoscolus texanus*. *Journal of Chemical Ecology* 17: 1909–1916.
Lopo, A.C. (1983). Sperm-egg interactions in invertebrates. In: *Mechanism and Control of Animal Fertilization* (Ed. J.F. Hartmann). New York, London: Academic Press, pp. 269–324.
Lovenberg, W., Weissbach, H., and Udenfriend, S. (1962). Aromatic L-amino acid decarboxylase. *Journal of Biological Chemistry* 237: 89–93.
Low, M.G., Futerman, A.H., Ackermann, K.E., Sherman, W.R., and Silman, I. (1987). Removal of covalentty bound inositol from Torpedo acetylcholinesterase and mammalian alkaline phosphatases by deamination with nitrous acid. Evidence for a common nembrane-anchoring structure. *Biochemical Journal* 241: 615–619.
Lozeron, H., and Maggiora, A. (1965). A propos of chemical radioprotection. *Dermatologica* 131: 28–40.
Luh, B.S., and Phithakpol, B. (1972). Characteristics of polyphenol oxidase related to browning in cling peaches. *Journal of Food Science* 37: 264.
Lukasiewicz-Rutkowska, H., Tretyn, A., Cymerski, M., and Kopcewicz, J. (1997). The effect of exogenous acetylcholine and other cholinergic agents on photoperiodic flower induction of *Pharbitis nil*. *Acta Soc. Bot. Polon.* 66: 47-54.
Lukes, R. (1930a). O novem zpusobu pripravy kyselin mastnych. *Chemicke Listy* 24: 197-200.
Lukes, R. (1930b). O nekterych derivatech kyseliny homolevulove. *Chemicke Listy* 24: 297-300.
Lundström, J. (1970). Biosynthesis of mescaline and 3,4-dimethoxyphenethylamine in *Trichocereus pachanoi*. *Acta Pharmacologica Suecica* 7: 651–666.
Lundström, J. (1971a). Biosynthesis of mescaline and tetrahydroisoquinoline alkaloids in *Lophophora williamsii*. Occurrence and biosynthesis of catecholamine and other intermediates. *Acta Chemica Scandinavica* 25: 3489–3499.
Lundström, J. (1971b). Biosynthetic studies on mescaline and related cactus alkaloids. *Acta Pharmacologica Suecica* 8: 275–302.
Lundström, J. (1971c). Biosynthesis of tetrahydroisoquinoline alkaloids in *Lophophora williamsii*. *Acta Pharmacologica Suecica* 8: 485–496.
Lundström J., and Agurell, S. (1971). Biosynthesis of mescaline and tetrahydroisoquinoline alkaloids in *Lophophora williamsii* (Lem.). Coult. *Acta Pharmacologica Suecica* 8: 261–274.
Lundström, J., and Agurell, S. (1972). Biosynthesis of mescaline and other peyote alkaloids. In: *Biochemie und Physiologie der Alkaloide* (Eds. K. Mothes, K. Schreiber and H.R. Schütte) (Int. Symp. Halle (Saale) 25–28 June 1969). Berlin: Academie-Verlag, pp. 89–94.
Lunevskii, V.Z., Zherelova, O.M., Vostrikova, I.Y., and Berestovsky, G.N. (1983) Excitation of Characeae cell membranes as a result of activation of calcium and chloride channels. *Journal of Membrane Biology* 72: 43–58.
Lynn, K.R. (1989). Four lysozymes from latex of *Asclepias syriaca*. *Phytochemistry* 28: 1345–1348.
Lysenko, G.G., Gins, V.K., Piskunova, N.P., Pukhalsky, B.A., and Mukhin, E.N. (1987). Photoreduction of $NADP^+$ and oxygen in chloroplasts from leaves of spring wheat at tillering and shooting phases. *Fiziologia Rastenii (USSR)* 34: 461–468.
Lysov, A.V. (1982). System acetylcholine-acetylcholinesterase in tissues of various organs. In: *Comparative Biochemistry of Metabolism Changes in Animals*. (Ed. A.V. Lysov). Kuibyshev: Kuibyshev State University, pp. 68–71.
Lyte, M. (1993). The role of microbial endocrinology in infectious disease. *Journal of Endocrinology*. 137: 343-345.
MacFarlane, W.V. (1963). The stinging properties of *Laportea*. *Economical Botany* 17: 303–311.
Macnab, R. (1985). Bacterial chemotaxis. In: *Molecular Aspects of Cellular Regulation* (Eds. P. Cohe and M.D. Houslay). Vol. 4. Molecular Mechanisms of Transmembrane Signalling. pp. 455–487. Amsterdam: Elsevier.
Madhavan, S., Sarath, G., Lee, B.H., and Pegden, R.S. (1995). Guard cell protoplasts contain acetylcholinesterase activity. *Plant Science* 109: 119–127.

Maelicke, A. (1991). Acetylcholine esterase: the structure. *Trends in Biochemical Science* 16: 355–356.

Maheshwari, S.C., Gupta, R., and Gharyal, P.K. (1982). Cholinesterase in plants. In: *Recent Development in Plant Science: SSSMMM Sircar Memorial Volume.* (Ed. S.P. Sen). New Delhi: Today and Tomorrows Printers and Publishers, pp. 145–160.

Malborn, C.C., Bahouth, W., Brandwein, H.J., George, S.T., Graciano, M.P., Moxham, C.P., and Rapiejko, P.J. (1987). The structure and biology of beta-adrenergic receptors. In: *Synaptic Transmitters and Receptors* (Ed. S. Tucek). Praha: Academia, pp. 239–247.

Major, R.T., and Dürsch, F. (1958). N^α, N^α-Dimethylhistamine, a hypotensive principle in *Casimiroa edulis* Llave et Lex. *Journal of Organic Chemistry* 23: 1564-1565.

Mamedov, M.D., Hayashi, H., Wada, H., Mohanty, P.S., Papageorgiou, G.C., and Murata, N. (1991). Glycinebetaine enhances and stabilizes the evoluation of oxygen and the synthesis of ATP by cyanobacterial thylakoid membranes. *FEBS Letters* 294: 271–274.

Mann, P.J.G. (1955). Purification and properties of the amine oxidase of pea seedlings. *Biochemcial Journal* 59: 609–620.

Mann, P.J.G. (1961). Further purification and properties of the amine oxidase of pea seedlings. *Biochemical Journal* 79: 623–631.

Mann, J.D., Fales, H.M., and Mudd, S.H. (1963). Alkaloids and plant metabolsim. VI, O-Methylation in vitro of norbellandine, a precursor of Amaryllidaceae alkaloids. *Journal of Biological Chemistry* 238: 3820–3823.

Mansfield, D.H., Webb, G., Clark, D.G., and Taylor, I.E.P. (1978). Partial purification and some properties of a cholinesterase from bush bean (*Phaseolus vulgaris* L.) roots. *Biochemical Journal* 175: 769–777.

Marchi, B., Trielli, F., Falugi, C., Corre, M.C., and Fenaux, L. (1996). Cholinomimetic drugs may affect growth and metamorphosis of the sea urchin larva. *Oceanologica Acta* 19: 287–291.

Marme, D., and Dieter, P. (1983). Role of Ca^{2+} and calmodulin in plants. In: *Calcium and Cell Function.* (Ed. W.J. Cheung), Vol. 4. New York, London, Paris, San Diego, San Francisco, Sao Paulo, Sydney Tokyo Toronto: Academic Press, pp. 263–311.

Marshall P.B. (1959). Catechols and tryptamines in the "matoke" banana (*Musa paradisiaca*). *Journal of Pharmacy and Pharmacology* 40: 639.

Marquardt, P., and Falk, H. (1957). Vorkommen und Syntheses von Acetylcholine in Pflanzen and Bakterien. *Arzneimittelforschung* 7: 203–211.

Marquardt, P., and Spitznagel, G. (1959). Bakterielle Acetylcholine Bildung in Kunstlichen Nahrboden. *Arzneimittelforschung* 9: 456–465.

Marquardt, P., and Vogg, G. (1952). Pharmakologische and chemische Untersuchungen uber Wirkstoffe in Bienenpollen. *Arzneimittelforschung* 2(353): 267–271.

Marquardt, P., Schumacher H., and Vogg, G. (1952). Die chemische Konstitution des blutdrucksenkenden Factors in der Kartoffel. *Arzneimittelforschung* 2: 301–304.

Martin, F.W. (1972). In vitro measurement of pollen tube growth inhibition. *Plant Physiology* 49: 924–925.

Martin, Y.R.L., Carmona, A.J., Frances, I.F.P., and Blesa, A.C. (1987). Estudio de la actividad citoquinina en flores de platanera. Efecto de L-dopamina sobre el bioensayo del callo de Soja. *Anales de Edafologiay Agrobiologia* 46: 475–484.

Mason, H.S. (1948). The chemistry of melanin. III. Mechanism of the oxidation of dihydroxyphenylalanine by tyrosinase. *Journal of Biological Chemistry* 172: 83–99.

Massoulie, J. (1986). Polymorphism of cholinesterase: possible insertion of the various molecular forms in cellular structures. In: *Dynamics of Cholinergic Function* (Ed. I. Hanin). New York, London: Plenum Press, pp. 727–732.

Massoulie, Y., and Bon, S. (1982). The molecular forms of cholinesterase and acetylcholinesterase. *Annual Review of Neuroscience* 5: 57–106.

Massoulie, Y., Sussman, J.L., Doctor, B.P., Soreq, H., Velan, B., Cygler, M., Rotundo, R., Schafferman, A., Silman, I., and Taylor, P. (1992). Recommendation for nomenclature in cholinesterases. In: *Multidisciplinary Approaches to Cholinesterase Functions* (Eds. A. Shafferman and B. Velan). New York: Plenum Press, pp. 285–288.

Massoulie, Y., Pezzementi, L., Bon, S., Krejci, E., and Vallette, F. (1993). Molecular and cellular biology of cholinesterases. *Progress in Neurobiology* 41: 31–91.

Mayer, A.M. (1987). Polyphenoloxidases in plant—recent progress. *Phytochemistry* 26: 11–20.

Mayer, A.M., and Harel, E. (1979). Polyphenol oxidases in plant. *Phytochemistry* 18: 193–215.

McKeand, J.B., Knox, D.P., Duncan, J.L., and Kennedy, M.W. (1992). Antibody binding and inhibition of secreted enzymes of *Dictyocaulus viviparus*. *Journal of Cellular Biochemistry* (Supplement) 16A: 149.

McQueen, J.K. (1987). Classical transmitters and neuromodulators. In: *Basic and Clinical Aspects of Neuroscience* (Eds. E. Fluckiger, E.E. Muller and M.O. Thorner), Vol. 2. Transmitter Molecules in the Brain. Berlin, Heidelberg etc.: Springer-Verlag, pp. 7–16.

Mechoulam, R., Sondheimer, F., Melera, A., and Kincl, F.A. (1961). The structure of zapotidine. *Journal of American Chemical Society* 83: 2022–2023.

Medda, R., Padiglia, A., and Floris, G. (1995). Plant copper-amine oxidases. *Phytochemistry* 39: 1–9.

Melin, J.M., Sommarin, M., Sandelius, A.S., and Jergil, B. (1987). Identification of Ca^{2+}-stimulated polyphosphoinositide phospholipase C in isolated plant plasma membranes. *FEBS Letters* 223: 87–91.

Meskes-Lozoya, M., Lozoya, X., Marles, R.J., Socy-Breau, C., Sen, A., and Arnason, J.T. (1990). N,N-dimethyltryptamine alkaloid in *Mimosa tenuiflora* bark (tepescohuite). *Archives of Investigator Medicine* 21: 175–177.

Metzler, D.E. (1977). *Biochemistry. The Chemical Reactions of Living Cells*. Vol. 3. New York, San Francisco: Academic Press, 100 pp.

Mikhelson, M.J. and Zeimal, E.V. (1970) *Acetylcholine*. Moscow: Nauka. 280 pp.

Mikhelson, M.J., and Zeimal, E.V. (1973). *Acetylcholine. An Approach to the Molecular Mechanism of Action*. Oxford, New York: Pergamon Press, 241 pp.

Miller, D.D., Harrold, M., Wallace, R.A., Wallace, L.J., and Uretsky, N.J. (1988). Dopaminergic drugs in the cationic form interact with D_2 dopamine receptors. *Trends in Pharmacological Science* 9: 282–284.

Minneman, K.P., and Johnson, R.D. (1987). α_1-adrenergic receptors linked to inositol phosphate and cyclic AMP accumulation in rat brain. In: *Synaptic Transmitters and Receptors* (Ed. S. Tucek). Praha: Academia, pp. 310–314.

Misra, S., Mohapatra, T.M., and Rathaur, S. (1993). *Wuchereria bancrofti* identification of parasitic acetylcholinesterase in microfilariae infected human serum. *Tropical Medicine and Parasitology* 44: 75–78.

Miura, G.A., and Shih, T.M. (1984a). Cholinergic constituents in plants: characterization and distribution of acetylcholine and choline: *Physiologia Plantarum* 61: 417–421.

Miura, G.A., and Shih, T.M. (1984b). Identification of propionylcholine in higher plants. *Physiologia Plantarum* 62: 341–343.

Miura, G.A., Broomfield, C.A., Lawson, M.A., and Worthley, E.G. (1982). Widespread occurrence of cholinesterase activity in plant leaves. *Physiologia Plantarum* 56: 28–32.

Molyneux, D.E. and McKinlay, R.G. (1989). Observation on germination and acetylcholinesterase activity in wheat seeds. *Annals of Botany* 63: 81–86.

Momonoki, Y.S. (1992). Occurrence of acetylcholine-hydrolyzing activity at the stele-cortex interface. *Plant Physiology* 99: 130–133.

Momonoki, Y.S., and Momonoki, T. (1991). Changes in acetylcholine levels following leaf wilting and leaf recovery by heat stress in plant cultivars. *Nippon Sakumotsu Gakkai Kiji. Japanese Journal of Crop Science* 60: 283–290.

Momonoki, Y.S., and Momonoki, T. (1992). The influence of heat stress on acetylcholine content and its hydrolyzing activity in *Macroptilium atropurpureum* cv. Siratro. *Japanese Journal of Crop Science* 61: 112–118.

Momonoki, Y.S., and Momonoki, T. (1993a). Changes in acetylcholine-hydrolyzing activity in heat-stressed plant cultivars. *Japanese Journal of Crop Science* 62: 438–448.

Momonoki, Y.S., and Momonoki, T. (1993b). Histochemical localization of acetylcholinesterase in leguminous plant, siratro (*Macroptilium atropurpureum*). *Japanese Journal of Crop Science* 62: 571–576.

Monkovic, I., and Spenser, I.D. (1965a). Biosynthesis of berberastine. *Journal of American Chemical Society* 87: 1137–1138.

Monkovic, I., and Spenser, I.D. (1965b). Biosynthesis of berberine and berberastine. Incorporation of catecholamines. *Canadian Journal of Chemistry* 43: 2017–2026.

Morgan, J.I., Wigham, C.G., and Perris, A.D. (1984). The promotion of mitosis in cultured thymic lymphocytes by acetylcholine and catecholamines. *Journal of Pharmacy and Pharmacology* 36: 511–515.

Mors, W.B., Cascon, S.C., and Moussatche, H. (1959). Presenca, nos aculeos da faveleira (*Cnidosculus*) de histamina e de uma outra substancia ativa na musculatura lisa. *Ciencia Cultura* 11: 136 (Supplement).

Mukherjee, I. (1980). The effect of acetylcholine on hypocotyl elongation in soybean. *Plant and Cell Physiology* 21: 1657–1660.

Munnik, T., Irvine, R.F., and Musgrave, A. (1998). Phospholipid signalling in plants. *Biochimica et Biophysica Acta* 1389: 222-272.

Munoz-Delgado, E., and Vidal, C.J. (1987). Solubilization and partial characterization of acetylcholinesterase from the sarcotubular system of skeletal musle. *Neurochemical Research* 12: 597–605.

Nachmanson, D., and Neumann, E. (1975). *Chemical and Molecular Basis of Nerve Activity*. New York, San Francisco, London: Academic Press, 404 pp.

Nakajima, H., and Hatano, S. (1962). Acetylcholinesterase in the plasmodium of the myxomycete, *Physarium polycephalum*. *Journal of Cellular Comparative Physiology* 59: 259–264.

Nalbandyan, R.M. (1986). Copper-containing proteins of brain and their significance in etiology of schizophrenia. *Neurokhimiya (USSR)* 5: 74–84.

Nederkoorn, P.H.J., Timmerman, H., and Donne-Op den Kelder, G.M. (1997). *Signal Transduction by G Protein-Coupled Receptors*. New York, Berlin: Springer-Verlag, 151 pp.

Nelson, L. (1978). Chemistry and neurochemistry of sperm motility control. *Federal Proceedings of Federation of American Soiety of Experimental Biology* 37: 2543–2549.

Nettleship, L., and Slaytor, M. (1974). Limitation of feeding experiments in studying alkaloid biosynthesis in *Peganum harmala* callus cultures. *Phytochemistry* 13: 735–742.

Neuwald, F. (1952). Choline in the leaves of *Digitalis purpurea* and *Digitalis lanata*. *Archives of Pharmacology* 285: 459–461.

Niaussat, P., Laborit, H., Dubois, C., and Hiaussat, M. (1958). Action de la serotonine sur la croissance des jeunes plantules d'Avoine. *Comptes Rendus de la Societe de Biologie* 152: 945–947.

Nicolova, R., Chvoika, L., Hadacova, V., and Schejbalova, V. (1986). The effect of phenylacetic acid on cholinesterase activity and on some isoenzymes of esterases and cholinesterases of pea in vitro. *Biologia Plantarum* 28: 145–148.

Niyazova, M.M., and Bulychev, A.A. (1989). The reversible decrease in the fluorescence yield of chlorophyll *a* at the generation of the action potential in moss *Anthoceros* cells. *Biofizika* 34: 272–274.

North, J.A. (1986). Muscarinic receptors and membrane ion conductances. *Trends in Pharmacological Science*, Supplement: 19-22.

North, J.A. (1994). Handbook of Receptors and Channels. Boca Raton: CRC Press. 384 pp.
Nourse, A., Schabort, J.C., Dirr, H.W., and Dubery, I.A. (1989). Purification and properties of an esterase from *Cucurbita maxima* fruit tissue. *Phytochemistry* 28: 379–383.
Nurmann, G. and Strack, D. (1979). Sinapine esterase, 1. Characterization of sinapine esterase from cotyledons of *Raphanus sativus. Zeitschrift für Naturforschung* 34 C: 715–720.
Oata, Y. (1974). Removal of the sugar inhibition of flowering in *Lemna gibba* G-B by catecholamines. *Plant and Cell Physiology* 15: 63–68.
Oata, Y., and Hoshino, T. (1974). Diurnal change in temperature sensitivity of *Lemna gibba* G-3 induced by acetylcholine in continuous light. *Plant and Cell Physiology* 15: 1063–1072.
Oata, Y., and Nakashima, H. (1978) Photoperiodic flowering in *Lemna gibba* G-3: Time measurement. *Botanical Magazine,* Tokyo (Special issue) 1: 177–198.
Oberthür, W., Lottspeich, F., and Hucho, F. (1987). The ion channel of the nicotinic acetylcholine receptor. In: *Receptors and Ion Channels* (Eds. Y.A. Ovchinnikov and F. Huge). Berlin, New York: Walter de Gruyter, pp. 3–11.
Oelrich, P.B., and Robertson, P.A. (1970). Purification of pain-producing substances from *Dendrocnide (Laportea) moroides. Toxicon* 8: 89–90.
Ohnishi, J. (1964). Le changement de volume du chloroplaste, accompagne de photophosphorylation et les proteins ressemblantes a l'active et a la myosine extractes du chloroplasts. *Journal of Biochemistry (Tokyo)* 55: 494–503.
O'Neill, S.D., and van Tassel, D.L. (1994). Methods for controlling flowering in plants. United States Patent, N 5, 300, 481. Date of Patent 5 April 1994. pp. 1–10.
Oniani, D.A. (1975). The Study of Effects of Physiologically Active Substances on the Regulation of the Rate of Protoplasma Movement and Membrane Potential in Cells of *Nitella.* Ph.D. Thesis. Tbilisi State University, Tbilisi, 24 pp.
Oniani, D.A., Vorob'ev, L.N., and Kudrin, A.N. (1973). Influence of noradrenaline, adrenaline and izadrin on the motility of protoplasm of cells of *Nitella syncarpa.* In: *Characean Algae and Their Use in the Studies of the Biological Cellular Processes.* (Ed. K.K. Yankyavichus). Vilnus: Institute of Botany of Lit SSR, pp. 423–432.
Oniani, D.A., Vorob'ev, L.N., and Kudrin, A.N. (1974). Influence of noradrenaline on the motility of protoplasm of cells of *Nitella. Soobshcheniya Grusinian SSR (Proceedings of Grusinian SSR)* 73: 457–460.
Oniani, D.A., Kudrin, A.N., Lomsadze, B.A., and Vorob'ev, L.N. (1977a). The study of bioelectric potential and rate of the protoplasm movement in cells of *Nitella syncarpa* under adrenaline and noradrenaline. *Soobshcheniya Grusinian SSR (Proceedings of Grusinian SSR)* 86: 457–460.
Oniani, D.A., Kudrin, A.N., Lomsadze, B.A., and Vorob'ev, L.N. (1977b). The study of regulation of intracellular processes with participation of the mediatory receptors in cells of *Nitella syncarpa. Soobshcheniya Grusinian SSR (Proceedings of Grusinian SSR)* 87: 173–176.
Opritov, V.A., Pyatygyn, S.S., and Retivin, V.G. (1991). *Bioelectrogenesis in Higher Plants.* Moscow: Nauka, 216 pp.
Ott, P. (1985). Membrane acetylcholinesterases: Purification, molecular properties and interactions with amphiphilic environments. *Biochimica et Biophysica Acta* 822: 375–392.
Oury, A., and Bacq, Z.M. (1938). Ester instable de la choline sans cholinesterase dans la pomme de terre et un Champignon. *Archives of International Physiology* 47: 92–101.
Pagani, F., and Romussi, G. (1969). Sui constituenti dei semi di *Cleome pungens* Willd. Nota II. Isolamento del cloruro di feruloicolina. *Il Farmaco-Ed.Sc.* 24: 257–262.
Palmer, J.K. (1963). Banana polyphenoloxidase. Preparation and properties. *Plant Physiology* 38: 508–513.
Park, C.H., Kim, S.H., Choi, W., Lee, Y.J., Kim, J.S., Kang, S.S., and Suh, J.H. (1996). Novel anticholinesterase and antiamnesic activities of dehydroevodiamine, a constituent of *Evodia rutaecarpa. Planta Medica* 62: 405–409.

Parups, E.V. (1976). Acetylcholine and synthesis of ethylene in etiolated bean tissue. *Physiologia Plantarum* 36: 154–156.
Patil, T.M. and Kulkarni, K.K. (1989). Stomatal response to organophosphoropesticide spray in *Lycopersicon esculentum* Mill. *Geobios* (Jodhpur) 16: 82-65.
Paton, W.D.M. (1958). The release of histamine. In: *Progress in Allergy* (Ed. P. Karlos). Vol. 5. pp. 79–148. Basel, New York: Karger S.
Paul, A.G. (1973). Biosynthesis of the peyote alkaloids. *Lloydia* 36: 36–45.
Pec, P., and Frebort, I. (1991). Competition of homologous substrates, putrescine and cadaverine, in the reaction catalyzed by pea diamine oxidase. *Biochemistry International* 24: 633–640.
Penel, C., Darimont, E., Greppin, H., and Gaspar, Th. (1976). Effect of acetylcholine on growth and isoperoxidases of the lentil (*Lens culinaris*) root. *Biologia Plantarum* 18: 293–298.
Pelletier, S.W. (1983). The nature and definition of alkaloids. In: *Alkaloids: Chemical and Biological Perspectives* (Ed. S.W. Pelletier). Vol. 1. New York etc.: J. Wiley & Sons, pp. 1–31.
Pertseva, M.N. (1988). Modern views on the evolution of system hormone-hormonal receptor. *Uspekhi. Sovremennoi Biologii (Advantage of Modern Biology, USSR)* 105: 439–449.
Pertseva, M.N. (1989). *Molecular Base of the Development of the Hormone-Competency*. Leningard: Nauka, 310 pp.
Pertseva, M.N. (1990a). The path of the evolution of the hormonal signal realization system. *Sechenov Physiological Journal of the USSR* 76: 1126–1137.
Pertseva, M.N. (1990b). Is the evolution similarity between chemosignalling systems of eukaryotes and prokaryotes? *Journal of Evolutionary Biochemistry and Physiology* 26: 505–513.
Peters, J.E., Wu, P.H.L., Sharp, W.R., and Paddock, E.F. (1974). Rooting and the metabolism of nicotine in tobacco callus cultures. *Physiologia Plantarum* 31: 97–100.
Pickard, B.G. (1973). Action potentials in higher plants. *Botanical Review* 39: 172–201.
Pickles, V.R., and Sutcliffe, J.F. (1955). The effects of 5-hydroxytryptamine, indole-3-acetic acid and some other substances, on pigment effusion, sodium uptake, and potassium efflux, by slices of red beet-root in vitro. *Biochimica et Biophysica Acta* 17: 244–251.
Polevoi, V.V. (1975). Regulatory systems of organisms. *Vestnik LGU* (Russia) 15: 104–108.
Polevoi, V.V. (1985). Evolution of plant hormonal system. In: *Evolution of Functions in Plant World* (Eds. V.V. Polevoi and Yu.I. Maslov) Leningrad: Leningrad State University Press, pp. 216–229.
Potopalskii, A.I., Petlichnaya, L.I., and Ivasivka, S.V. (1989). *Berberis and Its Preparations in Biology and Medicine*. Kiev: Naukova Dumka, 288 pp.
Pottosin, I.I. (1992). Single channel recording in the chloroplast envelope. *FEBS Letters* 308: 87–90.
Poupat, C., and Sevent, T. (1975). Cinnamoylhistamine, et hordenine alcaloides *d'Acacia spirorbis*. *Tetrahydron Letters* 14: 1881–1882.
Prosser, C.L. (1986). *Adaptational Biology: Molecules to Organisms*. New York etc.: J. Wiley and Sons, 784 pp.
Protacio, C.M., Dai, Y.R., Lewis, E.F., and Flores, H.E. (1992). Growth stimulation by catecholamines in plant tissue/organ cultures. *Plant Physiology* 98: 89-96.
Raineri, M., and Modenesi, P. (1986). Preliminary evidence for a cholinergic-like system in lichen morphogenesis. *Histochemical Journal* 18: 647–654.
Ramwani, J., and Mishra, R.K. (1986). Purification of bovine striatal dopamine D_2 receptor by affinity chromatography. *Journal of Biological Chemistry* 261: 8894–8898.
Ranjeva, R., and Boudet, A.M. (1987). Phosphorylation of proteins in plants: regulatory effects and potential involvement in stimulus. Response coupling. *Annual Review of Plant Physiology* 38: 73–93.

Rapport, M.M., Green., A.A., and Pages, I.H. (1948). Crystalline serotonin. *Science* 108: 329–330.
Reed, D.J. (1968). Tryptophan decarboxylation in cell-free extracts of Alaska pea epicotyls. In: *Biochemistry and Physiology of Plant Growth Substances.* Proc. 6th International Conference on Plant Growth Substances. July 24–29, 1967, Ottawa (Eds., F. Wightman and G. Setterfield). pp 243–258. Ottawa: The Runge Press Ltd.
Reed, W.A., and Bonner, B.A. (1971). Investigations on the rapid phytochrome induced inhibition of *Tropasolum* stem elongation. *Plant Physiology (Suppl.)* 47: 2.
Regula, I. (1970). 5–Hidroksitriptamin u ljutoj Koprivi (*Urtica pilulifera* L.). *Acta Botanica Croatica* 29: 69–74.
Regula, I. (1972). Identifikacija serotonina u vrste *Elaeagnus umbellata* Thunb. *Acta Botanica Croatica* 31: 105–108.
Regula, I. (1974). Kromatografska identifekacija serotonina u Koprivi *Urtica pilulifera* var. *dodartii* (L.) Aschers. *Acta Botanica Croatica* 33: 89–91.
Regula, I. (1981). Serotonin in the tissues of *Loasa vulcanica* ed. Andre. *Acta Botanica Croatica* 40: 91–94.
Regula, I. (1985). The presence of serotonin in the embryo of *Juglans mandshurica* Maxim. *Acta Botanica Croatica* 44: 19–22.
Regula, I. (1986). The presence of serotonin in embryo of black walnut (*Juglans nigra*). *Acta Botanica Croatica* 45: 91–95.
Regula, I. (1987). The presence of serotonin in *Blumenbachia contorta* Hook Fil. *Acta Botanica Croatica* 46: 45–48.
Regula, I., and Devide, Z. (1979). Occurrence of some indoles in *Shepherdia argentea* (Pursh) Nutt. *Acta Botanica Croatica* 38: 41–44.
Regula, I., and Devide, Z. (1980). The presence of serotonin in some species of genus *Urtica*. *Acta Botanica Croatica* 39: 47–50.
Regula, I., Kolevska-Pletikapic, B., Kranik-Rasol, M., and Cusin, B. (1988). Presence of serotonin in *Juglans ailanthifolia* var. *ailanthifolia* and its physiological effects in plants. *Abstracts of the 6th Congress of the Fed. Eur. Soc. on Plant Physiology.* Split, Yugoslavia, 4-10 September, 1988, p. 749.
Regula, I., Kolevska-Pleticapic, B., and Krsnik-Rasol, M. (1989). Presence of serotonin in *Juglans ailanthifolia* var. *ailanthifolia* Carr and its physiological effects on plants. *Acta Botanica Croatica* 48: 57–62.
Reynolds, J.D., Kimbrough, T.D., and Weekley, L.B. (1985). The effect of light quality on 5-hydroxyindole metabolism in leaves of *Sedum morganianum* (Crassulaceae). *Biochemie und Biophysics der Pflanzen* 180: 345–351.
Reynolds, T. (1989). Comparative effects of heterocyclic compounds in inhibition of lettuce fruit germination. *Journal of Experimental Botany* 40: 391–404.
Riov, J., and Jaffe, M.J. (1973a). Cholinesterase from mung bean roots and its inhibition by plant growth retardants. *Experientia* 29: 264–265.
Riov, J., and Jaffe, M.J. (1973b). Cholinesterases from plant tissues. I. Purification and characterization of a cholinesterase from mung bean roots. *Plant Physiology* 51: 520–528.
Riov, J., and Jaffe, M.J. (1973c). Cholinesterases from plant tissues. II. Inhibition of bean cholinesterase by 2-isopropyl-4-dimethyl-amino-5-methyl-phenyl-1-peperidine carboxylate methyl chloride (AMO-1618). *Plant Physiology* 52: 233–235.
Rizvi, S.J.H., and Rizvi, V. (1987). Improving crop productivity in India: role of allelochemicals. In: *Allelochemicals. Role in Agriculture and Forestry. ACS Symposium Ser.* (Eds. G.R. Waller), Vol. 330. Washington: American Chemical Society, pp. 69–75.
Robb, D.A., Swain, T., and Mapson, L.W. (1966). Substrates and inhibitors of the activated tyrosinase of broad bean (*Vicia faba* L.). *Phytochemistry* 5: 665–675.
Roberts, D.M., Zielinski, R.E., Schleicher, M., and Watterson, D.M. (1983). Analysis of suborganellar fractions from spinach and pea chloroplasts for calmodulin-binding proteins. *Journal of Cell Biology* 97(N5, part 1): 1644–1647.

Roberts, M.F. (1986). Paper latex and alkaloid storage vacuoles. In: *Plant Vacuoles* (Ed. B. Martin). New York, London: Plenum Press, pp. 513–528.

Robertson, P.A., and Macfarlane, W.V. (1957). Pain-producing substances from the stinging bush *Laportea moroides*. *Australian Journal of Experimental Biology* 35: 381–394.

Robinson, J.D., and Pratap, P.R. (1991). Effect of choline on Na^+ and K^+-interactions with the Na^+/K^+-ATPase. *Biochimica et Biophysica Acta* 1069: 281–287.

Robinson, S.P., and Downton, W.J.S. (1984). Potassium, sodium, and chloride content of isolated intact chloroplasts in relation to ionic compartmentation in leaves. *Archives of Biochemistry and Biophysics* 228: 197–206.

Roddick, J.G. (1989). The acetylcholinesterase-inbihitory activity of steroidal glycoalkaloids and their aglycones. *Phytochemistry* 28: 2631–2634.

Rojas, E., Pollard, H.B., and Heldman, E. (1985). Real-time measurements of acetylcholine-induced release of ATP from bovine medullary chromaffin cells. *FEBS Letters* 185: 323–327.

Rosenberg, H., and Paul, A.G. (1970). The isolation and synthesis of dolichotheline. *Phytochemistry* 9: 655–657.

Rosenberg, H., and Stohs, S.J. (1976). Effect of histidine decarboxylase inhibitors on the production of an aberrant alkaloid in *Dolichothele sphaerica*. *Phytochemistry* 15: 501–503.

Roshchina, V.D., and Roshchina, V.V. (1989). *The Excretory Function of Higher Plants*. Moscow: Nauka, 214 pp.

Roshchina, V.V. (1986). Cholinesterase from pea chloroplasts. *Doklady AN SSSR* 290: 486–489.

Roshchina, V.V. (1987a). Action of acetylcholine agoinsts and antagonists on reactions of photosynthetic membranes. *Photosynthetica* 21: 296–300.

Roshchina, V.V. (1987b). Cholinergic system in chloroplasts of higher plants. *Abstracts 14th Int. Bot. Cong.*, Berlin-Dahlem, Botanical Museum, West Berlin, p. 482.

Roshchina, V.V. (1988a). Characterization of pea chloroplast cholinesterase: Effect of inhibitors of animal enzymes. *Photosynthetica* 22: 20–26.

Roshchina, V.V. (1988b). Cholinesterases from chloroplasts of higher plants. *Fiziologiya Rastenii (USSR) (Soviet Plant Physiology)* 35: 899–906.

Roshchina, V.V. (1989a). Biomediators in chloroplasts of higher plants. 1. The interaction with photosynthetic membranes. *Photosynthetica* 23: 197–206.

Roshchina, V.V. (1989b). Reaction of chloroplast membranes with biomediators. *Biophysika (USSR)* 24: 602–605.

Roshchina, V.V. (1990a). Biomediators in chloroplasts of higher plants. 2. The acetylcholine-hydrolyzing proteins. *Photosynthetica* 24: 110–116.

Roshchina, V.V. (1990b). Biomediators in chloroplasts of higher plants. 3. Effect of dopamine on photochemical activity. *Photosynthetica* 24: 117–121.

Roshchina, V.V. (1990c). Biomediators in chloroplasts of higher plants. 4. Reception by photosynthetic membranes. *Photosynthetica* 24: 539–549.

Roshchina, V.V. (1990d). Regulation of chloroplast's reactions by secondary metabolites acetylcholine and biogenic amines. *Acta Botanica Croatica* 49: 29–35.

Roshchina, V.V. (1991a). *Biomediators in Plants. Acetylcholine and Biogenic Amines. Acetylcholine and Biogenic Amines*. Pushchino: Biological Center AN SSSR, 192 pp.

Roshchina, V.V. (1991b). Nerotransmitters catecholamines and serotonin in plants. *Uspekhi Sovremennoi Biologii (Advances in Modern Biology, Russia)* 111: 622–636.

Roshchina, V.V. (1991c). *Reactions of Chloroplasts on Biomediators*. Theses of Doctor Science Report, Moscow, Institute of Plant Physiology, 48 pp.

Roshchina, V.V. (1992). The action of neurotransmitters on the seed germination. *Biologicheskie nauki* 9: 124–129.

Roshchina, V.V. (1994). Chemosensory mechanisms in allelopathy. In: *Allelopathy in Agriculture and Forestry* (Eds. S.S. Narwal and P. Tauro). Jodhpur (India): Scientific Publishers, pp. 273–285.

Roshchina, V.V. (1996). Volatile plant excretions as natural antiozonants and origin of free radicals. In: *Allelopathy. Field Observation and Methodology* (Eds. S.S. Narwal and P. Tauro). Jodhpur: Scientific Publishers, pp. 233-241.

Roshchina, V.V. (1998). Cholinergic mechanisms in plant fertilization. *Proceedings of 10th Intern. Symposium on Cholinergic Mechanisms*. Amsterdam: Elsevier Publ., pp. 484-485.

Roshchina, V.V. (1999a). Mechanisms of cell-cell communication. In: *Allelopathy Update* (Ed. S.S. Narwal), Vol. 2. Enfield, New Hampshire: Science Publishers, pp. 3-25.

Roshchina, V.V. (1999b). Chemosignalization at pollen. *Uspekhi Soveremennoi Biologii (Trends in Modern Biology, Russia)* 119: 557-566.

Roshchina, V.V. (2000a). The neurotransmitter substances function in plants. *Russian Journal of Physiology* 86: 1300-1307.

Roshchina, V.V. (2000b) Plant reactions as sensitive tests on toxins—natural antineurotransmitter drugs. *Abstracts of 13th World Congress on Animal, Plant and Microbial Toxins*, Paris, 18-22, September 2000. Paris: International Society of Toxinology.

Roshchina, V.V. (2001) Molecular-cellular mechanisms in pollen allelopathy. *Allellopathy Jounral* 8: 11-28.

Roshchina, V.V., and Alexandrova, I.Ph. (1991). Enzyme from fungus *Aspergillus niger*, hydrolyzed cholinic esters. *Biologicheskie Nauki* 12: 50-54.

Roshchina, V.V., and Karpilova, I.F. (1993). Biomediators in chloroplasts of higher plants. 5. The interaction of acetylcholine-hydrolyzing protein with other water-soluble proteins of chloroplasts. *Photosynthetica* 28: 273-279.

Roshchina, V.V., and Melnikova, E.V. (1995). Spectral analysis of intact secretory cells and excretions of plants. *Allelopathy Journal* 2: 179-188.

Roshchina, V.V., and Melnikova, E.V. (1996). Microspectrofluorometry: A new technique to study pollen allelopathy. *Allelopathy Journal* 3: 51-58.

Roshchina, V.V., and Melnikova, E.V. (1998a). Allelopathy and plant reproductive cells: participation of acetylcholine and histamine in signalling in the interactions of pollen and pistil. *Allelopathy Journal* 5: 171-182.

Roshchina, V.V., and Melnikova, E.V. (1998b). Pollen-pistil interaction: response to chemical signals. *Biology Bulletin* 25: 557-563.

Roshchina, V., and Melnikova, E.V. (1999). Microspectrofluorimetry of intact secreting cells with applications to the study of allelopathy. In: *Principles and Practices in Plant Ecology: Allelochemical Interactions* (Eds. Inderjit, K.M.M. Dakshini and C.L. Foy). Boca Raton, USA: CRC Press, pp. 99-126.

Roshchina, V.V., and Mukhin, E.N. (1984). The acetylcholinesterase activity in chloroplasts of higher plants. *Doklady AN SSSR* 278: 754-757.

Roshchina, V.V., and Mukhin, E.N. (1985a). Acetylcholinesterase activity in chloroplasts and acetylcholine effects on photochemical reactions. *Photosynthetica* 19: 164-171.

Roshchina, V.V., and Mukhin, E.N. (1985b). Acetylcholine action on the photochemical reactions in chloroplasts. *Plant Science* 42: 95-98.

Roshchina, V.V., and Mukhin, E.N. (1986). Acetylcholine, its role in plant life. *Uspekhi Sovremennoi Biologii (Trends in Modern Biology—Reviews, Russia)* 101: 265-274.

Roshchina, V.V., and Mukhin, E.N. (1987a). Acetylcholine-acetylcholinesterase system in isolated pea chloroplasts and regulation of some photosynthetic reactions. *Fiziologiya Rastenii (USSR)* 34: 67-73.

Roshchina, V.V., and Mukhin, E.N. (1987b). Acetylcholine changes photochemical reactions and stimulates Na^+ and K^+ efflux from chloroplasts. *Fiziologiya Rastenii (USSR)* 34: 907-911.

Roshchina, V.V., and Mukhin, E.N. (1989). Reactions of chloroplasts on allelopathically active alkaloids and polyacetylenes. In: *Methodological Problems of Allelopathy* (Ed. A.M. Grodzinsky). Kiev: Naukova Dumka, pp. 128-131.

Roshchina, V.V., and Roshchina, V.D. (1993). *The Excretory Function of Higher Plants.* Berlin, Heidelberg, New York, London, Tokyo: Springer-Verlag, 314 pp.

Roshchina, V.V., and Semenova, M.N. (1990). Plant cholinesterases: activity and substrate-inhibitory specificity. *Journal of Evolutionary Biochemistry and Physiology (USSR)* 26: 644–651.

Roshchina, V.V., and Semenova, M.N. (1995). Neurotransmitter systems in plants. Cholinesterase in excreta from flowers and secretory cells of vegetative organs in some species. In: *Proceedings of the Plant Growth Regulation Society of America* (Eds. D. Greene and G. Cutler), 22: Annual Meeting, 18–20 July 1995. Minneapolis: Fritz C.D., pp. 353–357.

Roshchina V.V., Melnikova E.V. and Kovaleva, L.V. (1996). Autofluorescence in the pollen-pistil system of *Hippeastrum hybridum. Doklady Biological Science* 349: 403–405.

Roshchina, V.V., Melnikova, E.V., Kovaleva, L.V. (1997). The changes in autofluorescence during the male gametophyte development. *Russian Journal of Plant Physiology* 44: 45–53.

Roshchina, V.V., Melnikova, E.V., Kovaleva, L.V. and Spiridonov, N.A. (1994). Cholinesterase in plant pollen grains. *Doklady Biological Sciences* 337: 424–427.

Roshchina, V.V., Melnikova, E.V, Mit'kovskaya, L.I. and Karnaukhov, V.N. (1998a). Microspectrofluorimetry for the study of intact plant secretory cells. *Journal of General Biology (Russia)* 59: 531–554.

Roshchina, V.V., Melnikova, E.V, Popov, V.I., Novoselov, V.I., Peshenko, I.V., Khutsyan, S.S., and Fesenko, E.E. (1998b). Modelling of the chemosignal transduction at pollen. In: *Reception and Intracellular Signalization* (Ed. V.I. Zinchenko). Pushchino: Biological Center of Russian Academy of Sciences, pp. 244–247.

Roshchina, V.V., Popov, V.I., Novoselov, V.I., Melnikova, E.V., Gordon, R. Ya., Peshenko, I.V., and Fesenko, E.E. (1998c). Transduction of chemosignal in pollen. *Tsitologiya (Cytology, Russia)* 40: 964–971.

Roshchina, V.V., Golovkin, B.N., Melnikova, E.V., Novoselov, V.I. and Gordon R.Ya. (2001). Microanalysis of pollen from hothouse plants. *Bulletin of Central Botanical Garden of Russian Academy of Sciences,* 179: in press.

Rowatt, E. (1948). The relation of pantotenic acid to acetylcholine formation by a strain of *Lactobacillus plantarum. Journal of General Microbiology* 2: 25–30.

Rozen, V.B. (1986). Hormone Receptors, Their Structure, Characteristics and Modes of Functioning in Cells. In: *Physiology of Hormonal Reception* (Ed. V.G. Shalyapina). pp. 5–33. Leningrad: Nauka.

Rubin, L.L., Chalfin, N.A., Adamo, A., and Klymkowsky, M.W. (1985). Cellular and secreted forms of acetylcholinesterase in mouse muscle cultures. *Journal of Neurochemistry* 45: 1932–1938.

Rueffer, M. (1988). Biosynthetic studies of protoberberines and related alkaloids in plant cell culture. In: *Chemistry and Biotechnology of Biologically Active Natural Products,* 4th Int. Conf. Budapest, 10–14 August 1987 (Ed. Cs. Szantay). Budapest: Academiai Kiado, pp. 237–245.

Rügenhagen, C., Dietze, P., Fecker, L.F., Goddijn, O.J.N., and Hoge, J.H.C. (1993). Increased production of serotonin by suspension and root cultures of *Peganum harmala* transformed with a tryptophan decarboxylase cDNA clone from *Catharanthus roseus. Transgenic Research* 2: 336–344.

Saha, J.C., and Kasinathan, S. (1963). Histamine action of the latex of *Calotropis gigantea* Ait. *Archives International Pharmacodynamical Therapy* 143: 78–89.

Saimi, Y., Martinac, B., Gustin, M.C., Culbertson, M.R., Adler, J., and Kung C. (1988). Ion channels in *Paramecium,* yeast and *Escherichia coli. Trends in Biochemical Sciences* 13: 304–309.

Sajner, J., and Veris, O. (1958a). Über das Vorkommen von Histamin in *Viscum album*. *Pharmazie* 13: 170.
Sajner, J., and Veris, O. (1958b). Histaminwirkungen von Weissdorninfius. *Pharmazie* 13: 52–54.
Sakamura, S., Terayama, Y., Kawakatsu, S., Ichihara, A., and Saito, H. (1980). Conjugated serotonins and phenolic constituents in safflower seed (*Carthamus tinctorius*). *Agricultural and Biological Chemistry* 44: 2951–2954.
Salleo, A., Salleo, S., and Urna, G. (1977). Impiego di segmenti di fusto di *Cucurbita* sp. per la identificazione dei tessuti dotati di attivita bioelectrica in plante superiori. *Giornale Botanico Italiano* 111: 298–299.
Sasse, F., Heckenberg U., and Berlin, J. (1982). Accumulation of β-carboline alkaloids and serotonin by cell cultures of *Peganum harmala*. II. Interrelationship between accumulation of serotonin and activities of related enzymes. *Zeitschrift für Pflanzenphysiologie* 105: 315–322.
Sastry, B.V.R., and Sadavongvivad, C. (1979). Cholinergic systems in non-nervous tissues. *Pharmacological Reviews* 30: 65–132.
Satter, R.L., and Galston, A.W. (1981). Mechanism of control of leaf movements. *Annual Review of Plant Physiology* 32: 83–110.
Satter, R.L., and Moran, N. (1988). Ionic channels in plant cell membranes. *Physiologia Plantarum* 72: 816–820.
Satter, R.L., Applewhite, P.B., and Galston, A.W. (1972). Phytochrome-controlled nyctinasty in *Albizzia julibrissin*. V. Evidence against acetylcholine participation. *Plant Physiology* 50: 523–525.
Saunders, J.A., and McClure, J.W. (1973). Acetylcholine inhibition of phytochrome-mediated increases in a flavonoid and in phenylalanine ammonia-lyase activity of etiolated barley plumules. *Plant Physiology* 51: 407–408.
Saxena, P.R., Pant, M.C., Kishor, K., and Bhargava, K.P. (1965). Identification of pharmacologically active substances in the Indian stinging nettle *Urtica parviflora* Roxb. *Canadian Journal Physiology and Pharmacology* 43: 869–876.
Saxena, P.R., Tangri, K.K., and Bhargava, K.P. (1966). Identification of acetylcholine, histamine and 5-hydroxytryptamine in *Girardinia heterophylla* (Decne). *Canadian Journal of Physiology and Pharmacology* 44: 621–627.
Schäffner, A.R. (1998). Aquaporin function, structure, and expression: are there more surprises to surface in water relations? *Planta* 204: 131–139.
Scheiber, G., Kolar, H., Foisner, R., and Kubicek, C.P. (1986). Choline stimulates synthesis of extracellular proteins in *Trichoderma reesei* QM9414. *Archives of Microbiology* 144: 41–47.
Schiffmann, E., and Gallin, J.I. (1979). Biochemistry of phagocyte chemotaxis. *Current Topics in Cell Regulation* 15: 203–261.
Schloss, P., Mayser, W., and Betz, H. (1992). Neurotransmitter transporters. A novel family of integral plasma membrane proteins. *FEBS Letters* 307: 76–80.
Schönbohm, E., and Meyer-Wegener, J. (1989). Actin polymerization as an essential process in light- and dark-controlled chloroplast anchorage. *Biochemie und Physiologie der Pflanzen* 185: 337–342.
Schütte, H.R. (1989). Secondary plant substances. Benzylisoquinoline alkaloids. In: *Progress in Botany* (Eds. H.D. Behnke, K. Esser, K. Kubitzki, M. Runge and H. Ziegler), Vol. 51. Berlin, Heidelberg, New York, London, Paris, Tokyo, Hong Kong: Springer-Verlag, pp. 113–133.
Schütte, H.R. (1991). Secondary plant substances: Monoterpenoid indole alkaloids. In: *Progress in Botany* (Eds. H.D. Behnke, K. Esser, K. Kubitzki, M. Runge and H. Ziegler), Vol. 52. Berlin, Heidelberg, New York, London, Paris, Tokyo, Hong Kong, Barcelona, Budapest: Springer-Verlag, pp. 84–96.

Schütte, H.R., Knofel, D., and Hyer, O. (1966). Biosynthese von Lupinealkaloiden. 19. Mitt. Anreichrerung and charakterisierung der Diaminoxydase aus Keimlingen von *Lupinus luteus*. *Zeitschrift für Pflanzenphysiologie* 55: 110–118.
Selbie, L.A., and Hill, S.J. (1998). G protein-coupled receptor cross-talk: the fine-tuning of multiple receptor-signalling pathways. *Trends in Pharmacological Sciences* 19: 87–93.
Semenova, M.N., and Roshchina, V.V. (1993). Cholinesterase in anthers of higher plants. *Russian Plant Physiology* 40: 255–259.
Serrano, E.E., and Zeiger, E. (1989). Sensory transaction and electrical signaling in guard cells. *Plant Physiology* 91: 795–799.
Seuwen, K., Magnaldo, I., and Pouyssegur, J. (1988). Serotonin stimulates DNA synthesis in fibroblasts acting through $S-HT_1\beta$ receptors coupled to a G_1-protein. *Nature* 335: 254–256.
Shiina, T., and Tazawa, M. (1986). Regulation of membrane excitation by protein phosphorylation in *Nitellopsis obtusa*. *Protoplasma* 134: 60–61.
Shostakovskaya, I.V., and Babskii, A.M. (1984). The influence of adrenaline on calcium transport and oxidative phosphorylation in mitochondria. *Ukrainean Biokhimicheskii Zhurnal (USSR)* 56: 57–62.
Sillard, R.G., Jarv, J.L., and Bartfai, T. (1985). Kinetic evidence for cooperativity of quinuclidinyl benzylate interaction with muscarinic receptor from rat brain. *Biologicheskie Membrany (Biological Membranes, USSR)* 2: 426–432.
Silver, A. (1974). *The Biology of Cholinesterases*. Amsterdam: North Holland Publ. Co., 230 pp.
Simon, P., Bonzon, M., Greppin, H., and Marme, D. (1984). Subchloroplasts localization of NAD kinase activity: evidence for a Ca^{2+} calmodulin-dependent activity at the envelope and for a Ca^{2+}, calmodulin-independent activity in the stroma of pea chloroplasts. *FEBS Letters* 167: 332–337.
Singh, V.K., and Singh, D.K. (1996). Enzyme inhibition by allicin, the molluscidal agent of *Allium sativum* L. (garlic). *Phytotherapy Research* 10: 383–386.
Sinyukhin, A.M. (1973). Functional activity of action potential in ferns and mosses at fertilization. *Biophyzika (USSR)* 18: 477–482.
Sinyukhin, A.M., and Britikov, E.A. (1967). Action potential in the reproductive system of plants. *Nature* 215: 1278–1279.
Skulachev, V.P. (1989). *Energetics of Biological Membranes*. Moscow: Nauka, 564 pp.
Slezak, E. (1984). Acetylcholine bioregulator of biochemical and physiological processes in plants. *Abstracts of 16th Conference of FEBS*. Moscow: Beckman Instrument, p. 422.
Slezak, E. (1987). Chemiczne mediatory pobudzeina u roslin. *Sprawozdania Wroclawskiego Towarzystwa Naukowego* 42B: 69–70.
Smack, D.P., and Colombini, I. (1985). Voltage-dependent channels found in the membrane fraction of corn mitochondria. *Plant Physiology* 79: 1094–1097.
Small, D.H. (1988). Serum acetylcholine possesses trypsin-like and carboxypeptidase B-like activity. *Neuroscience Letters* 95: 307–312.
Small, D.H. (1989). Acetylcholinesterase zymogens of neuropeptide processing enzymes? *Neuroscience* 29: 1–9.
Small, D.H., and Murtman, R.J. (1985). Association of serotonin, dopamine, or noradrenaline with actine-like component in pheochromosytoma (PC12) cells. *Journal of Neurochemistry* 45: 825–831.
Small, D.H., and Simpson, R.J. (1988). Acetylcholinesterase undergoes autolysis to generate trypsin-like activity. *Neuroscience Letters* 89: 223–228.
Small, D.H., Ismael, Z., and Chubb, I.W. (1986). Acetylcholinesterase hydrolyses chromogranin A to yield low molecular weight peptides. *Neuroscience* 19: 289–295.
Smallman, B.N., and Maneckjee, A. (1981). The synthesis of acetylcholine by plants. *Biochemical Journal* 194: 361–364.

Smith, B.N., and Meeuse, B.J.D. (1966). Production of volatile amines and skatole at anthesis in some *Arum* lily species. *Plant Physiology* 41: 343–347.

Smith, J.W., and Kirshner, N. (1960). Enzymatic formation of noradrenaline by the banana plant. *Journal of Biological Chemistry* 235: 3589–3591.

Smith, T.A. (1977). Phenethylamine and related compounds in plants. *Phytochemistry* 16: 9–18.

Smith, T.A. (1980). Plant amines. In: *Secondary Plant Products. Encyclopedia of Plant Physiology* (Eds. E.A. Bell and B.V. Charlwood), Vol. 8. Berlin, Heidelberg: Springer-Verlag, pp. 433–460.

Smith, T.A. (1981). Amines. In: *The Biochemistry of Plants* (Eds. P.K. Stumpf and E.E. Conn), Vol. 7. New York, London: Academic Press, pp. 249–268.

Soll, I. (1988). Purification and characterization of a chloroplast outer-envelope-bound, ATP-dependent protein kinase. *Plant Physiology* 87: 898–903.

Solyakov, L.S., Sablin, S.O., Kuusk, V.V., and Agabekyan, R.S. (1989). Acetylcholinesterase from pea roots: Isolation, purification, characterization of soluble and membrane-bonding forms. *Biokhimiya (USSR)* 54: 88–94.

Soreq, H., Gnatt, A., Loewenstein, Y., and Neville, L.F. (1992). Excavations into the active-site gorge of cholinesterase. *Trends in Biochemical Science* 17: 353–358.

Spanswick, R.M. (1972). Electrical coupling between cells of higher plants. A direct demonstration of intercellular communication. *Planta* 102: 215–227.

Spencer, R., and Gear, I. (1962). Inclusion of radioactive dopamine-^{14}C in C_3-hydrastine. *Journal of American Chemical Society* 84: 1059–1061.

Staehelin, L.A., and Arntzen, C.J. (1983). Regulation of chloroplast membrane function: protein phosphorylation changes the spatial organization of membrane components. *Journal of Cell Biology* 97: 1327–1337.

Stanley R.G., and Linskens, H.F. (1974). *Pollen Biology, Biochemistry and Management*. Berlin: Springer-Verlag, 307 pp.

Steelink, C., Yeung, M., and Caldwell, R.L. (1967). Phenolic constituents of healthy and wound tissues in the giant cactus (*Carnegiea gigantea*). *Phytochemistry* 6: 1435–1440.

Steffens, P., Nagakura, N., and Zenk, M.H. (1985). Purification and characterization of the berberine bridge enzyme from *Berberis beaniana* cell cultures. *Phytochemistry* 24: 2577–2583.

Stephenson, M., and Rowatt, E. (1947). The production of acetylcholine by a strain of *Lactobacillus plantarum*. *Journal of General Microbiology* 1: 279–298.

Stone, T.W. (Ed.) (1996). *CNS Neurotransmitters and Neuromodulators. Dopamine*. Boca Raton: CRC Press. 272 pp.

Strack, D., Tkotz, N., and Klug, M. (1978). Phenylpropanoid metabolism in cotyledons of *Raphanus sativus* and the effect of competitive in vivo inhibition of L-phenylalanine ammonia-lyase (PAL:) by hydroxylamine derivatives. *Zeitschrift für Pflanzenphysiologie* 89: 343–353.

Strack, D., Nurmann, G., and Sachs, G. (1980). Sinapine esterase. II. Specificity and change of sinapine esterase activity during germination of *Raphanus sativus*. *Zeitschrift für Naturforschung* 35c: 963–966.

Strakhovskaya, M.G., Fraikin, G.Ya., and Rubin, L.B. (1982). Role of serotonin in the release of effects of photodefense and short wavelength ultra-violet light on yeast *Candida quilliermondii*. *Biological Bulletin (Izvestiya AN SSSR, ser. Biology)* 4: 624–629.

Strakhovskaya, M.G., Ivanova, E.V., Belenikina, N.S., and Fraikin, G.Ya. (1989). Photoactivation of the enzymatic synthesis of serotonin from 5-hydroxytryptophan in yeast extracts. *Biologicheskie Nauki (USSR)* 1: 42–44.

Süss K.H. (1981). Identification of chloroplast thylakoid phosphoproteins. Evidence for the absence of phosphoryl-polypeptide intermediates in the ATPase complex. *Biochemistry and Biophysics Research Communication* 102: 724–729.

Sutton, H.C., and Winterbourn, C.C. (1989). On the participation of higher oxidation states of iron and copper in Fenton reactions. *Free Radical Biology and Medicine* 6: 53–60.
Suvorov, N.I., and Shashkov, V.S. (1975). *Chemistry and Pharmacology of the Substances for the Profilactics of the Radiationary Injuries*. Moscow: Atomizdat, 224 pp.
Suzuki, Y. (1966). Possibility of pyridoxal phosphate as a coenzyme of pea amine oxidase. *Naturwissenschaften* 53: 504.
Takimoto, A. (1992). A catecholamine-derived flower-inducing substance. *Shokubutsu no Kagaku Chosetsu* 27: 56–66. *Japan* (*Chem Abstr*. 117: 402, 248523w).
Talapatra, B., Dutta, S., Maiti, B.C., Pradhan, D.K., and Talapatra, S.K. (1974). Terpenoid constituents of Indian Glochidion species: *Glochidion acuminatum* and *Glochidion thomsoni*. Partial synthesis of Glochidone. *Australian Journal of Chemistry* 27: 2711–2714.
Tanada, T. (1972). On the involvement of acetylcholine in phytochrome action. *Plant Physiology* 49: 860–861.
Tester, M. (1990). Plant ion channels: whole cell and single-channel studies. *New Phytologist* 114: 305–340.
Tester, M. and Blatt, M.R. (1989). Direct measurement of K^+ channels in thylakoid membranes by incorporation of vesicles into plant lipid bilayers. Plant ion channels: whole cell and single-channel studies. *Plant Physiology* 91: 249–252.
Thimann, K.V. and Grochowska, M. (1968). The role of tryptophan and tryptamine as IAA precursors. In: *Biochemistry and Physiology of Plant Growth Substances*. Proc. 6th International Conference on Plant Growth Substances. July 24–29, 1967. Ottawa (Eds. F. Whightman and G. Setterfield). pp. 231–242. Ottawa: The Runge Press Ltd.
Thomas, P., and Janave, M.T. (1973). Polyphenol oxidase activity and browning of mango fruits induced by gamma irradiation. *Journal of Food Science* 38: 1149.
Thomas, P., and Nair, P.M. (1971). Effect of gamma irradiation on polyphenol oxidase activity and its relation to skin browning in bananas. *Phytochemistry* 10: 771–777.
Tocher, R.D., and Tocher, C.S. (1969). Biosynthesis of 3-hydroxytyramine in plants. *Abstract of 11th International Botanical Congress*. p. 219.
Tocher, R.D., and Tocher, C.S. (1972). DOPA decarboxylase in *Cytisus scoparius*. *Phytochemistry* 11: 1661–1667.
Toriyama, H. (1975). The effects of acetylcholine and noradrenaline upon the immature sieve tube of root tip. *Abstracts of 12th International Botanical Congress*. Moscow: Nauka, 374 pp.
Toriyama, H. (1978). Observational and experimental studies of meristem of leguminous plants. I. Effects of acetylcholine, red light and far-red light on protoplasts of root tip meristem. *Cytologia* 43: 325–337.
Toriyama, H., and Jaffe, M.J. (1972). Migration of calcium and its role in the regulation of seismonasty in the motor cell of *Mimosa pudica* L. *Plant Physiology* 49: 72–81.
Toutant, J.P., Massoulie, J., and Bon, S. (1985). Polymorphism of pseudocholinesterase in *Torpedo marmorata* tissues. Comparative study of the catalytic molecular properties of this enzyme with acetylcholinesterase. *Journal of Neurochemistry* 44: 580–592.
Tretyak, T.M., and Arkhipova, L.V. (1992). Intracellular activity of mediators. *Uspekhi Soveremennoi Biologii (Trends in Modern Biology)* 11: 265–272.
Tretyn, A. (1987). Influence of red light and acetylcholine on $^{43}Ca^{2+}$ uptake by oat coleoptile cells. *Cell Biology International Reports* 11: 887–896.
Tretyn, A. (1990). Mechanisms of the acetylcholine action in plants. *Postepy Biologie Kommorki* 17: 97–11.
Tretyn, A., and Bossen, M.E. (1990). The effect of a calcium-channels antagonist nifedipine and agonist Bay K-8644 on the phytochrome-controlled swelling of etiolated wheat protoplasts. *Physiologia Plantarum* 78: 230–235.
Tretyn, A., and Kendrick, R.E. (1990). Induction of leaf unrolling by phytochrome and acetylcholine in etiolated wheat seedlings. *Photochemistry and Photobiology* 52: 123–129.

Tretyn, A., and Kendrick, R.E. (1991). Acetylcholine in plants: presence, metabolism and mechanism of action. *Botanical Reviews* 57: 33–73.
Tretyn, A., and Kwiatkowska, K. (1987). Characterization of plant cholinergic system. Plant cholinesterases. *Postepy Biologie Kommorki* 14: 83–96.
Tretyn, A., and Tretyn, M. (1988a). Diurnal acetylcholine oscillation in green oat seedlings. *Acta Physiologia Plantarum* 10: 243–246.
Tretyn, A., and Tretyn, M. (1988b). Charakterystka roslinnego systemu cholinergicznego acetylcholina. *Postepy Biologii Kommorki* 15: 477–494.
Tretyn, A., and Tretyn, M. (1990). Oscillation of acetylcholine in oat seedlings. *Chronobiologia* 17: 45–52.
Tretyn, A., Slesak, E., and Andersz, A. (1985). Interaction of light and the cholinergic system in the process of seed germination. In: *Light and Hormonal Interactions of Plants* Symposium, Occasion 175 Anniversary Humbold University, Berlin 2–7 Sept. 1985, Abstr. Berlin, pp. 93–94.
Tretyn, A., Slesak, E., and Kwiatkowska, K. (1986). Localization of AChE activity in plant cells in LM/TEM/SEM. *Folia Histochemistry and Cytobiology* 4: 328–329.
Tretyn, A., Bobkiewicz, W., Tretyn, M., and Michalski, L. (1987a). The identification of acetylcholine and choline in oat seedlings by gas chromatography and nuclear magnetic resonance (NMR). *Acta Societatis Botanicorum Poloniae* 56: 499–511.
Tretyn, A., Kado, R.T., and Kendrick, R.E. (1987b) Loading and localization of Fluo-3 and fluo-3/AM calcium indicators in *Sinapis alba* root tissue. *Folia Histochemistry and Cytobiology* 35: 41–51.
Tretyn, A., Kopcewicz, J., and Slezak, E. (1988). Interaction of light and the cholinergic system in the regulation of seed germination. *Biologia Plantarum* 30: 338–342.
Tretyn, A., Lukasiewicz-Rutkowska, H., and Kopcewicz, J. (1997). Isolation, purification and identification of acetylcholine in *Pharbitis nil* seedlings. *Acta Physiologiae Plantarum* (Poland) 19: 303-309.
Tretyn, A., Michalski, L., Tretyn, M., and Bobkiewicz, W. (1987c). Changes in the endogenous acetylcholine during deetiolation of *Avena sativa* seedlings. *Plant Growth Regulators*. Procedings 4th Inst. Symp., Pamporovo, 1986, Part 1, Sofia, pp. 737–741.
Tretyn, A., Bossen, M.E., and Kendrick, R.E. (1990a). The influence of acetylcholine on the swelling of wheat (*Triticum aestivum* L.) protoplasts. *Journal of Plant Physiology* 136: 24–29.
Tretyn, A., Kendrick, R.E., Bossen, M.E., and Vredenberg, W.J. (1990b). Influence of acetylcholine agonists and antagonists on the swelling of etiolated wheat (*Triticum aestivum* L.) mesophyll protoplasts. *Planta* 182: 473–479.
Tsavkelova, E.A., Botvinko, I.V., Kudrin, V.S., and Oleskin, A.V. (2000) Detection of neurotransmitter amines in microorganisms, using high performance liquid chromatography. *Doklady of Russian Academy of Sciences* 372: 840–842.
Tucek, S. (1983). The synthesis of acetylcholine. In: *Handbook of Neurochemistry. Enzymes of the Nervous System* (Ed. A. Lajtha), Vol. 4. New York and London: Plenum Press, pp. 219–249.
Tukey, H.B., Jr. (1966). Leaching of metabolites from above-ground plant parts and its implications. *Bulletin of Torrey Botanical Club* 93: 385–401.
Tulus, M.R., Ulubelen A., and Öser, F. (1961). Choline und Acetylcholin in den Blättern von *Digitalis ferruginea* L. *Archiv der Pharmazie und Berichte der Deutschen Pharmazentischen Geseleschaft* 294: 11–17.
Tung, H.F., and Raghavan, V. (1969). Effects of growth retardants on the growth of excised roots of *Dolichos lablab* L. in culture. *Annals of Botany* 32: 509–519.
Tyler, V.E., Jr. (1958). Occurrence of serotonin in a hallucinogenic mushroom. *Science* 128: 718.
Tyler, V.E., Jr., and Groger, D. (1964a). Occurrence of 5-hydroxytryptamine and 5-hydroxytryptophan in *Paneolus sphinctrinus*. *Journal of Pharmacological Science* 4: 462–463.

Tyler, V.E., Jr., and Groger, D. (1964b). Investigation of the alkaloids of *Amanita* species. I. *Amanita muscaria*. *Planta Medica* 12: 334–339.

Tzagolov, A. (1963a). Metabolism of sinapine in mustard plants. 1. Degradation of sinapine into sinapic acid and choline. *Plant Physiology* 38: 202–206.

Tzagolov, A. (1963b). Metabolism of sinapine in mustard plants. 2. Purification and some properties of sinapine esterase. *Plant Physiology* 38: 207–213.

Tzendina, M.B., Frishmann, D.I., Levchenko, V.F., and Berman, A.L. (1988). The similarity of primary structure and homology of rhodopsin, β-adrenoreceptor and muscarinic receptor. *Journal of Evolutionary Biochemistry and Physiology (USSR)* 24: 797–806.

Udenfriend, S., Lovenberg, W., and Sjoerdsma, A. (1959). Physiologically active amines in common fruits and vegetables. *Archives in Biochemistry and Biophysics* 85: 487–490.

Umrath, K., and Thaler, I. (1980). Auslosung von Blattbewegungen bei *Mimosa* und von Krummungen von *Lupinus*-Hypokotylen, gedeutet duch Freisetzung von Erregungssubstanz und Auxin. *Phyton* 20: 333–348.

Umrath, K., and Thaler, I. (1981). 5-Hydroxyindolyl-3-essigsaüre als ein Auxin. *Berichte Deutsche Botanische Gesicht* 94: 143–149.

Urich, K. (1994). Comparative Animal Biochemistry. Berlin, Heidelberg: Springer-Verlag. 784 pp.

Uspenskaya, V.D., and Goryachenkova, E.V. (1958). The purification of diamino oxidase by electrophoresis. *Biokhimiya (USSR)* 23: 212–219.

Üstünes, L., Özer, A., Laekeman, G.M., Corthout, J., Pieters, L.A.C., Baeten W., Herman, A.G., Claeys, M. and Vlietinck., A.J. (1991) Chemical characterization and pharmacological activity of nazlinin, a novel indole alkaloid from *Nitraria schoberi*. *Journal of Natural Products* 54 N 4: 959–966.

Vackova, K., Kutachek, M., de Almeida, P.M., Eder, J., Hoffmann, I., Aslam, M., Vassilev, G., and Kefeli, V.I. (1982a). Enzymes of acetylcholine and hydroxycinnamoylcholine ester metabolism. *Plant Physiology (Bulgaria)* 8: 3–12.

Vackova, K., Kutachek, M., de Almeida, R.M., Eder, Y., Hoffman, U., Aslam, M., Vassilev, G., and Kefeli, V.I. (1982b). Metabolism of cholinic esters on hydroxyknelenite and acidity in plants and enzymes. *Bulgarian Plant Physiology* 8: 3–11.

Vackova, K., Kutucek, M., and de Almeida, R.M. (1984). Some properties of pea cholinesterase and its activity in plant parts at different growth stages. *Biologia Plantarum* 26: 275–284.

Vaisfeld, I.L., and Kassil, G.N. (1981). *Histamine in Biochemistry and Physiology*. Moscow: Nauka, 278 pp.

Valles, K.L.M., Proudlove, M.O., Beechey, R.B., and Moore, A.L. (1984). Membrane potential measurements in wheat-leaf mesophyll protoplasts. *Biochemical Society Transactions* 12: 851–852.

Vamos-Vigyazo, L. (1981). Polyphenol oxidase and peroxidase in fruits and vegetables. *Critical Review of Food Science and Nutrition* 15: 49–127.

Varsheney, K.A., Gangwar, L.P., and Goel, N. (1988). Choline and betaine accumulation in *Trifolium alexandrinum* L. during salt stress. *Egypt Journal of Botany* 31: 81–86.

Vassilev, G.N. (1987). New biologically active substances with growth-regulating activity. *Plant Growth Regulators: Proceedings of 4th International Symposium Pamporovo*, September 28 to October, 4 1986, Sofia, Part 2, pp. 319–336.

Vaughn, K.C., and Wilson, K.J. (1981). Improved visualization of plastid fine structure: plastid microtubules. *Protoplasma* 108: 21–27.

Vdovichenko, L.M. (1966). Acetylcholine and oxidative processes in mitochondria. *Ukrainian Biokhimicheskii Zhurnal (USSR)* 38: 185–193.

Vdovichenko, L.M., and Demin, N.N. (1965). Acetylcholine and respiration of mitochondria of brain cells. *Doklady AN SSR* 162: 1434–1436.

Verbeek, M., and Vendrig, J.C. (1977). Are acetylcholine-like cotyledon factors involved in the growth of cucumber hypocotyl? *Zeitschrift für Pflanzenphysiologie* 83: 335–340.

Viale, A., Ngernprasirtsiri, J., and Akazawa, T. (1991). Characterization and intraorganellar distribution of protein kinases in amyloplasts isolated from cultured cells of sycamore (*Acer pseudoplatanus* L.). *Plant Physiology* 96: 1142–1149.

Vialli, D.M., Barbetta, F., Zanotti, L., and Mihalyi, K. (1973). Estendibilita del concettodi systema cellulare enterocromaffine ai vegetali. *Archives of Histochemistry* 45: 270–282.

Vidgren, J., Svensson, L.A., and Liljas, A. (1994). Crystal structure of catechol-o-methyltransferase. *Nature* 368: 354.

Villalobos-Peetrini, R., and Laguarda-Figueras, A. (1967). Radioprotection of *Vicia faba* by serotonin-creatinine sulfate complex. *Radiation Botany* 7: 369–373.

Villalobos, J., Ramirez, F., and Moussatche, H. (1974). Some observations on the presence of acetylcholine and histamine in plants. *Cience e Culture* 26: 690–693.

Vizi, E.S., and Lendvai, B. (1999). Modulatory role of presynaptic nicotinic receptors in synaptic and non-synaptic chemical communications in the central nervous system. *Brain Research Reviews* 30: 219–235.

Vogel, J., Iff, E.T., and Moussatche, H. (1964). Sobre o teorem acetylcholina na semente e latex de jaca (*Actocarpus integraem*) condicoes variadas de desen volvimento. *Annais XIII Congress Society of Botanica Brasilia Recife*, p. 84.

Volinskii, B.G., Bender, K.I., Freidman, S.L., Bogoslovskaya, S.I., Voronina, K.V., Glazirina, G.A., Kaprelova, T.S., Koloskova, I.T., Kuznetzova, S.G., and Martinov, L.A. (1983). *Medicinal Plants in Scientific and Folk Medicine*. Saratov: Saratov University, 360 pp.

Waalkes, T.P., Sjoerdsma, A., Greveling, C.R., Weissbach, H., and Udenfriend, S. (1958). Serotonin, norepinephrine, and related compounds in bananas. *Science* 127: 648–650.

Wagner, H. (1988). Non-steroid, cardioactive plant constituents. In: *Economic and Medicinal Plant Research* (Eds. H. Wagner, H. Hikino and N.R. Farnsworth). London: Academic Press, pp. 17–38.

Walaas, E., Walaas, O., and Haavaldsen, S. (1963). Spectrophotometric and electron-spin resonance studies of complexes of catecholamines with Cu (II) ions and the interaction of ceruloplasmin with catecholamines. *Archives of Biochemistry and Biophysics* 100: 97–100.

Webb, L.J. (1948). *Guide to the Medicinal and Poisonous Plants of Queensland*. Council for Scientific and Industrial Research, Commonwealth of Australian Bulletin No. 232. Melbourne: Gourley JJ, Government Press, 202 pp.

Weigel, P., Lerma, C., and Hanson, A.D. (1988). Choline oxidation by intact spinach chloroplasts. *Plant Physiology* 86: 54–60.

Wenthold, R.J., Mahler, H.R., and Moore, W.J. (1974). Half-life of acetylcholinesterase in mature rat brain. *Journal of Neurochemistry* 22: 941–943.

Werle, E. (1955). Amine und Betaine. In: *Modern Methods of Plant Analysis* (Eds. K. Paech and M.V. Tracey), Vol. 4. Berlin: Springer-Verlag, pp. 517–523.

Werle, E., and Pechmann, E. (1949). Über die Diamin-oxydase der Pflanzen und ihre adaptative Bildung durch Bakterien. *Liebig Annal. Chemie* 562: 44–60.

Werle, E., and Raub, A. (1948). Über Vorkommen, Bildung und Abbau biogener Amine bei Pflazen unter besonderer Beruck-sichtigung des Histamines. *Biochemische Zeitschrift* 318: 538–553.

Werle, E., and Roewer, F. (1950). Monoaminoxydase in Pflanzen. *Biochemische Zeitschrift* 320: 298–301.

Werle, E., and Zabel, A. (1948). Über die Verbreitung der Histaminase im Pflanzenreich. *Biochemische Zeitschrift* 318: 554–559.

Werle, E., Trautschold, I., and Aures, D. (1961). 13Reinigung und Characterisierung der Amino-Oxydase ause Erbsen. *Hoppe-Seyler's Zeitschrift für Physiologie Chemie* 326: 200–211.

West, G.B. (1958). Tryptamine in edible fruits. *Journal of Pharmacy and Pharmacology* 10: 589–590.

West, G.B. (1959a). Indole derivatives in tomatoes. *Journal of Pharmacy and Pharmacology* 11: 275–277.
West, G.B (1959b). Tryptamines in tomatoes. *Journal of Pharmacy and Pharmacology* 11: 319–320.
West, G.B. (1960). Carcinoid tumours and pineapples. *Journal of Pharmacy and Pharmacology* 12: 768–769.
Whatley, J.M., John, P., and Whatley, F.R. (1979). From extracellular to intracellular: the establishment of mitochondria and chloroplasts. *Proceedings of Royal Society, London* B204: 165–187.
Wichers, H.J., Prass, N., and Huizing, H.J. (1989). XIX. *Mucuna pruriens:* In vitro production of L-DOPA. In: *Biotechnology in Agriculture and Forestry*, Vol. 7. Medicinal and Aromatic Plants. II. (Ed. Y.P.S. Bajaj). pp. 349–366. Berlin, Heidelberg. Springer-Verlag.
Wichers, H.J., Visser, J.F., Huizing, H.J., and Pras, N. (1993). Occurrence of L-DOPA and dopamine in plants and cell cultures of *Mucuna pruriens* and effects of 2,4-D and NaCl on these compounds. *Plant Cell, Tissue and Organ Culture* 33: 259–264.
Wierenga, J.M., and Hollingworth, R.M. (1992). Inhibition of insect acetylcholinesterase by the potato glycoalkaloid alpha-chaconine. *Natural Toxins* 1: 96–99.
Williams, C.H., Lawson, J., and Backwell, F.R.C. (1992). Inhibition and inactivation of monoamine oxidase by 3-amino-1-phenyl-prop-1-enes. *Biochimica et Biophysica Acta* 1119: 111–112.
White, H.L., and Cavallito, C.J. (1970). Inhibition of bacterial and mammalia choline acetyltransferase by steryl pyridine analogs. *Journal of Neurochemistry* 17: 1579–1589.
White, J.M., and Pike, C.S. (1974). Rapid phytochrome-mediated changes in adenosine 5'-triphosphate content of etiolated bean buds. *Plant Physiology* 53: 76–79.
Whittaker, V.P. (1963). Identification of acetylcholine and related esters of biological origin. *Handbuch der Experimentellen Pharmakologie*, Berlin: Springer-Verlag. 15: 1–39.
Whittaker, V.P. (1987). Cholinergic transmission: past adventures and future prospects. In: *Cellular and Molecular Basis of Cholinergic Functions* (Eds. M.J. Dawdall and J.N. Hawthorne). Chichester, Weinheim: Ellis Horwood and VCH Verlagsgesellschaft, pp. 495–512.
Wieland, T., and Motzel, W. (1953). Über das Vorkommen von Bufotenin im gelben Knollenblatterpilz. *Annuals der Chemie* 581: 10–16.
Wisniewska, J. and Tretyn, A. (1999). Effect of light on the level of acetylcholine in seedlings of the wild—type and phytochrome mutants of tomato (*Lycopersicon esulentum* Mill). *Acta Physiologiae Plantarum* (Poland) 21: 221-230.
Wong, P.K., and Chang, L. (1988). The effects of 2,4-D herbicide and organophosphorus insecticide on growth, photosynthesis and chlorophyll synthesis of *Chlamydomonas reinhardtii* (mit-positive). *Environment Pollution* 55: 179–190.
Wood, H.N., Lin, M.C., and Braun, A.C. (1972). The inhibition of plant and animal adenosine 3',5'Cyclic monophosphate phosphodiesterases by a cell-division-promoting substance from tissues of higher plant species. *Proceedings of National Academy of Sciences of USA* 69: 403–406.
Wu, J.Y. (1983). Decarboxylases. Brain glutamate decarboxylase as a model. In: *Handbook of Neurochemistry. Enzymes in the Nervous System* (Ed. A. Laitha), Vol. 4. New York, London: Plenum Press, pp. 111-131.
Youguchi, R., Okuzumi, M., and Fujii, T. (1960). Seasonal variation in numbers of mesophilic and halophilic histamine-forming bacteria inshore of Tokyo Bay and Sagamni Bay. *Nippon Suisan Gakkaishi* 56: 1467–1472.
Young, R.J., and Laing, J.C. (1990). Biogenic amine binding sites in rabbit spermatozoa. *Biochemistry International* 21: 781–787.
Yunghans, H., and Jaffe, M.J. (1972). Rapid respiratory changes due to red light or acetylcholine during the early events of phytochrome-mediated photomorphogenesis. *Plant Physiology* 49: 1–7.

Yurin, V.M. (1990). Electrophysiological aspects of enobiology of plant cell. 3. Non-state ionic flows. *Vestnik AN BSSR Ser Biology* 5: 7–10.
Yurin, V.M., Ivanchenko, V.M., and Galactionov, S.G. (1979a). *Regulation of Functions of Plant Cell Membranes.* Minsk: Nauka i Technika, 200 pp.
Yurin, V.M., Gusev, V.V., Kudryashov, A.P., Matusov, G.D., and Safronova, N.N. (1979b). Testing of membrane-tropic effect of physiologically active substances I. Qualitative estimations. *Vestnik AN Belorussian SSR, ser. Chemical Sciences* 1: 97–100.
Zhang, J., Blusztajn, J.K., and Zeisel, S.H. (1992). Measurement of the formation of betaine aldehyde and betaine in rat liver mitochondria by a high pressure liquid chromatography-radioenzymatic assay. *Biochimica et Biophysica Acta* 1117: 333–339.
Zieher, L.M., and de Robertis, E. (1963). Subcellular localization of 5-hydroxytryptamine in rat brain. *Biochemical Pharmacology* 12: 596–598.
Zieher, L.M., and de Robertis, E. (1964). Distribution subcellular de noradrenaline y dopamine en el cerebro de raba. *VI Congress of Association of Latinoamerican cie Fisiologie Vifia Mar.*, Chile, p. 23.
Zholkevich, V.N. (1981). On the nature of root pressure. In: *Structure and Function of Plant Roots* (Eds. R. Browwer, O. Gasparicova, J. Kolek and B.C. Louhman). (Ser. Developments in Plant and Soil Sciences, Vol. 4. Proceedings of the 2nd Int. Symp., Bratislava, Czechoslovakia, September 1–5, 1980). The Hague, Boston, London: Martinus Nijhoff/ Dr. W. Junk Publishers, pp. 157–158.
Zholkevich, V.N., and Chugunova, T.V. (1995). Ineraction of cytoskeleton proteins, biomediators and phytohormones in plant water transport regulation. *Doklady of Russian AN* 341: 122–125.
Zholkevich, V.N., and Chugunova, T.V. (1997). Effect of neuromediators on root pumping activity. *Doklady of Russian AN* 356: 122–125.
Zholkevich, V.N., Sinitsyna, Z.A., and Peisakhzon, B.I. (1979). On the nature of root pressure. *Fiziologia Rastenii (Soviet Plant Physiology)* 26: 978–993.
Zholkevich, V.N., Chugunova, T.V., and Korolev, A.V. (1990). New data on the nature of root pressure. *Studia Biophysica* 130: 209–216.
Zholkevich, V.N., Gusev, N.A., Kaplya, A.V., Pakhomova, G.I., Pilshchikova, N.V., Samuilov, F.D., Slavnyi, P.S., and Shmat'ko, I.G. (1989). *Plant Water Exchange.* Moscow: Nauka, 256 pp.
Zholkevich, V.N., Zubkova, N.K., and Korolev, A.V. (1998). Effect of colchicine and noradrenaline on exudate secretion of *Helianthus annuus* L. roots in the absense of water uptake from an environment. *Doklady of Russian Academy of Sciences* 359: 551–553.
Zukovskii, S.G., and Evstigneeva, T.A. (1983). Action of organophosphorus insecticides on pea cholinesterase. *Bulletin of All-Union Science Institute of Plant Defence* 56: 26–31.

Subject Index

Abscisic acid   163
Absorbance changes
    light-induced at 520nm   67, 68
ACC = 1-Aminocyclopropane-1-carboxylic
    acid   14
Acetaldehyde (Acetal)   28
Acetate   13, 14, 19, 98, 99
Acetate uptake   88, 89
Acetic acid   13, 98, 99, 138, 147
Acetylcholine (ACh)   84, 86, 88-94, 97-100, 125, 130, 146, 179, 180, 187, 197, 201, 207-210, 213
    agonists   85, 86, 89, 90, 93, 94, 97, 192
    antagonists   66, 85, 86, 88-91, 94, 97, 98, 119, 192
    as antioxidant   80
    as cation   195
    as chemosignal   149
    as defensive agent   196
    as intercellular mediator or hormone   202
    as intracellular mediator   202
    as possible hormone   197
    as protector element   149
    as radioprotector   199
    as regulator of energetic and metabolic processes   202
    as substrate in metabolism   195
    biosynthesis   5, 11-14, 17, 18, 98, 99
    catabolism   14, 99, 100
    concentration in stomata   148, 203, 208
    contents in animals   120
    content in plants   5-12, 17, 18, 61, 148
    content in chloroplasts   12, 91, 180
    control of Na^+/K^+ channels   188
    dependence on light   13
    determination   4
    effects of age   13
    effects of heat stress   11
    enzymes of biosynthesis   84, 98, 99, 203
    enzymes of catabolism   81, 99, 100, 203
    hydrolysis enzymic   114, 99-101, 113, 125, 126, 132, 133, 137, 138, 149, 177, 195, 204
    hydrolysis non-enzymic   19
    identification   4, 5
    in bacteria   4, 202
    in chloroplasts   12, 91
    in cytoplasm   91
    in flagellum   203
    in food and medicinal plants   211, 212
    in fungi   4, 187
    in honey   11
    in insects   149
    in interactions between organisms   62, 63
    in male cells (spermatozoa)   185
    in motile organs   123
    in pistil   149
    in pollen   11, 77, 90, 149, 185
    in pollen allelopathy   206
    in reproductive organs   11
    in secretions   60, 185, 196
    in silage   4
    in stinging trihomes(hairs)   5
    light influence   13, 202
    localization in cell   12, 14, 120-126, 148, 191
    localization in chloroplasts   19
    metabolism   13
    methods of determination
    modes of mediation   187
    non-mediatory function   192, 193
    occurrence in plants   5-13, 113
    pharmacological effects   211
    physiological activity   207

physiological roles 54-80, 189
precursors 13
protective role 197
regulatory function 192, 193, 203
regulatory role
  in photosynthesis 203
  in respiration 203
regulation of ion channel opening 88
role in electric events 189
Acetylcholine effects on
  ATPase activity in chloroplasts 72
  ATP level 70, 71, 173
  ATP synthesis 71, 192
  autofluorescence of intact cell 79, 80
  bioelectric reactions 198
  Ca^{2+} efflux (output) 66
  Ca^{2+} 66
  chemotaxis of bacteria 62
  chloroplast reactions 193
  contraction of protoplasts 61
  development 56, 57
  elongation 52-54, 60
  elongation of pollen tubes 54, 55
  energetic reactions 69, 89
  enzymic activity 196
  ethylene emission 80
  exudation of xylem sap 62
  flowering 57, 88
  flower opening 62
  fluorescence of chlorophyll 80
  formation of secondary roots 52, 53, 57
  germination of fungi spores 54, 57
  germination of pollen 54, 55, 176
  germination of seeds 54, 55, 57, 58
  germination of vegetative microspores 54
  growth 52-57, 197
  growth of seedlings 88
  H^+ conductivity 190
  H^+ efflux 198
  hormone fluxes 65
  influence of environmental factors 57
  influence of light 57, 58
  incorporation of radioactive phosphorus 69
  ion fluxes 65
  ion permeability 63, 64, 66, 184
  K^+ efflux 67
    from intact chloroplasts 69
  K^+ exchange 66, 69, 190
  K^+ uptake 190
  leaf shrinking of pericycle cells 88
  leaf unrolling 65
  liberation of mediators and hormones 204
  light 52, 53, 56
  membrane potential 63-65, 189, 190
  membrane potential of chloroplasts 67, 68
  $\Delta\mu H$ 71
  metabolic reactions 69, 80
  movement of cytoplasm 61
  mitosis 59
  Na^+ efflux 67
  Na^+ efflux from chloroplasts 69
  Na^+ exchange 69
  Na^+/K^+ permeability of chloroplasts 64, 67, 192
  $NADP^+$ photoreduction 71-74
  nasty movements 62
  opening of ion channels 65, 190, 191
  oxidative phosphorylation 194
  oxygen uptake 71, 198
  peroxidase activity 88
  pH dependence 58
  pH gradient 68
  photochemical reactions in chloroplasts 71-76, 184
  photophosphorylation 71, 195
  phytochrome-controlled processes 198
  ^{32}P incorporation in phospholipids 80
  pressure of roots 62, 88
  proton efflux 63
  proton exchange 63
  protone uptake 71
  release of ATP 195
  root pressure 148
  shrinking of pericycle cells 88
  sporulation 198
  substrate-linked phosphorylation 194
  swelling of mitochondria 67
  swelling of pollen 188
  swelling of protoplasts 62, 65
  unfolding of leaves 53
  utilization of inorganic phosphate 71
Acetylcholine hydrolysis 19, 99, 100, 135, 136, 147
  enzymic 19, 99, 100, 126-130, 135-138
  nonenzymic 19
Acetylcholine regulation
  of membrane permeability 66

Subject Index

Na^+ and K^+ output 64, 66
Acetylcholinesterase (true cholinesterase) 18, 19, 101, 113, 114, 130, 132, 134, 136, 142, 143, 147, 148, 187, 204
   activity 18, 19, 101, 114, 123, 137
   active center 100, 143, 148
   allosteric centers 100
   animals 133, 135, 136, 142, 143, 146, 149
   as trypsine-like enzyme 135
   at development 114
   carboxypeptidase activity 135
   catalytic center 148
   content in plants 100-115
   effects on CO_2 fixation by leaves 148
   excretion by oocytes 149
   excretion by parasites 149
   free (weak-bounded) forms 127-132
   functions 62
   globular 133
   half-time of life 130
   in animals 100, 116, 119
   in chemorecognition 149
   in chloroplasts 120, 130
   in defensive system 149
   in erythrocytes 139
   inhibition of ribulose-1, 5-bisphosphate carboxylase 147, 148
   inhibition by organophosphate pesticides 147
   inhibition by substrate 99, 100, 119, 128-130
   in human brain 113
   in latex 131
   in pistil 116
   in plasmodium of myxomycete 149
   in pollen 185
   in protoplasts 123
   in symbiosis 63
   interaction with catecholamines 81
   interaction with dopamine 81, 147, 148
   in stomata 65, 148
   interface between stele-cortex 148
   in traps 123, 124
   localization in cell 120-126
   membrane-bound forms 133
   molecular forms 132-134
   occurrence in plants 100-115
   plural forms 129, 130
   regulation of the activity 80
   rate of hydrolysis 136
   role in energetic processes 195
   substrate specificity 100, 132.
Acetyl~CoA 13, 14, 16, 18, 98
Acetyl~CoA synthetase 13
N-Acetylhistamine 47-50
^{14}C-Acetyl-β-methylcholine 101, 114, 116, 117
N-Acetylserotonin 42
Acetylthiocholine 99, 101, 116-120, 125, 130, 133, 136-139, 141, 142, 150, 216-218
   as substrate 99, 127-129, 139
   colored reaction 99, 216-218
ACHE see acetylcholinesterase
Acid
   acetic 16, 138, 147
   amino 179
   aromatic 179
   ascorbinic 195
   caffeic 163
   carbaminic 138-141
   chlorogenic 163
   cinnamic 15, 60, 163
   p-coumaric 163
   dehydroxymandelic 27, 28
   dihydroascorbinic 195
   3, 4-dihydrophenylacetic acid 28
   dioxymandelic 27, 28
   ferulic 163
   gibberellic 145
   glyoxalic 219, 220
   5-hydroxyindole acetic 41, 42, 53
   p-hydroxyphenylacetic 167
   imidazole acetic acid 50, 51
   indoleacetic 39, 41, 42, 53, 59, 60, 145, 163
   isovaleric 48
   lysergic 41, 42
   malic 119
   methylimidazole acetic 49, 51
   α-naphthylacetic 138
   oxycinnamic 15
   sinapic 15
   vanillic 27, 28, 33, 196
Acidation of internal medium 123
Actin 61, 158, 182
Actin-like components (proteins) 61, 158
Actin microfilaments 61
Action potential 61, 66, 85, 87, 149, 187, 188, 200, 201, 207, 209
   in reproductive organs 188

mechanism of formation 189
of chloroplasts 191
spreading 189
Active forms of oxygen 33, 34, 76, 176, 186
Active oxygen 76, 176
Active oxygen species (Reactive oxygen species) 76
Active sites of enzymes
Actomyosine filaments 158
Adenosine-5'-monophosphate 169
Adenosine triphosphate 138
S-Adenosylmethionine (SAM) 163, 164
Adenylate cyclase 83, 88, 150, 169-173, 179, 191, 192, 201, 207, 210, 220
   connected with receptor 173
   location 170, 173
   pathway 171
Adenylate cyclase system 169, 175, 187, 201, 207, 210, 220
ADP 171, 198
ADP formation 198
Adrenaline 24, 29, 33, 73, 151, 161, 166, 168-175
   agonist 33, 151
   antagonist 151
   as hormone 196, 200
   as substrate 166-168
   at stress 74
   biosynthesis 25, 26, 161, 164
   catabolism 27, 28, 166
   constants of binding 158
   content in plants 24
   effects 32
   enzymes of catabolism 33, 168
   enzymes of synthesis 26, 164
   in chloroplasts 25
   localization in cell 25
   occurrence in plants 24
   oxidation 34
   pharmacological effects 211
   precursors of alkaloids 29, 32, 33
   redox reactions 71-76, 195, 196
Adrenaline effects on 32
   bioelectric processes 65
   Ca^{2+} efflux (release) from intact chloroplasts 152, 175, 196
   Ca^{2+}/Mg^{2+} permeability in chloroplasts 156
   cyclosis of cytoplasm 61, 63, 152
   cytoplasm movement 61, 152
   elongation of seedlings 53, 60
   exudation of root sap 62
   flowering 152, 153
   fluorescence of phycoerythrine 76
   germination of seeds 55
   membrane potential 65, 152
   Mg^{2+} release(efflux) from chloroplasts 69, 70, 152, 190
   $NADP^+$ photoreduction 73
   oxidative phosphorylation 194
   photophosphorylation 73, 152, 156, 195
   photoreactions in chloroplasts 73-76
   substrate-linked phosphorylation 194
Adrenergic system of regulation 60, 191, 204
Adrenoblockers 42, 153, 156, 196
Adrenoreception 178
Adrenoreceptors 32, 150-153, 158, 202
   α-type 150, 151, 158, 211
   β-type 150-153, 200, 211
Adrenochrome 73-76, 198
Adrenolutin 73, 74
Affinity chromatography on
   1-methyl-9[N-β-(ε-aminohexanoyl)-β-amino-propylamino]acridinium bromide ligand-binding Sepharose 2B 126
   Sepharose 4B immobilized by n-aminophenyltrimethyl ammonium iodide 126
   Sepharose 4B~CoA SH 98
Affinity to neurotransmitters 84
Affinity to their antagonists 84, 94
Agents
   Antitumor 42
   Antiviral 42
Agonists
   of acetylcholine 85, 86, 89, 90, 93, 94, 97, 192
   of biogenic amines 152
   of dopamine 150, 151, 153, 156
   of histamine 151, 157-159
   of noradrenaline 151, 155, 156
   of serotonin 151, 155, 156
Agriculture 205
Alanine 136
β-Alanylhistamine 46
Albumine 137
Alcohols aliphatic C_2-C_4 179
Aldehydes 27

Subject Index

of betaine 16
vanillic 27, 28, 196
hydroxyindoleacetic 41
dihydrocinnamic 30
dioxyphenylacetaldehyde 34
Algae 65, 100, 101, 121, 180
Alkaloids 27, 28, 33, 196, 208
   anticholinesterase activity 196, 213
   as antimediator drugs 213
   as blockers of cholinoreceptor 196
   as agonists of acetylcholine 196
   at wounding 22
   biological effects 32, 146
   biosynthesis from biogenic amines 196
   formation at stress 22
   glycoalkaloids 213
   in plant culture of tissue 22
   originated from dopamine 22, 28, 196
   steroidal glycoalkaloids 146
   synthesis 208
Alkaloids derived from dopamine 27-32
Alkaloids derived from noradrenaline 29
Alkaloids derived from serotonin 41, 42
Alkaloids derived from histamine 48-51
Allelochemicals 41
Allelopathic action 60
Allelopathic recognition 178
Allelopathic relations 187
Allelopathy 60
Allergic reactions 47, 205
Allylcholine 16
Allylesterase 14
Amides of D-lysergic acid 42
Aminergic system 207
Amine oxidases 168
Amines 179
   as stressory compounds 196
   quarternary 20, 48
Amino acids 179
Amino acid analysis 136
γ-Aminobutyric acid 204
Aminochromes 33, 34, 74-76, 208
1-Aminocyclopropane-1-carboxylic acid (ACC) 14
2-Amino-6, 7-dihydroxy-1, 2, 3, 4-tetrahydronaphthalene 152
Aminoergic system of plants as biosensor 205
Aminoergic system of regulation 176, 207
Aminooxidase 27, 28, 164, 166, 177, 208
   copper-containing 27, 28

n-Aminophenyltrimethyl ammonium iodide 126
Aminoreceptors 151, 208
Amitryptiline 156
Ammonium detoxication 166
AMO-1618 20, 59, 140, 144, 145
AMP 171
Anabasine 213
Analog of Ellman reagent (Red analog) 114, 118, 119, 123, 124, 142, 215-218
Angiosperms 103
Anhalamine 29, 32
Anhalonidine 29, 32
Antagonism 97
Antagonism concept 97
Antagonists 84
   effects 88, 89
   of acetylcholine 65, 91
   of biogenic amines 151, 155
   of dopamine 150, 154
   of histamine 151, 152, 157-159
   of noradrenaline 151-153, 155, 158, 159
   of serotonin 151, 156, 158
Anthers 103, 106, 108-110, 115-119, 144, 215
Antibody against acetylcholinesterase 100
Anticholinergic drugs 113
Anticholinesterase activity of retardants 147
Anticholinesterase agents (inhibitors) 145-147, 185
Anticholinoreceptor drugs 85, 90, 113
Antimediatory agents 206
Antioxidants 187
Apomorphine 150, 153
Apoplast 166
Arabinose 137
Arecoline 86, 89, 93, 94
Arnifolin 213
Aquiporins 188
Aromatic acids 16, 179
Ascorbinic acid (ascorbate) 195, 196
Asparagine 136
ATCH and AthCh - *see* acetylthiocholine
Artificial phytocenosis 206
Atomic absorbance spectrophotometer 66
ATP 83, 168, 171, 173, 195
ATPase 180, 195
   in chloroplasts 72, 80
ATP level 88, 70
ATP synthesis 16, 71, 88, 184, 192, 195
   in chloroplasts 71, 156, 192

ATP synthetase   174, 188
ATP utilization   88
Atropine   33, 60, 65, 66, 85, 86, 88-93, 97, 113, 173, 196
Atropine effects on
  ATP level   173
  bacterial motility   88
  ethylene release   89
  incorporation of ^{32}P into phospholipids   88
  K$^+$ efflux   66
  Na$^+$ efflux   66
  nasty movements   62
  protoplast contraction   66
Attractants   179
Autofluorescence
  effect of acetylcholine   91, 92
  effect of atropine   91, 92
  effect of d-tubocurarine   91, 92
  of germination organs   149
  of pistil stigma   77, 78, 89, 91, 92, 149, 184, 185
  of pollen   77, 91, 149, 185
  of secretory hairs   159, 162
  of vegetative microspores   79, 80
Automnaline   30
Auxin   53, 60
Axes of leaves   24, 62
Azulenes   76

Bacteria   4, 62, 63, 101, 201, 204
  blue-green   5, 14
  photosynthesizing   62
Bacteriorhodopsin   200
Barley   16, 41, 53
Bay K-8644   65
Beet   16
Benadryl   157
Benperidol   156
Benzoylcholine   16, 146
Benzoylthiocholine   117, 216
Benzylisoquinoline alkaloids   32
Berberastine   28, 29, 208
Berberine   28, 29, 32, 213
Berberine bridge enzyme   32
Berberine synthesis   32
Betacyanin   26, 27
Betacyanin efflux   66
Betaine   16, 196, 197
Betaine aldehyde   16
Betazole   157

Biogenic amines   150, 155, 175, 177, 179, 180, 206, 208-210
  agonists   152
  antagonists   155
  as defensive agents   196
  as cations   195
  as chemical signals   178, 183
  as intracellular mediators   202
  as germination stimulators   206
  at stress   206
  as substrates   196
  biosynthesis   159-164, 203
  catabolism   164-169, 177, 203
  effects on ATP synthesis in chloroplasts   73, 192
  effects on NADP$^+$ photoreduction   73
  effects on photochemical reactions   73, 184
  in nitrogen exchange   202
  in pollen allelopathy   206
  in secretions   196
  interaction with membranes   150
  localization in cell   25
  modes of mediatory control   187
  non-mediatory functions   192-194
  protector function   199
  reception   152
  regulation of membrane potential   65
  regulatory function   192, 194
  role in electric events   188, 189
Biocenosis   178
Biomediators (terminology)   3, 53, 59, 60
  relationship with phytohormones   60
Biosensors   205, 210
Biotests   4
Biopotential   187
Blue→Red transition of autofluorescence   78, 79
  at the development of vegetative *Equisetum* microspores   79, 80
Borreline   42
Breeding   90, 115, 205, 206
Brucine   213
Bufotenine   41, 42, 151, 156
Bufotenidine   41, 42
α-Bungarotoxin   66, 85, 86, 89, 93-95
Burimamide   157
Burns   47
BuThCh or BTCh (see butyrylthiocholine)
Butyrylcholine   12, 14, 15, 100, 119, 125, 132
  effects   65

hydrolysis 119
occurrence in plants 15
Butyrylcholinesterase (pseudocholinesterase) 100, 132, 134, 142, 143, 147, 148, 204
Butyrylthiocholine (BuThCh) 116-120, 128, 129, 133, 138, 216

Cabbage conserved 4
Cactus 23, 32, 33, 48, 198
Cadaverine 48, 164
Caffeic acid 163
Caffeine 174
Caffe(o)ylcholine 16, 146
Calcium (Ca^{2+}) 174, 175, 181, 182, 191, 207, 208
   as secondary messenger 83, 150, 151, 168, 171, 174-177, 191
   channels 62, 188, 191
   concentration in cell 171
   -containing-proteins 171
   dependent processes 159
   effects 65
   efflux 66, 171, 174
   fluxes 188
   Mg^{2+}-activated enzymes 173
   permeability 65, 156
   regulation 181, 182
   sensor 182
   uptake 64
Callus 8, 18, 21-24, 26, 38, 53, 59, 60, 121, 122, 198
Calmodulin 65, 66, 158, 171, 173-175
Calmodulin-dependent proteins 171
Calmodulin inhibitor 65, 66, 158
Calvin cycle 17
cAMP 83, 88, 150, 151, 168-176, 179, 187, 191, 195, 196, 207, 208
   activators and inhibitors 175, 176
   synthesis 171, 195
   system 175, 176
Canadine 30-32
Capsaicin 29, 33, 145, 196
Carbam(o)ylcholine 53, 86, 89, 93
Carbaminic acid (carbonate) 19, 138, 139, 141, 142
Carbaminic inhibitors 101, 142, 144
Carbon dioxide fixation 148
β-Carboline alkaloids 42
Carboxylase 162
Carnegine 29, 33

Carnivorous plants 115, 123, 183
Carotenoids 184
Casimiroedine 50
Catabolic ways for the catecholamines 27-34
Catalase 148
Catechol 163, 164, 168,
Catecholamines 147, 150, 151, 156, 167, 168, 179, 207, 208, 210, 213
   agonists 156, 158
   aminochrome formation 33, 34
   antagonists 156
   antioxidant features 199
   antiviral and antitumor functions 199
   as chemical signals 179
   as oxidants 185
   as protectors 197
   as radioprotectors 199
   assays 219-220
   autooxidation 71
   biosynthesis 25, 26
   catabolism 27-29, 33, 164, 166, 167
   complexes with metals 73, 74, 205
   dependence on external factors 26
   in isolated chloroplasts 25
   in male cells (spermatozoa) 185
   in pollen 185
   interaction with serotonin alkaloids 196
   liberation 204
   localization in cell 25, 26, 39
   metabolism 25
   methods determination 20, 219, 220
   microspectrofluorimetric assay 219, 220
   occurrence in plants 20
   oxidation 33, 34, 166, 185, 208
   pharmacological effects 205, 211
   physiological activity 207
   precursors of alkaloids 28, 29, 32, 33, 196
   radicals 34
   reactions with electron carriers 74-76
   redox reactions 71-76, 208
   sensitive ion channels 201
   standard potentials 195
Catecholamine effects on
   ATP synthesis in chloroplasts 71
   Ca^{2+}/Mg^{2+} permeability in chloroplasts 192
   chemotaxis 63
   development 57
   ethylene production 80

flowering 57, 153
flower opening 62
fluorescence of phycoerythrin 75
growth 53
morphogenesis 57, 153
mitosis 59
nasty movement 62
secretory cells 204
Catecholamine radical 34
Catecholamine quinone 34
Catechol-o-methyltransferases 27, 41, 151, 164, 168
Catecholamine-o-transferase 27, 28
Catecholoxidase 168
CCC (chlorocholine chloride) 140, 144
Cell-cell communication 91
Cell-cell interaction 90
Cell-cell joints 122
Cell-cell recognition 182, 187
Cellular contacts 182
Cellular pores 189
Cell cultures 167
Cell wall 58, 120-122, 175
Cellular contacts 182
Cerulloplasmin 74
CF_o-factor 188
cGMP 83, 88, 150, 151, 168, 172-174, 176, 187, 191, 196, 207, 208
α-Chaconine 113, 213
Channels for ions 87, 183, 201
K^+/Na^+ 62
K^+ 85
Na^+ 87
ChE -see cholinesterase
Chelating agents 167
Chelerythrine 213
Chelidonine 28
Chemical signalization 179, 184, 208
Chemical stimulus 185
Chemoreception 178
Chemosignal spreading 185
Chemotaxis 62, 63, 178
Chilling 16
Chill-tolerance 16
Chlorine (Cl^-)channels 168, 188, 189
Chlorobenproprit 158
Chlor(o)choline chloride 140, 145, 147
Chloroethyl trimethyl ammonium 197
Chlorogenic acid 163
Chloromercurium benzoate 163

Chlorophyll 16, 20, 79, 80, 97, 202
fluorescence 80
synthesis 19, 79, 80, 147
Chloroplast 12, 17, 19, 61, 64, 67, 80-91, 93-98, 122, 125, 129, 158, 173-175, 179, 183, 184, 188, 190, 192-195, 201, 202, 207
as model 91, 93-98, 122, 125
cholinesterase 127, 129
envelope 68, 91, 125, 173, 184, 188, 190, 192-194
ion channels 188-191, 201
ion permeability 67, 184
isolated 90
K^+ efflux 190
membrane potential 67, 68, 190, 191
microtubules 180
movement 175
Na^+ efflux 190
phototaxis 175
receptors 158-201
Chlorpheniramine (chlortrimeton) 157
Chlorpromazine 150, 151, 154, 156, 158, 159
dimethylsulfonium 154
trimethylsulfonium 154
Choline 13, 14, 16-18, 98, 99, 138, 196, 197
cation 146
derivatives 17
effects 16
Cholineacetylase 14, 15, 18-20, 84, 98, 99, 203
Choline acetyltransferase (choliacetylase) 14, 15, 18-20, 84, 98, 99, 176, 203, 208, 209
activity 98, 99
evolution 203
in bacteria 99, 203
in brain 98, 99
in flagellum 203
in mammalia 99
localization in cell 14, 99
purification 98, 99
Choline kinase 88
Cholinergic regulatory system 66, 90, 176, 191, 203, 207-209
Cholinesterase 14, 18, 20, 84, 99-150, 177, 196, 208-210, 215-219
active center 100, 130, 142, 143, 149
activity 14, 15, 19, 20, 99, 100, 113, 114, 120
activity determination 101, 115

anionic center   138
arylacetylase activity   204
as hydrolase   204
as indicator of the acetylcholine presence   113
as protease   149
assay   116, 118, 119
at fertilization   149
defensive role   139
depression   59
esteratic center   138, 139
evolution   132, 203, 204
in algae   101
in animals   133, 134, 204
in anthers   117
in bacteria   99, 101, 204
in cell wall   122
in chloroplasts   19, 122, 127, 129, 130, 141, 143
in chloroplast envelope   125
in contractile and motile reactions   148
in flagellum   203
in flowers   215
in fungi   99, 101, 204
in generative organs   115
in germinating pollen grains   122
inhibition   117, 208
inhibition by substrate   118-120
inhibitors   20, 59, 100, 101, 113, 114, 116, 117, 123, 125
in honey   119
in mammals   139
in motile organs   62
in pistil   115
in pollen   115, 119, 123, 139, 142, 149, 185, 215
interaction with electron carriers   136, 137, 147
interaction with celluar components and enzymes   136, 137, 147
in thylakoids   125, 142
in vesicles   122
kinetic characteristics   115-120, 139, 141
K_m 217
localization in cell   58, 120-126
microspectrofluorimetric assay   217, 218
occurrence in plants   100-115
of pests   146, 205
of weeds   205
peptidase activity   204
physiological roles   139

plant excretions   149
rates of hydrolysis   204, 217
special roles in plants   147-150
stomata   115
substrate specificity   116, 118, 119, 125, 204
synthesis regulation   145
taxon and tissue -specific   100
true   100
types   100, 117, 119, 126, 132
Cholinesterase activity
cycle   114
during development   114, 115
effect of temperature   145
in aleuronic layer   114
in anthers   117,
in carnivorous plants   115
in chloroplasts   125, 141-143, 184
in compartments   122
inhibition by substrate   118, 119
in germinating pollen grain   122
in pistil   116
in plant excretions and extracts   14-19, 115
in flower   116
in pollen   149
in stomata   115
ion regulation   145
kinetic parameters   139, 141, 204
methods of determination   99, 100
oscillation   114
optimums pH and temperature   128, 129, 145
regulation   139-146, 148
Cholinesterase activity determination   99, 100, 215-219
biochemical methods   101, 115, 215-217
color reaction   99, 215, 216
histochemical methods   101, 115, 217, 218
radioisotope method   11
Cholinesterase isolated
aminoacid composition   136
aggregates   132, 135, 136
autoproteolysis   138
carboxypeptidase activity   135
electrophoresis   131-135
free form   126-132, 135, 136
from animals   133-135, 139
from chloroplasts   129-132, 134, 136, 137

from latex 131, 135, 137, 139, 141
inhibition by substrate 128-130, 132
inclusion of sugars 137
isoforms 130-132
kinetic parameters 126, 128, 129, 137
membrane-bound 126, 127, 129-132, 135, 136
methods of purification 126-139, 218, 219
molecular masses (Mr) 126, 131-136, 141
peptidase activity 138
plural forms 129-136
rates of hydrolysis 126, 127, 130, 136
Stokes radius 137
substrate specificity 128, 129, 132, 137, 138
subunits 126, 131-134, 141
Cholinesterase inhibitors 20, 59, 100, 101, 114, 116, 123, 125, 140, 146, 147, 149, 185 212
carbonate 101, 138-141, 143, 145
competitive 126, 138, 139
constants of inhibition (Ki) 139, 141
constants of the rate of inhibition (K_2) 139, 141, 143
derivatives of limonene 140, 145
effects on plant growth 145, 147
organophosphates (organophosphorus) 101, 138-140, 142-145
quaternary nitrogen-containing 138, 140
Cholinic esters 18, 60, 100, 116, 125, 130
hydrolysis 116-120, 130, 137, 138, 204
synthesis 85-88, 91
Cholinolytics 204
Cholinoxidase 196
Cholinoreception 85-88, 91, 93-98, 178
Cholinoreceptor 85-88, 91, 149, 200-202, 204, 208
as ionophore 87
blockers 85
coupling with adenylate cyclase 173
methods of the study 85
muscarinic type 35, 85, 98, 200
nicotinic type 33, 85, 87, 93, 200, 201
Chromatography methods 4, 5, 20, 34, 39, 98, 99, 126, 127, 131-135
Cicutotoxin 145
Cimetidine 157
Cinnamic acid 60
Cinnamoylcholine 167

Cinnamoylhistamine 50, 51
Citric acetylesterase 113
Clathrin 12, 180, 201
Clathrin-coated vesicles 180
Clemastin (tavegyl) 152, 155, 157, 159
Clock biological 62
Clones self-incompatible 115
Coclaurine 31, 32
Colchicine 27, 30, 61, 158
Connexons 182
Contacts
between cell wall and plasmalemma 122
between cells 183
between pollen and pistil 91, 92, 122, 149, 183, 184
between organelles within cell 125, 183, 200
intercellular 66, 206
intracellular 66, 206
intramitochondrial 191
non-structured 182
sites 122
structured 182
types 182
Contractile elements and systems 60, 62, 90, 148, 158, 180, 203
Contractile proteins 61-63
Copper-containing enzymes 147, 161, 164, 166-168
Coumaroylcholine 16
Coumaroylserotonin 34
Crinine 33
Cryptopine 28
Cultured cells and tissues 23, 28, 32, 36, 42, 57, 80
Curves of the concentration dependency 117, 119, 120, 125, 133, 156, 161
bell-shaped 94, 100, 116, 119, 120, 130, 133
dose-effect 97, 161
Cyanide 166
Cyclic AMP see c AMP
Cyclic GMP (see cGMP)
Cyclic nucleotides 88, 150, 160, 168, 174
biosynthesis 88, 169
content in plant cell 172, 173
effects on plant reactions 173, 174, 176
enzymes of their biosynthesis and catabolism 88, 170-173
Cyclosis of cytoplasm 61
Cytochalasin B 61, 158

Cytochromes 71, 73-76, 136, 137, 174, 195, 196
Cytokinin 59, 60
Cytoplasmic movement (stream) 175, 189

Deamination of DOPA and its aldehyde 27, 28
Decarboxylase of aromatic amino acids 25, 26, 40, 160-164, 177
Decarboxylase of dioxyphenylalanine 25, 26
Decarboxylases 151, 160, 162-164, 177
Dehydroevodiamine 146, 213
Depolarization of membranes 64, 65, 175, 189, 199, 201
Diamine oxidase 27, 41, 48, 49, 164, 166, 187
Diamines 48, 164, 165
Diacylglycerol 83, 88, 151, 171, 174
Dibutyryl cAMP 175, 176
Dichlorophenolindophenol 195
Dichlorvos 139, 140, 144
2, 3-Dihydroindole-5, 6-quinone 34
5, 6-Dihydroxyindole 34
3, 4-Dihydroxy-5-methoxyphenethylamine 32
3, 4-Dihydroxyphehylalanine (DOPA) 22, 25, 26, 29, 60, 162
3, 4-Dihydroxyphenylacetaldehyde 32
3, 4-Dihydroxyphenylalanine decarboxylase 162
Diisopropyl phosphofluoridate (diisopropyl fluorophosphate) 20, 101, 113, 139-141, 204
Dimaprit 157
Dimedone 51
Dimethoate 140, 144
Dimethylhistamine 47-50, 157
5, 5-Dithiobis-2-nitrobenzoic acid (Ellman reagent) 99, 118, 119
6, 7-diOHATN or 2-amino-6, 7-dihydroxy-1, 2, 3, 4-tetrahydronaphthalene 152-156, 159
Dioxybenzaldehyde 27
Diphenhydramine (benadril) 157
Diuron 75, 77
DNA synthesis 59, 199
Dolichotheline 48-50
Dopamine 147, 148, 155, 164, 168, 170, 219, 220
  agonists 150, 153-156
  antagonists 150, 151, 154, 156
  as chemosignal 178
  as protective agent 197
  as oxidant 75-76, 185
  at wounding stress 198
  biosynthesis 25-27, 159-164
  catabolism 27-33, 166, 167
  constants of binding 156
  content in plants 21, 22, 59, 212
  interaction with amino acids 33
  localization in cell 25
  metabolism 25
  oxidation 33, 34, 198, 199
  precursor of alkaloids 27-30, 32, 33
  redox reactions 74, 195
DOPA-decarboxylase 26, 163, 167
DOPA-formation 25, 26
Dopamine effects on
  ATPase activity in chloroplasts 71
  ATP synthesis 71,
  betacyanin efflux 66
  Ca^{2+} efflux (output) 152
  Ca^{2+}/Mg^{2+} permeability 152, 156
  chloroplast reactions 193
  development 57
  elongation of seedlings 60
  elongation of pollen tubes 54, 55
  energetic reactions 69, 89
  fluorescence of phycoerythryn 70
  germination of pollen 56, 152, 154
  germination of seeds 55
  growth 53, 60
  $NADP^+$ photoreduction 73-75
  photophosphorylation 71, 73, 152, 156, 195
Dopamine-b-hydroxylase 151, 161
Dopamine quinone 34, 147, 148
Dopaminic receptor 150, 151

Electric impulse 87, 187
Electric potentials 187
Electrophoresis 4, 5, 131-135, 166
Ellman method 114
Ellman reagent 99, 114, 116, 118, 119, 123, 124, 142, 215
Ellman reagent red analog 114, 118, 119, 123, 124, 142, 215-218
Endocytosis 180, 183, 202
Endoplasmic reticulum (ER) 12-14, 17, 18, 48, 120
Enzymatic oxidation of catecholamines to melanin 33

Environmental monitoring 205, 206
Ephedrine 29, 33, 155, 213
Epinine 26
Eschscholtzine 31, 33
Eserine (physostigmine) 100, 145, 149
Esterase 122, 136, 185
    non-specific 113, 122, 125, 126, 136, 149
    specific (see cholinesterase)
Etiolated seedlings 108, 121, 122
Etioplasts 127, 130
Ethylene 14, 60, 88, 89
    biosynthesis 14
    interaction with biomediators 60
    release 88, 89
Ethylene-forming enzyme 16
Evolution of chemoreception 178
Excretions 120, 122, 123, 149, 177, 179, 184, 187
Exine of pollen 121-123, 217, 218
Exocytosis 180, 181
Exudation of xylem sap 62

Far red light effects 145
Fatty acids 179
Fensulfothion 144, 147
Fenton equition 74
Ferredoxin 16, 74, 136, 137
Ferredoxin-NADP$^+$ reductase 74
Ferruloylcholine 15, 16
Ferruloylhistamine 47
Ferruloylserotonin 34
Fertility 101, 149
Fertilization 78, 89-92, 149, 177, 184, 185, 188, 205, 206
Flagellum 203, 204
Flavin 27, 166
Flavin adenine dinucleotide 166, 167
Flower 90-92, 116, 188
Flowering 18, 153
Fluorescence
    histochemical reactions 76, 218-220
    of chlorophyll 78-80
    of intact cells 78-80
    of pistil 77, 78, 90-92
    of pollen 77, 90-92, 217-220
Formation of fruits and seeds 91, 92, 155, 159, 185
Formylcholine 119
Forscolin 175, 176
Free radical reactions 76, 80, 185, 187

Galactose 137

Galanthamine 33, 213
Gallic tumors 53
Gametophytes 115
Gap junctions 182, 183, 217
GD-7 143, 144
GD -42 143, 144, 217
Gibberellic acid 145, 163
Gibberellins 53, 60, 197
Gigantine 29, 32, 33
Girinimbine 42
Glutamine 136
Glutathione peroxidase 185
Glochidicine 50, 51
Glochidine 50, 51
Glochidone 50, 51
Glucose 137
Glycine 136
Glycine betaine 16, 197
Glycoalkaloids 146
Glycoprotein 137
Glyoxalic acid 186, 219, 220
Golgi apparatus 12, 26, 41, 48, 82, 120, 180
Golgi vesicles and cisternae 12, 26
G-proteins or transducins (GTP-binding proteins) 83, 170, 177
Gramine 40-42, 60
Growth regulators 41, 145, 146, 159
GTP 83, 168, 170
GTP binding proteins 170, 182
GTP sensitive ion channel 62
Guanylate cyclase 83, 150, 168, 170, 173, 179, 187, 191, 192
Guard cells 65, 108, 115, 122, 123, 148, 170, 188

H$^+$ channels 188
H$^+$ conductivity 190
H$^+$ efflux 198
H$^+$ ions 123, 147, 195
H$^+$ uptake 190, 191
Hairs 44, 115, 183
Hallucinating substances 32, 42, 151, 152
Heat stress 11, 19, 65
Heavy metals 205
Histamine 150, 151, 155, 158, 164-166, 170, 179, 197, 208
    agonists 151
    antagonists 151, 155, 157-159
    antioxidant features 199
    as chemosignal 178, 179
    at stress 47
    biosynthesis 47, 48, 162, 164

catabolism 48-51, 164
content in plants 43-47
enzymes of synthesis and catabolism 162, 164
in plant secretions 60
in pollen 77, 185, 205
in stinging trrichomes 43, 46
pharmacological effects 211
precursors of alkaloids 50, 196
protectory role 208
oxidative deamination 48
Histamine effects on
ATP synthesis in chloroplasts 71
betacyanin efflux 66
fluorescence of phycoerythryn 73
germination of pollen 56, 152, 153, 176
germination of seeds 55, 56
growth 152
in wounded tissue 54
lysozyme activity 80
$NADP^+$ photoreduction 73
Nucleic acid synthesis 54
photophosphorylation 71, 73
protein synthesis 54
redox reactions 71, 73
Histamine-N-methyltransferase 48
Histaminic receptors 150-152, 157-159
Histidine 47, 56, 136, 162, 164, 167
Histidine carboxylase 159
Histidine decarboxylase 47, 48, 151, 162, 164, 204
Homolycorine 33
Honey 11
Hordenine 33
Hormones 53, 171, 175, 180, 185, 200-203
Hydrastine 28, 31-33
Hydrogen peroxide 30-33, 74, 81, 147, 148, 161, 185-187
Hydroxylases 15, 160-162, 176, 208
β-Hydroxylase 3, 4-dioxyphenylethy-lamine 26
Hydroxyl radical 176
Hyperpolarization of membrane 64, 65, 199
Hyoscyamine 113

Illumination continious 18
Imidazole acetic acid 48, 49, 166
Imidazole alkaloids 48
N-Imidazolepropionylhistamine 48
Impulse 87, 88
Inderal (propranolol) 153

Indolamines 42, 167
Indole 33, 179
Indoleacetic (indolylacetic) acid 39, 40-42, 53, 59, 60, 88, 145, 163, 167, 197, 203
Indole acetaldehyde 40, 41
Indole polymers 33
Indole pyruvate (pyruvic acid) 40, 41
Indole-5, 6-quinone 34
Inhibitors of cholinesterase (see cholinesterase inhibitors)
Inmecarb 152, 158
Inositol triphosphate (system) 83, 88, 150, 168, 171-174, 177, 191, 207, 208
Insecticides 19, 20, 100, 144, 210
Insectivorous (carnivorous) plants 123
Intercellular spaces 189
Ion changes 90
Ion channels 62, 65, 85, 87, 170, 173, 187, 188, 190, 208-210
Ion channel
calcium Ca^{2+} 62, 65
closing 82
opening 82
potassium K^+ 62
potential-dependent 87
sodium Na^+ 62
Ionic strength 20
Ionophore 87
Ion permeability 85, 87
of chloroplasts 67-68
of organelles 67
of plasmalemma 63-67
of whole cells 63-67
Ion transport 90
across chloroplast membrane 67, 184
across plasmalemma 63-67
Ipraside 167
Iproniazid phosphate 163
Irritation 189, 190, 199
mechanical 62, 198
Iron-containing proteins 74, 167
Isadrin (isoproterenol, izadrin) 152, 153
Isobutylmethylxanthine 175, 176
Isocraugsodine 33
Isoergine 42
Iso-OMPA 143, 144, 217
Isoprenoid metabolism 16
Isoproterenol 152, 153, 155
Isoquinoline alkaloids 28, 168
Isovaleric acid 48
Isovaleric ~CoA 48

Joints between cells 122-124, 182
Juglone 145
Junction 123
    between stele and cortex 148
    between cell walls 123
    gap 182, 183
    non-gap 182, 183

K^+ efflux 66, 69, 87, 188
    in chloroplasts 67-69, 89
K^+ exchange 65, 66
K^+ uptake 64
Kinetin 145
K^+-uptake 64
Kur-14 152, 153, 156, 158

Latex 21, 22, 25, 43, 128, 130, 131, 135, 139, 141, 196
Laticifer 28, 128, 130, 131, 139
Leaf drop 11
Leaf recovery 11
Leaf unrolling 89, 90
Leaf wilt 11
Leucine 136, 179
Light absorbance changes at 520 nm 67, 190
Light effects
    blue-light 58, 64, 198
    far red light 14, 18
    on the acetylcholine biosynthesis 14, 17, 18
    on the catecholamine biosynthesis 26, 27
    on growth reactions 57
    period 25
    seed germination 57
    red light 14, 17, 18, 52, 53, 58, 59, 63, 64, 88
    white light 17, 18
Limonene derivatives 140
Liver 46
Long day conditions 18
Long-day plants 18, 57
Lophophorine 29, 32
LSD_{25} 156
Lycopine 33
Lysergic acid 41, 42, 151
Lysergol 12, 42
Lysine 33, 136
Lysozyme 80, 196

Magnesium Mg^{2+} 167

Magnolin(e) 213
Mannich reaction 27, 28
Mass-spectroscopy 5, 20
Melanin 33, 34
Melatonin 41, 42
Membrane
    depolarization 175
    post-synaptic 32, 33, 66
    potential 63-69, 85, 87, 188-190, 207
    potential of organelles 67-69, 177
    potential of whole cells 63-67
    processes 63-69
    shrinking 90
    swelling 90
Mescaline 29, 32
Messengers 82
secondary 82, 168-176
    synthesis 88
Metanephrine 27, 28, 60
Methods of the receptor study
    pharmacological 88
    radiactive ligands 88
5-Methoxy-N-acyl tryptamine 42
3-Methoxy-4-hydroxymandelic acid 60
3-Methoxytyramine 28
o-Methylandrocymbine 30
o-Methylation 27, 28
Methylcatecholamine 27, 28
Methylglyoxal-bis(guanylhydrazone) 57
N-Methylhistamine 48, 49, 50, 152, 157, 159, 167
1, 4-Methylimidazole acetic acid 48, 49
Methylparathion 147
Methyltransferases 151, 164, 176, 208
Methylxanthins 174
Metiamide 157
Mexamine 156
Microelectrod technique 64
Microelectrophoresis 122, 123
Microfilaments 180
Microsomes 39, 48, 120, 135, 173,
Microspores vegetative 78-80
Microspores generative (pollen) 115, 123, 149
Microtubules 158, 180
Milk 46
Mitochondria 16, 25, 27, 39, 41, 47, 48, 67, 70, 120, 167, 173, 174, 179, 188, 190-192, 194, 201, 207
Mitotic index 53
Monoamines 25, 164-167, 191, 204

Subject Index   271

Monoamine oxidase   27, 28, 41, 48, 49, 151, 164-168
Models of   90
    cell-cell communications   90-92
    chloroplast-cytoplasm communications   91, 93-98
    cytoplasm-organelle communications   91, 93-98
Morphine   22, 27-29, 32, 213
Morpholine   213
Morphogenesis   89, 206
Motile reactions   60
Motor activity   24
Motor organs   24, 61
Movement
    nastic   62
    of cytoplasm   61
    of leaves   61, 63
Motility of bacteria   88, 89
Muscarine   65, 85, 86, 89, 93, 94
Mycelium of fungi   10, 58, 132
Mycorhiza   180

Na^+
    in cytoplasm   68
    within chloroplasts   68
Na^+-efflux   66, 69
    from intact chloroplasts   67, 68, 89, 93
Na^+/K^+- permeability   64
Na^+-transport   65
Na^+-uptake   87
NAD^+ kinase   171
$NADP^+$-photoreduction   71-73, 77
α-Naphthylacetic acid   138
Nasty movements   62
Neostigmine (proserine)   20, 58, 59, 62, 100, 101, 116-118, 123, 125-127, 133, 138-142, 147-150, 204, 217
Nerve-muscle preparations   4
Nervous system   100
Nettles   11, 14
Nicotine   65, 85, 86, 89, 90
Nicotinic cholinoreceptors   93
Nifedipine   65
Nitrogen metabolism   25
NMR- spectroscopy   5
Noradrenaline (norepinephrine)   57, 155, 158, 164, 167, 170, 175, 219
    agonists   151-153, 155
    antagonists   66, 151-155, 158, 159
    as generator of superoxide radical   187
    as hormone   196

    as oxidant   185
    as precursors of alkaloids   27-29
    at fruit ripening   24
    at stress   24
    biosynthesis   25, 26, 160-164
    catabolism   27-28, 164-166
    constants of binding   158
    content in plants   22-24
    in axes   62
    in chloroplasts   25
    in pulvini   62, 63
    localization in cell   25
    occurrence in plants   22-24
    in oxidative processes   24
    pharmacological effects   211
    redox reactions   71-76, 195, 196, 199
    release   204
    stimulation of accumulation   25
Noradrenaline effects on   32
    Ca^{2+} efflux from chloroplasts   70, 152, 156, 175
    chloroplasts   156
    CO_2 fixation   148
    cyclosis of cytoplasm   61, 63
    elongation   53, 60
    elongation of hypocotyles   60
    exudation of xylem sap   62
    flowering   57, 152, 153
    fluorescence   76
    membrane potential   65, 152, 190
    Mg^{2+} efflux from chloroplasts   69, 152
    movement of cytoplasm   61
    $NADP^+$ -photoreduction   73, 74
    plant growth   53
    photophosphorylation   73, 152, 195
    pollen germination   56, 152
    seed germination   55
Norbellandine   164, 168
Norcarnegine   29, 31, 32,
Norcoclaurine   27, 32
Norcoclaurine synthase   32
Norepinephrine (see noradrenaline)
Norlaudanosoline   32
Normetanephrine   27, 28, 60
Norreticuline   31
Norsecurinin   33
Nucleolus   120, 122, 125
Nucleus   39, 48, 59, 173, 174, 179
Nyctinasty   39, 47, 59

Organic acid   18
Organophosphates   100, 138, 142-144, 204

effects on acetylcholine metabolism 19, 20
Organophosphate inhibitors 20, 100, 143, 144, 217
Oscillation in cholinesterase activity 114
Oxidative deaminations 27, 28, 164
N^{α}-4'-oxodecanoylhistamine 50
Oxygen effects on acetylcholine content 18
Oxygen release 16
   at photosynthesis 16
Oxygen uptake 198
Oxytryptophan 40

Pain 47
Papaverine 27-29, 32
Paper chromatography 5, 20
Papillae 115
Paraoxon 140, 144
Paspalicine 42
Pectinesterase 14, 113
Pellotine 29, 32
Permeability of cell and organelles 60, 64, 66
Permethrine 20, 145
Peroxidase 80, 148, 166, 185, 187
Peroxides 73-76
   formation 71
Pesticides 20, 58, 148, 176, 185, 187
pH effects on acetylcholine metabolism 20
pH gradient in chloroplasts 68
Phenol 76
Phenylalanine 25, 26, 136
Phenylalanine hydroxylase 25, 26
Phenylalanine monoxidase 25, 26
Phenylethanolamine-N-methyltransferase 26, 164
Phenylethylamines 25, 26, 161
Phosphatidylcholine 13
Phosphatidylinositol phosphate 88
Phosphatidyl inositol phosphodiesterase (phospholipase C) 83
Phosphodiesterase 169, 171, 174, 176
Phospholipase C 83, 172
Phospholipid 16
Phosphon D 139, 140, 144, 147
Phosphorylation
   oxidative and substrate 71, 173, 194
   of proteins 83, 160, 170-174, 191, 201, 202
   photosynthetic 61, 72, 73, 89, 90, 93, 94, 171

Photomorphogenesis 59, 197
Photoperiod 18, 25, 153
Photoperiodism 19
Photophosphorylation 71-73, 89, 90, 93, 94, 158, 160, 161, 173
Photosynthesis 13, 17, 18, 130, 147, 175, 195, 202
Photosystems 18, 74, 173, 196
Phototaxis of chloroplasts 175
Phthallophos 20
Phycoerythrin 75, 76, 199
Phyllandin 33
Phylogenesis 200, 201
Physostigmine (eserine) 5, 20, 58, 59, 100, 101, 113, 116-118, 125, 126, 133, 138, 139, 140-142, 145, 147-150, 196, 204, 217
Phytocenosis 91, 206
Phytochrome 14, 18, 29, 53, 59, 61, 65, 197, 198, 202, 203
Phytohormones 59
Peridoles fluoridated 151
Pirilamine (neoantergan) 157
Pistil 20, 90-92, 105, 106, 109, 111, 112, 115, 116, 122, 149, 150, 154, 155, 159
Pistil-pollen interaction 90-92, 149, 206
Plasmadesmata 66, 183
Plasmalemma (plasmic membrane) 58, 62, 65, 66, 120-122, 153, 170-174, 179, 183, 185, 188, 192, 199, 207
Plastids 39, 125, 136, 173, 183, 184
Plastocyanin 71, 74-76, 136, 137, 147, 195, 196
Plumules 70
Pollen 8, 9, 44-47, 54, 55, 76-78, 90, 91, 101, 103-106, 108-111, 115, 117-123, 125, 141, 147, 149, 150, 155, 176, 179, 180, 184-186, 188, 205, 215, 218-220
Pollen exine 122, 184
Pollen fertility 101
Pollen germination 54-56, 122, 147, 149, 152-155, 176, 180, 184-187
Pollen loads 115
Pollen-pistil contacts and recognition 91, 92, 122, 149, 183, 184
Pollen-pollen interaction 185, 187
Pollination 91, 92, 120, 159
Pollutants 206
Polyphenoloxidase 15, 33, 34, 166, 168, 199
Potential of rest 190
Potential (voltage)–dependent ion channels 201

Priming   168
Proline   136
Propionylcholine   14, 15, 65, 100, 119, 130
Propionylthiocholine   128, 129, 138, 216
Propranolol   152, 153
Proserine (neostigmine)   100, 149
Protein kinase   170-174, 191, 192, 201
Protein phosphatase   171
Protein phosphorylation   30
Protonema   53
Pseudocholinesterase (butyrylcholinesterase)   100, 132, 143, 204
Psilocin   41, 42
Psilocybin   41, 42
PTCh-see propionylthiochopline
Pulvini   8, 9, 11, 19, 23, 61-63, 65, 123
Putrescine   48, 164
Pyrethrine   27, 113
Pyridoxal phospate   27, 163, 166
Pyruvate   13

Q_{76} limonene derivative   140, 144, 145
Q_{80} limonene derivative   140, 144, 145
Quercetin   163
Quarternary compounds   139, 140, 147
Quinone   33
Quinuclidinyl benzylate   85, 86, 89, 93, 94, 96
Quipasine   156

Radiation (γ and x)   198, 199
Reaction centers of photosystems   67, 173, 174, 184
Red analog of Ellman reagent   114
Red light effects   145, 197
Reserpine   41, 42, 196
Respiration   203
Retardants   58, 144, 145, 205
Reticuline   27, 28, 30-32
Rhizobia and rhizogenesis   53, 57, 63
Rhodopsin   200
Rhoeadine   31, 33
Ribulose-1, 5-bisphosphate carboxylase   136, 137, 147, 148, 196
Rigor   147
Root pressure   62, 88-90, 148

Salsoline   213
Salsolonol   28
Salynity   197, 208
Sanguinarine   213
Sanguirythrine   213

Schiff base   29
Scopolamine   113
Scoulerine   30-32
Secondary messengers   150, 168-177, 179, 187, 191-193, 196, 201, 207, 208
Secretion   82, 123, 150, 175, 181, 182, 219, 220
Secretory vesicles   12, 82, 181, 192
Securinin   33
Self-compatible clone   149, 150
Self-incompatible clone   149, 150, 205
Semicarbazide   166
Serine   13, 136
Serine-containing proteins   100, 141
Serine protease   141
Serotonin(5-oxytryptamine, 5-hydroxytryptamine)   34, 150, 156, 158, 159, 170, 171, 175, 208, 213
 agonists   151, 152
 antagonists   65, 66, 151, 152, 156, 158
  as hormone an growth signal   53, 54, 196, 197, 200, 201, 203
 as precursors of alkaloids   41, 42, 196
 as protector   198, 199
 binding with DNA   199
 biosynthesis   39-41, 160, 162, 163
 catabolism   39, 41, 42, 166
 content in plants   34-39
 in bacteria   202
  in stinging trichomes   35
 localization in cell   39
 metabolism   39
 methods of determination   34
 occurrence in plants   34-39, 212
 oxidation   41, 198
Serotonin effects on
 betacyanin efflux   66
 Ca^{2+} efflux   66, 175
 Ca^{2+}/Mg^{2+} permeability in chloroplasts   192
 DNA synthesis   59
 formation of rhizospere   53
 germination of pollen   56, 152, 153, 176
 germination of seeds   55, 152, 153, 158, 159, 167
 growth   53, 54, 203
 membrane potential   65
 mitotic index   53
 morphogenesis   57
 Na^+ transport   65, 66
 pharmacological effects   211
 RNA synthesis   59

Serotonin receptors 150, 151, 202
Sesquiquinones 33, 179
Shepherdine 41, 42
Shikimate pathway 39
Short-day plants 18, 57
Short red light 18
Sinapine choline (sinapoylcholine) 14-16, 113, 138, 196
Sinapinesterase 14, 113, 146
Slime and slime hairs 107, 123, 124, 149, 183
Solanine 113, 213
Spermidine 48, 57, 164, 166
Spermine 166
Spiroperidol 156
Sporoderm 122
Standard potentials 195
Sterol 16
Stigma of pistil 91, 115
Stinging trichomes (hairs) 4, 10, 11, 14, 37, 39, 43, 46, 47, 98
Stomata 16, 65, 108, 115, 121, 123, 147, 148, 188, 190
Stress 197, 206, 208, 209
Stress reactions 47, 195
Strychnine 213
Sugar 179
Sulpiride 150, 151, 154
Superoxide dismutase 33, 148, 185, 187
Superoxide anion radical 33, 34, 76, 77, 81, 147, 148, 176, 185-187
Swelling of cells and organells 62, 67, 159, 188
Symbiosis 63, 121
Symplast 189
Synapse 66, 82, 83, 120, 178, 179, 181, 189, 191, 195, 200, 207
Synaptic contacts 66, 187, 207
Synaptic membranes 11, 66, 82
Synaptic vesicles 181
Tabun 204
Tallus 121
Tanada effect 63
Tavegyl (clemastin) 152, 155, 157, 159, 162
Tazettine 33
Tendrills 23, 37
Tetraethylammonium 85, 86, 89
Tetrahydrobiopterin 160, 161
Tetrahydrocolumbamine 31, 32
Tetrahydroprotoberberinic alkaloids 32

Thebaine 32
Theophylline 174-176
Thioperamide 158
Thylakoids 68, 91, 93-97, 125, 142, 173, 184, 190-194
Tonoplast 66, 173, 188, 192
Transducins 177
Transferases 208
Transmethylase of phenylethanolamines 26
Trap of carnivorous plant 107, 109, 115, 123, 124
Threonine 136
Trifluoroperazine 65, 66, 152, 156, 158, 159
Trimethylammonium 149, 153
3, 4, 5-Trihydroxyphenethylamine 32
Trypsin 160
Tryptamine 39, 40, 162, 163, 165-167
Tryptophan 39-41, 59, 160, 161, 163, 179
Tryptophan decarboxylase 40, 163
Tryptophan-5-hydrohylase 40, 160, 162
d-Tubocurarine 33, 65, 88-93, 188, 196, 208
Tubulin 61
Tyramine 25-27, 29, 161, 165-167
Tyramine hydroxylase 26, 151
Tyrosinase 166, 168
Tyrosine 25, 26, 136, 160-162, 167, 168
Tyrosine carboxylyase 162
Tyrosine decarboxylase 162, 163
Tyrosine hydroxylase 25, 26, 160, 161
Tyrosine-3-monoxidase 25, 26

Unicellular organisms 67, 172, 178, 179, 183, 199-203, 207

Vacuoles 25, 39, 188
Valine 136
Vanillic acid 27, 28, 33, 166
Vanillic aldehyde 27, 28
Vanillin 33
Verapamil 65
Vesicles 11, 122, 180-182, 184
  clathrine-coated 25, 39, 180
  secretory 25, 92, 181, 182
Vinblastine 41, 42

Water
  deficit 206

Subject Index 275

destruction at photosynthesis 175
 channels 188
 transport 188
 uptake 89
Weed-like species 206
Wilting 65
Wind-pollinated plants 139
Wounding 166

Yeast 172, 174, 179, 181, 198, 200, 201

Yohimbane 41, 42
Yohimbine 41, 42, 66, 151-153, 155, 156, 158, 159, 161, 213

Xanthine oxidase 49
Xylose 137

Zapotidine 50, 51
Zeatin 66

Latin Index

PLANTS
Acacia argentea 51
Acacia polystacha 51
Acacia spirorbis 51
Acalipha indica 107
Acanthaceae 103
Acer negundo 103
Aceraceae 103
Aconitum napellus 21-23
Aconitum paniculatum 22, 23
Adiantaceae 102
Adiantum capillus-veneris 102
Aeschynomene indica 107
Aegopodium 165
Aesculus hippocastanum 108, 115, 118-120, 142
Agavaceae 103
Ageratum conyzoides 104
Agropyron repens 54, 55, 58, 198
Agrostis alba 44, 47
Aizoaceae 103
Alangiaceae 6
Alangium lamarckii 6
Albizzia sp. 19
Albizzia julibrissin 8, 17, 23, 24, 36, 61, 63, 109, 113, 142, 212
Alliaceae 103, 118, 128
Allium altaicum 14, 98, 103, 128, 130, 132
Allium cepa 118-120, 139, 142, 174
Allium sativum 113
Alopecurus pratensis 44, 47
Altingiaceae 6, 103
Amaranthaceae 6
Amaranthus caudatus 6, 17
Amaryllidaceae 103, 116, 118, 128, 142
Amsonia angustifolia 6
Anabasis aphylla 213
Anacardiaceae 6, 104

Ananas sp. 38
Ananas commosus 35, 38, 212
Anhalonium lewini 32
Annona reticulata 32
Annonaceae 104, 212
Anthoceros sp. 201
Anthocercis viscosa 110
Apiaceae 104
Apium graveolens 174
Apocynaceae 6, 104
Aquifoliaceae 6, 104
Araceae 35, 43, 104
Arachis hypogaea 54
Areca catechu 196
Arecaceae 104
Argemone mexicana 110
Argyrodendron peralatum 51
Arnica montana 213
Artemisia sp. 165
Artocarpus champeden 9, 11, 212
Artocarpus integra 9, 11, 212
Asclepiadaceae 43, 104
Asclepias syriaca 80, 104, 196
Asparagaceae 104
Asparagus racemosus 104
Aspidiaceae 104
Asteraceae 6, 21, 23, 24, 43, 104, 113, 118, 142, 212
Athyrium filix-femina 53, 60
Atropa belladonna 196
Avena sp. 64, 167
Avena fatua 8, 17
Avena sativa 8, 17, 44, 52, 108, 121, 122, 129, 141, 212

Balsaminaceae 105
Berberidaceae 28
Berberis sp. 167, 212

Latin Index 277

Berberis beaniana 32
Berberis stolonifera 32, 168
Beta sp. 65
Beta trigyna 43
Beta vulgaris 21, 197, 212
Beta vulgaris var *rapa* 43
Beta vulgaris f. *rubra* 66
Betulaceae 6, 105, 118, 119, 139, 142, 185
Betula pendula 6, 15, 105,
Betula verrucosa 105
Bidens biternata 104, 118, 142
Biota orientalis 103
Blumenbachia contorta 37
Boehmeria cylindrica 112
Boerhaavia diffusa 110
Boerhaavia sp. 110
Bombacaceae 105
Bombax ceiba 105
Bougainvillea glabra 110
Brassica campestris var *napobrassica* 6
Brassicaceae 6, 15, 43, 105, 113
Brassica kaber 54, 55, 198
Brassica oleracea 6, 105, 174
Brassica oleracea var. *botrytis* 14, 98
Brassica oleracea var. *gongylodes* 6
Bromeliaceae 35
Bromus erectus 44, 47
Brucea antidysenterica 213
Burceraceae 105

Cactaceae 21, 23, 32, 105, 118, 205, 212
Caesalpiniaceae 106
Caesalpinia pulcherrima 106
Cajanus cajan 55
Callistemon lanceolatus 110
Calocasia esculenta 104
Calotropis gigantea 43
Calotropis procera 104
Calystegia sepium 104, 106
Campanulaceae 106
Campanula hybrida 106
Canna indica 106
Cannabiaceae 106
Cannabis sativa 106
Cannaceae 106
Caprifoliaceae 6, 106, 113
Capsella bursa pastoris 7, 43
Capsicum annuum 196
Carica papaya 35, 106
Caricaceae 35, 106
Carissa carandus 104

Carnegiea gigantea 104
Carthamus tinctorius 35
Carum copticum 10
Caryophyllaceae 106
Caryota urens 104
Casimiroa edulis 47, 51
Cassia occidentalis 106
Cassia tora 106
Catharanthus roseus 40, 42, 104
Caulerpa racemosa 102
Caulerpa scalpelliformis 102
Caulerpaceae 102
Ceratophyllaceae 106
Ceratophyllum demersum 106
Chaetangiaceae 102
Chamaedorea elegans 104
Characeae 14, 102
Chelidonium album 213
Chelidonium majus 45, 110, 212
Chenopodiaceae 6, 16, 21, 43, 106, 113, 128, 131, 212
Chenopodium album 43, 54, 198
Chenopodium bonus-henricus 44
Chenopodium quinoa 44
Chlamydomonas reinhardtii 147
Chlorella sp. 74-76, 137
Chondodendron tomentosum 196
Chrysanthemum sp. 20
Chrysanthemum cinerariaefolium 104, 113
Cicer arietinum 52, 107, 115, 128, 131, 132, 141, 142, 144, 166, 204
Citisus laburnum 213
Cytisus scoparius 162, 163
Citrullus vulgaris 35, 46
Citrus aurantifolia 111
Citrus limettioides 172
Cladoforaceae 102
Cleome pungens 15
Clivia sp. 103, 116
Cnidoscolus oligandrus 35, 44
Cnidoscolus phyllacanthus 36, 44
Cnidoscolus texanus 7, 11, 36, 44
Coccinia cordifolia 107
Cocos nucifera 174
Codiaceae 102
Codiaeum variegatum 7, 15, 107, 212
Coleus blumei 109
Combretaceae 106
Commelina communis 106
Commelinaceae 106
Commiphora wightii 105

Compositae (Asteraceae) 6, 165
Convolvulaceae 6, 101, 106, 116, 128
Convolvulus arvensis 106, 116, 117, 119, 122, 127, 128, 130, 131, 134
Corchorus aestuans 112
Cornaceae 107
Cornus florida 107
Coronopus didymus 105
Corydalis pallida 23, 28
Corylus avellana 45, 47
Corynanthe yohimbe 42
Crassulaceae 23, 35, 107, 139
Crataegus oxyacantha 9, 212
Crataegus sp. 46
Crinum asiaticum 58
Crotalaria juncea 107
Cruciferae (Brassicaceae) 6, 15, 43, 105, 113
Cucumis anguria 7, 19
Cucumis sativus 7, 18, 19, 52
Cucurbita pepo 7, 17, 107, 136, 187, 212
Cucurbitaceae 7, 35, 107, 212
Cupressaceae 103
Cycadaceae 101, 103, 107
Cycas revoluta 103
Cyclamen sp. 46
Cynara sp. 174
Cynodon dactylon 110
Cynosurus cristatus 44, 47
Cyperaceae 107
Cyperus strigosus 107
Cytisus scoparius 21, 25

Dactylis glomerata 45, 47
Datura innoxia 101, 111
Daucus carota 9, 104
Delphinium sp. 46
Dennstaedtia punctilobula 107
Dennstaedtiaceae 107
Digitalis ferruginea 9, 212
Digitalis lanata 9
Digitalis purpurea 9
Dioon edule 103
Dioon spinulosum 103
Diospyros virginiana 107
Dolichos lablab 197
Dolichothele sp. 48, 50
Dolichothele sphaerica 50
Dracaena deremensis 103
Drosera capensis 107, 115, 149
Droseraceae 44, 107
Drosera sp. 44

Ebenaceae 107
Echinocereus pentalophus 105
Echinochloa crusgalli 54, 198
Eclipta prostrata 104
Eichhornia crassipes 110
Elaeagnaceae 35, 42, 212
Elaeagnus umbellata 35
Entada pursaetha 21, 22
Ephedraceae 103
Ephedra eguisetum 213
Ephedra foliata 103
Ephedra monosperma. 213
Ephedra sp. 33, 47
Epiphyllum sp. 118, 119
Epiphyllum hybridum 105, 115, 118, 119, 121
Equisetaceae 102
Equisetum arvense 54, 78-80
Equisetum ramosissimum 102
Erodium cicutarium 4, 212
Euglenaceae 44
Euglena sp. 44
Euphorbiaceae 7, 11, 35, 44, 47, 51, 101, 107, 113, 128, 212
Euphorbia hirta 107
Euphorbia milii 107
Euphorbia neriifolia 107
Euphorbia pulcherrima 107
Euphorbia viminalis 107
Evodia rutaecarpa 146, 213
Evolvulus nurninulurius 106

Fabaceae (Leguminosae) 7, 21-25, 34, 36, 44, 101, 107, 113, 128, 165, 166, 205, 212
Ficus bejamina 109
Ficus benghalensis 109
Ficus elastica 109
Ficus krishnae 109
Ficus racemosa 109
Ficus religiosa 110
Forsynthia sp. 110
Fragaria sp. 165
Fraxinus americana 110
Fumariaceae 23
Funariaceae 5, 8, 212
Funaria hygrometrica 18, 212
Funaria hygrometrica × *Physcomitrium piriforme* 8, 17, 18
Funkia (Hosta) sp. 109, 116

Galaxaura oblongata 102

Latin Index

Galanthus woronowi 213
Galega officinalis 213
Galinsoga quadriradiata 105
Gelsemium sempervirens 45, 47
Geraniaceae 8, 44, 118
Geranium pratense 118, 119,
Geranium thungbergii 212
Ginkgoaceae 101, 103, 113
Ginkgo biloba 103
Girardinia heterophylla 10, 37, 43, 46
Gladiolus sp. 108, 116, 117, 215
Gladiolus hybridum 116
Glechoma hederacea 109
Gleditsia triacanthos 107
Glochidon sp. 50, 51
Glochidion multiloculare 51
Glochidion philipicum 51
Glochidion thomsoni 50
Glochidion venulatum 51
Glycine max 52, 60, 88, 89, 107, 121, 165, 168, 174
Gnaphalium indicum 105
Gossypium sp. 164, 212
Gossypium herbaceum 109
Gossypium hirsutum 36, 45, 213
Gracilaria corticata 102
Gracilariaceae 102
Gramineae 8, 24, 44, 108, 129, 132, 212
Grateloupiaceae 102
Grevillea robusta 111
Griffonia simplicifolia 35, 36, 39, 40
Grossularia reclinata 212
Gymnocalicium castellanosii 105
Gymnocalicium zegarrae 106

Halymenia venista 102
Helianthus annuus 6, 17, 43, 62, 88, 98, 99, 105, 113, 212
Helminthocladia calvadosii 102
Helminthocladiaceae 102
Hemerocallis fulva 109, 115, 118, 119, 139, 142
Hermidium alipes 21
Hibiscus palustris 109
Hibiscus rosa-sinensis 109
Hidrocharitaceae 108
Hippeastrum hybridum 54-56, 77, 78, 90, 91, 103, 116-118, 121-123, 141, 142, 144, 149, 154, 155, 159, 176, 185-187, 215, 218
Hippocastanaceae 108, 118, 142
Hippophae rhamnoides 212

Hordeum vulgare 80, 162, 163, 213
Hosta (Funkia) sp. 109, 116
Hydrangea 165
Hydrangeaceae 116, 118
Hydrastis canadensis 28
Hydrocotylaceae 108
Hydrocotyle verticillata 108
Hydrophyllaceae 11
Hydrosme rivieri 43
Hydrylla verticulata 108

Ilex crenata 104
Ilex opaca 6, 104
Impatiens balsamina 105
Ipomoea sp. 188
Ipomoea abutiloides 6, 106
Ipomoea nil 106
Ipomoea tricolor 42
Iridaceae 108, 113, 116
Iris virginica 109
Iyengaria stellata 102

Jasminum sp. 165
Jatropha integerrima 107
Jatropha urens (Cnidosculus urens) 43, 44, 47, 212
Juglandaceae 34, 36, 212
Juglans sp. 163
Juglans ailanthifolia 36
Juglans mandshurica 36
Juglans nigra 36, 212
Juglans regia 36, 38, 39, 163, 212
Justicia gendarussa 103, 107

Kalanchoe pinnata 107
Kochia childsii 44

Labiatae (Lamiaceae) 45, 109, 165, 212
Lactuca sativa 21, 23, 24, 60
Lamiaceae 109
Lamium album 45, 212
Lantana camara 112
Laportea sp. 10, 46
Laportea moroides 5, 10, 37, 43, 46, 212
Loranthaceae 45
Lathyrus sp. 55
Lathyrus latifolius (latifolia) 58, 107
Lathyrus odoratus 107
Lathyrus sativus 54, 55, 107
Lauraceae 36, 109
Laurencia pediculariodes 102
Lavandula vera 165

Lavandula spica 165
Leguminosae 7
Lemna sp. 58
Lemna gibba 8, 19, 57, 60, 88, 172
Lemna paucicostata 57, 153
Lemna perpusilla 57
Lemnaceae 8
Lens culinaris 52, 108
Lentibulariaceae 45, 109
Lepidium sativum 43
Liliaceae 109, 116, 118, 142
Lilium tenuifolium 116
Liquidambar marginata 103
Liquidambar styraciflua 6
Liriodendron tulipifera 109
Livistona cinensis (chinensis) 104
Loasaceae 11, 36
Loasa vulcanica 36
Lolium perenne 45, 47
Lonicera japonica 6, 106
Lophophora williamsii 21, 25, 32
Loranthaceae 8
Lupinus albus 53
Lupinus luteus 165
Lycopersicon esculentum 18, 37, 38, 110, 159, 162, 163, 212
Lycopersicon sp. 46
Lygophyllaceae 34, 36

Macroptilium atropurpureum 9, 11, 109, 123
Magnolia grandiflora 109, 213
Magnoliaceae 109
Malus domestica 11, 118, 119, 139, 142
Malvaceae 36, 45, 109, 212
Mangifera sp. 172
Mangifera indica 104
Marchantia polymorpha 121
Matricaria chamomilla 105, 118, 142
Mazus pumilus 111
Medicago sativa 108, 165, 173
Melia azedarach 109
Meliaceae 109
Melilotus alba (albus) 108
Menispermaceae 28
Mesembryanthemum crystallinum 103
Mimosa pudica 19, 23, 24, 36, 61, 63, 109
Mimosa sp. 7, 45, 175, 212
Mimosaceae 8, 21-23, 36, 45, 109, 205, 212
Momordica charantia 35
Monostroma fuscum 22, 25

Monostromatadaceae 21
Moraceae 9, 109, 212
Morus alba 110
Mucuna pruriens 21, 23, 36, 39, 212
Musa acuminata 53, 59
Musa paradisiaca 110
Musa sapientum 36-38, 45, 161
Musa sp. 20, 21, 23, 24, 212
Musaceae 21, 23, 24, 36, 45, 110, 212
Myrtaceae 110

Narcissus pseudonarcissus 90, 104
Narcissus sp. 174
Nepenthaceae 45
Nepenthes sp. 45
Nephrolepis biserrata 102
Nerine bowdenii 164, 168
Nicotiana glauca 111, 148
Nicotiana plumbaginifolia 111
Nicotiana rustica 111
Nicotiana tabacum 89, 172
Nitella flexilis 64
Nitella sp. 14, 61, 63, 100, 102, 188-190
Nitella syncarpa 65
Nitrariaceae 37
Nitraria schoberi 37
Nyctaginaceae 21, 110
Nyctanthes arbor-tristis 112

Ocimum sanctum 109
Oleaceae 45, 110, 165
Oleandraceae 102
Osmundaceae 110
Osmunda regalis 110
Oxalidaceae 110
Oxalis corniculata 110

Panicum crusgalli 163
Papaveraceae 21, 23, 28, 45, 110, 118, 142, 212
Papaver bracteatum 21, 22, 25, 28
Papaver orientale 110, 115, 118, 142, 185
Papaver rhoeas 174
Papaver somniferum 21, 22, 25, 28, 32, 110, 214
Parmelia caperata 100, 121
Parthenium hysterophorus 105
Parthenocissus quenquefolia 112
Paspalum ciliatifolium 110
Passiflora foetida 37
Passiflora quadrangularis 23, 37

Passifloraceae 23, 37
Peganum harmala 36, 40, 42
Peltandra virginica 104
Peperomia metallica 190
Peperomia clusiaefolia 110
Peperomiaceae 110
Peristrophe paniculata 103
Persea sp. 21, 36
Persea americana 172
Petunia hybrida 111, 112, 115, 118-120, 142, 149, 150, 185
Pharbitis nil 6, 19, 115, 121, 122
Phaseolus aureus 7, 12, 15, 17, 23-25, 52, 53, 61, 63, 64, 70, 71, 88, 108, 121, 122, 126-128, 130, 131, 134, 141, 142, 144, 146, 172, 173, 198, 212
Phaseolus multiflorus 23, 36
Phaseolus mungo 163
Phaseolus vulgaris 7, 17, 60, 64, 70, 98, 99, 108, 114, 121, 122, 126, 128, 131, 132, 139, 141, 142
Philadelphus grandiflorus 111, 116, 118-120, 142
Phleum pratense 45, 47
Phyllanthus fraternus 107
Physalis minima 101, 112
Physostigma venenosum 20, 196
Pinaceae 5, 9, 103, 118, 142
Pinguicula sp. 45
Pinus abies 103
Pinus sylvestris 9, 101, 103, 115, 118, 142
Piper amalago 21
Piper betle 110
Piperaceae 21, 110, 190
Piptadenia columbina 42
Piptadenia macrocarpa 42
Piptadenia peregrina 36, 42
Piptadenia sp. 42
Pisum sativum 7, 8, 12, 14, 15, 17, 23-25, 36, 52, 53, 55, 57, 61, 64, 67, 68, 71-77, 80, 93-99, 101, 108, 114, 121, 122, 125-128, 130, 131, 133-137, 141, 142, 144-146, 148, 156, 160, 161, 163-166, 168, 172, 174, 190, 212
Plantaginaceae 9, 45, 110, 116, 118, 142
Plantago lanceolata 9, 17, 45
Plantago major 110, 116, 118, 139, 142
Plantago rugelli 9, 15, 110, 113
Platanaceae 110
Platanus occidentalis 110
Plectranthus australis 109

Poa pratensis 45, 47
Poaceae 110
Polyalthia longifolia 104
Polygonaceae 9, 110
Polygonum hydropiperoides 110
Pontederia cordata 111,
Pontederiaceae 110
Populaceae 9, 111
Populus balsamifera 111, 118, 139, 142
Populus grandidentata 9, 15, 111
Populus tremuloides × *P. tremula* 53
Porophyllum lanceolatum 6, 19, 57
Portulaca grandiflora 21, 24-27 ·
Portulaca oleracea 23, 212
Portulaca quadrifida 111
Portulacaceae 21, 23, 24, 111, 212
Primulaceae 46
Proteaceae 111
Prunus domestica 37, 212
Prunus serotina 9, 111
Psidium guajava 110
Psophocarpus tetragonolobus 108
Pteridaceae 101, 102
Pteris multifida 102
Punctariaceae 102
Punica granatum 111
Punicaceae 111
Putranjiva roxburghii 107

Quisqualis indica 106

Ranunculaceae 28, 46
Raphanus sativus 7, 14, 18, 19, 54, 55, 105, 146, 153, 158, 159, 167
Rauwolfia serpentina 42, 196, 213
Rhodomelaceae 101, 102
Rhumex obtusifolius 54
Rhus copallina 6, 104
Ricinus communis 107
Robinia pseudoacacia 8, 12, 15, 23, 25, 108, 122, 127, 129, 130, 132, 134, 212
Rosa sp. 111
Rosaceae 9, 37, 46, 111, 118, 142, 165, 212
Rosmarinus officinalis 165
Rubus sp. 165
Rumex obtusifolius 9, 17, 19, 54, 57
Ruscus aculeatus 104
Rutaceae 28, 46, 51, 111

Salicaceae 9, 111, 118, 142
Salix alba 9, 111, 118, 142

Salix caprea 9, 11, 111, 118, 139, 141, 142
Salsola kali 44
Salvia sp. 45
Salvinia natans 102
Salviniaceae 102
Samanea saman 23, 36, 61, 63
Sambucus nigra 106
Sarracenia sp. 46
Sarraceniaceae 46
Sassafras albidium 109
Saxifraga stolonifera 111
Saxifragaceae 111, 165, 212
Scrophulariaceae 9, 111, 212
Secale cereale 45, 47
Sedum morganianum 23, 25, 35, 39
Sedum pachyphyllum 23, 25, 39
Sesbania sesban 108
Setaria viridis 54, 198
Shepherdia argentea 35
Shepherdia canadensis 35, 42
Silybum marianum 43
Sinapis alba 7, 17, 19, 43, 52, 57
Sisymbrium irio 105
Smilacaceae 9
Smilax hispida 9
Solanaceae 9, 21-23, 37, 46, 101, 111, 113, 118, 129, 142, 205, 212
Solanum ambrosiacum 9
Solanum dulcamara 213
Solanum melongena 37, 112
Solanum nigrum 9, 112
Solanum tuberosum 9, 22, 24, 57, 112, 129, 132, 146, 213
Sonchus oleraceus 105
Sorghum vulgare 163
Spinacia oleracea 6, 17, 21, 43, 44, 47, 98, 99, 106, 128, 131, 142, 159, 162, 164, 190, 198, 212
Spongomorpha indica 102
Stachytarpheta jamaicensis 22
Stellaria media 106
Sterculariaceae 50
Stipa tenacissima 8, 212
Strychnos nux-vomica 213
Symplocarpus foetidus 35
Synadenium grantii 128, 131, 135, 137, 139, 141-145
Syringa vulgaris 45, 47

Tabernaemontana divaricata 104
Tagetes erecta 105
Taraxacum officinale 6, 11, 105, 165
Taxaceae 103, 113
Taxus baccata 103
Thermopsis lanceolata 213
Thevetia peruviana 104
Thymus vulgaris 165
Tilia cordata 46, 47, 112, 118, 119
Tilia platyphyllos 46, 47
Tiliaceae 46, 112, 118
Tradescantia virginiana 106
Trianthema portulacastrum 103
Tragopogon arvense 62
Trichocereus pachanoi 21, 23, 25, 32
Tridax procumbens 105
Trifolium alexandrinum 16
Trifolium incarnatum 165
Trifolium pratense 44, 108, 164, 165
Trifolium repens 44, 108, 164, 165
Trifolium sp. 8, 11
Trifolium subterraneum 165
Trigonella foenum-graecum 108
Triticum aestivum 53, 65, 163
Triticum vulgare 8, 17, 52, 53, 58
Tropaeolaceae 52
Tropaeolum majus 112
Tulipa sp. 109, 118, 139, 142

Udotea indica 102
Umbelliferae 10, 165
Urera sp. 46
Urticaceae 10, 12, 24, 25, 37, 43, 46, 47, 112, 129, 132, 209, 212
Urtica cubensis 37
Urtica dioica 10, 12, 14, 24, 25, 37, 39, 46, 98, 99, 112, 122, 125, 129, 130, 132, 134, 203, 212, 213
Urtica ferox 37
Urtica membranacea 37
Urtica parviflora 10, 46
Urtica pilulifera 37
Urtica thunbergiana 37
Urtica urens 10, 46
Urtica sp. 35, 37, 39
Utricularia sp. 109, 115, 123

Vaccinaceae 112, 113
Vaccinium corymbosum 112
Verbascum chinense 111
Verbenaceae 22, 112
Vernonia conyzoides 105
Vernonia elaeagnifolia 105

Viburnum dentatum 106
Viburnum dilatatum 6, 106
Vicia faba 8, 11, 108, 115, 121, 122, 148, 168, 170, 199
Vicia sp. 165
Vigna radiata 16
Vigna sesquipedalis 8, 52, 58, 88
Vigna sinensis 108
Vigna sp. 88
Vigna unguiculata 8, 18, 19
Vinca rosea 42, 172
Viola papilionacea 112
Violaceae 112
Viscum abietes 212
Viscum album 8, 45
Viscum austriacum 212
Vitidaceae 112
Vitis vinifera 111

Withania somnifera 112

Xanthium strumarium 6, 19, 57

Zamiaceae 101, 103
Zamia furfuracea 103
Zea mays 8, 12, 24, 47, 62, 108, 122, 123, 129, 130, 132, 134, 148, 163, 170, 172

FUNGI
Agaricaceae 5, 10, 37
Agaricus campestris 10, 50
Amanita mappa 37, 42
Ascomycetes (Hypocreaceae) 10, 46, 129
Aspergillus niger 101, 112, 129, 132

Basidiomycetae 46

Cercospora beticola 197, 198
Claviceps purpurea 4, 10, 43, 46
Coprinus comatus 47

Hypocreaceae (see Ascomycetes)

Lactarius blennius 10, 99

Panaeolus campanulatus 37
Phallus impudicus 46
Psalliota campestris 10
Psilocybe aztecorum 42

Physarium polycephalum 100

Saccharomyces cerevisiae 100, 172, 174, 198, 201

Trichoderma sp. 16
Trichoderma viride 54, 57, 198

ANIMALS
Apis 10
Aplysia californica 142

Canis familiaris (dog) 38, 47

Electrophorus electricus 123
Equus caballus (horse) 47

Geodia gigas (giant siliceous gigas) 47, 50

Invertebrata 47, 132, 141-143

Mammalia 38, 47, 142, 143
Mollusks 38
Musca domestica 14, 98, 99, 141, 142

Octopus 34, 38
Oryctolagus (Rabbit) 38

Rattus sp. (Rat) 10

Torpedo sp. 12, 85
Torpedo californica 87, 136, 200
Torpedo marmorata 87

Vertebrata 47, 130, 141, 142, 147

PROTOZOA
Paramecium sp. 201, 204

BACTERIA
Lactabacillus plantarum 203
Oscillatoria agardhii 5, 14, 98, 99
Pseudomonas fluorescens 62
Rhizobium 63
Rhodospirillaceae 62, 63, 88
Rhodospirillum rubrum 62, 63, 88
Thiospirillum rubrum 88
Thiospirillum jenense 63, 88